Graduate Texts in Mathematics 92

Graduate Texts in Mathematics

1 TAKEUTI/ZARING. Introduction to Axiomatic Set Theory. 2nd ed.
2 OXTOBY. Measure and Category. 2nd ed.
3 SCHAEFFER. Topological Vector Spaces.
4 HILTON/STAMMBACH. A Course in Homological Algebra.
5 MACLANE. Categories for the Working Mathematician.
6 HUGHES/PIPER. Projective Planes.
7 SERRE. A Course in Arithmetic.
8 TAKEUTI/ZARING. Axiometic Set Theory.
9 HUMPHREYS. Introduction to Lie Algebras and Representation Theory.
10 COHEN. A Course in Simple Homotopy Theory.
11 CONWAY. Functions of One Complex Variable. 2nd ed.
12 BEALS. Advanced Mathematical Analysis.
13 ANDERSON/FULLER. Rings and Categories of Modules.
14 GOLUBITSKY/GUILLEMIN. Stable Mappings and Their Singularities.
15 BERBERIAN. Lectures in Functional Analysis and Operator Theory.
16 WINTER. The Structure of Fields.
17 ROSENBLATT. Random Processes. 2nd ed.
18 HALMOS. Measure Theory.
19 HALMOS. A Hilbert Space Problem Book. 2nd ed., revised.
20 HUSEMOLLER. Fibre Bundles. 2nd ed.
21 HUMPHREYS. Linear Algebraic Groups.
22 BARNES/MACK. An Algebraic Introduction to Mathematical Logic.
23 GREUB. Linear Algebra. 4th ed.
24 HOLMES. Geometric Functional Analysis and its Applications.
25 HEWITT/STROMBERG. Real and Abstract Analysis.
26 MANES. Algebraic Theories.
27 KELLEY. General Topology.
28 ZARISKI/SAMUEL. Commutative Algebra. Vol. I.
29 ZARISKI/SAMUEL. Commutative Algebra. Vol. II.
30 JACOBSON. Lectures in Abstract Algebra I: Basic Concepts.
31 JACOBSON. Lectures in Abstract Algebra II: Linear Algebra.
32 JACOBSON. Lectures in Abstract Algebra III: Theory of Fields and Galois Theory.
33 HIRSCH. Differential Topology.
34 SPITZER. Principles of Random Walk. 2nd ed.
35 WERMER. Banach Algebras and Several Complex Variables. 2nd ed.
36 KELLEY/NAMIOKA et al. Linear Topological Spaces.
37 MONK. Mathematical Logic.
38 GRAUERT/FRITZSCHE. Several Complex Variables.
39 ARVESON. An Invitation to C^*-Algebras.
40 KEMENY/SNELL/KNAPP. Denumerable Markov Chains. 2nd ed.
41 APOSTOL. Modular Functions and Dirichlet Series in Number Theory.
42 SERRE. Linear Representations of Finite Groups.
43 GILLMAN/JERISON. Rings of Continuous Functions.
44 KENDIG. Elementary Algebraic Geometry.
45 LOÈVE. Probability Theory I. 4th ed.
46 LOÈVE. Probability Theory II. 4th ed.
47 MOISE. Geometric Topology in Dimensions 2 and 3.

continued after Index

Joseph Diestel

Sequences and Series in Banach Spaces

Springer Science+Business Media, LLC

Joseph Diestel
Department of Math Sciences
Kent State University
Kent, OH 44242
U.S.A.

Library of Congress Cataloging in Publication Data
Diestel, Joseph, 1943-
Sequences and series in Banach spaces.
(Graduate texts in mathematics; 92)
Includes bibliographies and index.
1. Banach spaces. 2. Sequences (Mathematics)
3. Series. I. Title. II. Series.
QA322.2.D53 1984 515.7'32 83-6795

Originally published by Springer-Verlag New York, Inc. in 1984
Softcover reprint of the hardcover 1st edition 1984

Typeset by Science Typographers, Medford, New York.

ISBN 978-1-4612-9734-5 ISBN 978-1-4612-5200-9 (eBook)
DOI 10.1007/978-1-4612-5200-9

Preface

This volume presents answers to some natural questions of a general analytic character that arise in the theory of Banach spaces. I believe that altogether too many of the results presented herein are unknown to the active abstract analysts, and this is *not* as it should be. Banach space theory has much to offer the practitioners of analysis; unfortunately, some of the general principles that motivate the theory and make accessible many of its stunning achievements are couched in the technical jargon of the area, thereby making it unapproachable to one unwilling to spend considerable time and effort in deciphering the jargon. With this in mind, I have concentrated on presenting what I believe are basic phenomena in Banach spaces that any analyst can appreciate, enjoy, and perhaps even use.

The topics covered have at least one serious omission: the beautiful and powerful theory of type and cotype. To be quite frank, I could not say what I wanted to say about this subject without increasing the length of the text by at least 75 percent. Even then, the words would not have done as much good as the advice to seek out the rich Seminaire Maurey-Schwartz lecture notes, wherein the theory's development can be traced from its conception. Again, the treasured volumes of Lindenstrauss and Tzafriri also present much of the theory of type and cotype and are must reading for those really interested in Banach space theory.

Notation is standard; the style is informal. Naturally, the editors have cleaned up my act considerably, and I wish to express my thanks for their efforts in my behalf. I wish to express particular gratitude to the staff of Springer-Verlag, whose encouragement and aid were so instrumental in bringing this volume to fruition.

Of course, there are many mathematicians who have played a role in shaping my ideas and prejudices about this subject matter. All that appears here has been the subject of seminars at many universities; at each I have received considerable feedback, all of which is reflected in this volume, be it in the obvious fashion of an improved proof or the intangible softening of a viewpoint. Particular gratitude goes to my colleagues at Kent State University and at University College, Dublin, who have listened so patiently to sermons on the topics of this volume. Special among these are Richard Aron, Tom Barton, Phil Boland, Jeff Connor, Joe Creekmore, Sean Dineen, Paddy Dowlong, Maurice Kennedy, Mark Leeney, Bob Lohman, Donal O'Donovan, and A. "KSU" Rajappa. I must also be sure to thank Julie Froble for her expert typing of the original manuscript.

Kent, Ohio
April, 1983

JOE DIESTEL

Contents

Some Standard Notations and Conventions

Throughout we try to let W, X, Y, Z be Banach spaces and denote by w, x, y, z elements of such. For a fixed Banach space X, with norm $\| \ \|$, we denote by B_X the closed unit ball of X,

$$B_X = \{x \varepsilon X: \| x \| \leq 1\},$$

and by S_X the closed unit sphere of X,

$$S_X = \{x \varepsilon X: \| x \| = 1\}.$$

Again, for a fixed X, the continuous dual is denoted by X^* and a typical member of X^* might be called x^*.

The Banach spaces c_0, l_p $(1 \leq p \leq \infty)$, $C(\Omega)$ and $L_p(\mu)$ $1 \leq p \leq \infty$ follow standard notations set forth, for example, in Royden's "Real Analysis" or Rudin's "Functional Analysis"; we call on only the most elementary properties of the spaces such as might be encountered in a first course in functional analysis. In general, we assume the reader knows the basics of functional analysis as might be found in either of the aforementioned texts.

Finally, we note that most of the main results carry over trivially from the case of real Banach spaces to that of complex Banach spaces. Therefore, we have concentrated on the former, adding the necessary comments on the latter when it seemed judicious to do so.

CHAPTER I

Riesz's Lemma and Compactness in Banach Spaces

In this chapter we deal with compactness in general normed linear spaces. The aim is to convey the notion that in normed linear spaces, norm-compact sets are small—both algebraically and topologically.

We start by considering the isomorphic structure of n-dimensional normed linear spaces. It is easy to see that all n-dimensional normed linear spaces are isomorphic (this is Theorem 1). After this, a basic lemma of F. Riesz is noted, and (in Theorem 4) we conclude from this that in order for each bounded sequence in the normed linear space X to have a norm convergent subsequence, it is necessary and sufficient that X be finite dimensional. Finally, we shown (in Theorem 5) that any norm-compact subset K of a normed linear space is contained in the closed convex hull of some null sequence.

Theorem 1. *If X and Y are finite-dimensional normed linear spaces of the same dimension, then they are isomorphic.*

PROOF. We show that if X has dimension n, the X is isomorphic to l_1^n. Recall that the norm of an n-tuple $(a_1, a_2, \ldots, a_n)$ in l_1^n is given by

$$\|(a_1, a_2, \ldots, a_n)\| = |a_1| + |a_2| + \cdots + |a_n|.$$

Let $x_1, x_2, \ldots, x_n$ be a Hamel basis for X. Define the linear map $I: l_1^n \to X$ by

$$I((a_1, a_2, \ldots, a_n)) = a_1 x_1 + a_2 x_2 + \cdots + a_n x_n.$$

I is a linear space isomorphism of l_1^n onto X. Moreover, for each $(a_1, a_2, \ldots, a_n)$ in l_1^n,

$$\|a_1 x_1 + a_2 x_2 + \cdots + a_n x_n\| \leq \left(\max_{1 \leq i \leq n} \|x_i\|\right)(|a_1| + |a_2| + \cdots + |a_n|),$$

thanks to the triangle inequality. Therefore, I is a bounded linear operator. (Now if we knew that X is a Banach space, then the open mapping theorem would come immediately to our rescue, letting us conclude that I is an open map and, therefore, an isomorphism—we don't know this though; so we continue). To prove I^{-1} is continuous, we need only show that I is bounded below by some $m > 0$ on the closed unit sphere $S_{l_1^n}$ of l_1^n; an easy normalization argument then shows that I^{-1} is bounded on the closed unit ball of X by $1/m$.

To the above end, we define the function $f: S_{l_1^n} \to \mathbb{R}$ by

$$f((a_1, a_2, \ldots, a_n)) = \|a_1x_1 + a_2x_2 + \cdots + a_nx_n\|.$$

The axioms of a norm quickly show that f is continuous on the *compact* subset $S_{l_1^n}$ of $\mathbb{R}^n$. Therefore, f attains a minimum value $m \geq 0$ at some $(a_1^0, a_2^0, \ldots, a_n^0)$ in $S_{l_1^n}$. Let us assume that $m = 0$. Then

$$\|a_1^0x_1 + a_2^0x_2 + \cdots + a_n^0x_n\| = 0$$

so that $a_1^0x_1 + a_2^0x_2 + \cdots + a_n^0x_n = 0$; since $x_1, \ldots, x_n$ constitute a Hamel basis for X, the only way this can happen is for $a_1^0 = a_2^0 = \cdots = a_n^0 = 0$, a hard task for any $(a_1^0, a_2^0, \ldots, a_n^0) \in S_{l_1^n}$. □

Some quick conclusions follow.

Corollary 2. *Finite-dimensional normed linear spaces are complete.*

In fact, a normed linear space isomorphism is Lipschitz continuous in each direction and so must preserve completeness; by Theorem 1 all n-dimensional spaces are isomorphic to the Banach space l_1^n.

Corollary 3. *If Y is a finite-dimensional linear subspace of the normed linear space X, then Y is a closed subspace of X.*

Our next lemma is widely used in functional analysis and will, in fact, be a point of demarcation for a later section of these notes. It is classical but still pretty. It is often called *Riesz's lemma.*

Lemma. *Let Y be a proper closed linear subspace of the normed linear space X and $0 < \theta < 1$. Then there is an $x_\theta \in S_X$ for which $\|x_\theta - y\| > \theta$ for every $y \in Y$.*

PROOF. Pick any $x \in X \setminus Y$. Since Y is closed, the distance from x to Y is positive, i.e.,

$$0 < d = \inf\{\|x - z\| : z \in Y\} < \frac{d}{\theta};$$

therefore, there is a $z \in Y$ such that

$$\|x - z\| < \frac{d}{\theta}.$$

Let

$$x_\theta = \frac{x - z}{\|x - z\|}.$$

Clearly $x_\theta \in S_X$. Furthermore, if $y \in Y$, then

$$\begin{aligned}\|x_\theta - y\| &= \left\| \frac{x-z}{\|x-z\|} - y \right\| \\ &= \left\| \frac{x}{\|x-z\|} - \frac{z}{\|x-z\|} - \frac{\|x-z\|y}{\|x-z\|} \right\| \\ &= \frac{1}{\|x-z\|} \Big\| x - \underbrace{(z + \|x-z\|y)}_{\text{a member of } Y} \Big\| \\ &> \frac{\theta}{d} d = \theta. \end{aligned}$$

□

An easy consequence of Riesz's lemma is the following theorem.

Theorem 4. *In order for each closed bounded subset of the normed linear space X to be compact, it is necessary and sufficient that X be finite dimensional.*

PROOF. Should the dimension of X be n, then X is isomorphic to l_2^n (Theorem 1); therefore, the compactness of closed bounded subsets of X follows from the classical Heine-Borel theorem.

Should X be infinite dimensional, then S_X is not compact, though it is closed and bounded. In fact, we show that there is a sequence (x_n) in S_X such that for any distinct m and n, $\|x_m - x_n\| \geq \frac{1}{2}$. To start, pick $x_1 \in S_X$. Then the linear span of x_1 is a proper closed linear subspace of X (proper because it is 1 dimensional and closed because of Corollary 3). So by Riesz's lemma there is an x_2 in S_X such that $\|x_2 - \alpha x_1\| \geq \frac{3}{4}$ for all $\alpha \in \mathbb{R}$. The linear span of x_1 and x_2 is a proper closed linear subspace of X (proper because it's 2-dimensional and closed because of Corollary 3). So by Riesz's lemma there is an x_3 in S_X such that $\|x_3 - \alpha x_1 - \beta x_2\| \geq \frac{3}{4}$ for all $\alpha, \beta \in \mathbb{R}$. Continue; the sequence so generated does all that is expected of it. □

A parting comment on the smallness of compact subsets in normed linear spaces follows.

Theorem 5. *If K is a compact subset of the normed linear space X, then there is a sequence (x_n) in X such that* $\lim_n \|x_n\| = 0$ *and K is contained in the closed convex hull of $\{x_n\}$.*

PROOF. K is compact; thus $2K$ is compact. Pick a finite $\frac{1}{4}$ net for $2K$, i.e., pick $x_1, \ldots, x_{n(1)}$ in $2K$ such that each point of $2K$ is within $\frac{1}{4}$ of an x_i, $1 \leq i \leq n(1)$. Denote by $B(x, \varepsilon)$ the set $\{y : \|x - y\| \leq \varepsilon\}$.

Look at the compact chunks of $2K$: $[2K \cap B(x_1, \frac{1}{4})], \ldots, [2K \cap B(x_{n(1)}, \frac{1}{4})]$. Move them to the origin: $[2K \cap B(x_1, \frac{1}{4})] - x_1, \ldots, [2K \cap B(x_{n(1)}, \frac{1}{4})] - x_{n(1)}$. Translation is continuous; so the chunks move to compact sets. Let K_2 be the union of the resultant chunks, i.e.,

$$K_2 = \{[2K \cap B(x_1, \tfrac{1}{4})] - x_1\} \cup \cdots \cup \{[2K \cap B(x_{n(1)}, \tfrac{1}{4})] - x_{n(1)}\}.$$

K_2 is compact, thus $2K_2$ is compact. Pick a finite $\frac{1}{16}$ net for $2K_2$, i.e., pick $x_{n(1)+1}, \ldots, x_{n(2)}$ in $2K_2$ such that each point of $2K_2$ is within $\frac{1}{16}$ of an x_i, $n(1)+1 \leq i \leq n(2)$.

Look at the compact chunks of $2K_2$: $[2K_2 \cap B(x_{n(1)+1}, \frac{1}{16})], \ldots, [2K_2 \cap B(x_{n(2)}, \frac{1}{16})]$. Move them to the origin:

$$[2K_2 \cap B(x_{n(1)+1}, \tfrac{1}{16})] - x_{n(1)+1}, \ldots, [2K_2 \cap B(x_{n(2)}, \tfrac{1}{16})] - x_{n(2)}.$$

Translation is still continuous; so the chunks, once moved, are still compact.

Let K_3 be the union of the replaced chunks:

$$K_3 = \{[2K_2 \cap B(x_{n(1)+1}, \tfrac{1}{16})] - x_{n(1)+1}\} \cup \cdots \cup \{[2K_2 \cap B(x_{n(2)}, \tfrac{1}{16})] - x_{n(2)}\}.$$

K_3 is compact, and we continue in a similar manner.

Observe that if

$$\begin{aligned}
x &\in K, \\
2x &\in 2K, \\
2x - x_{i(1)} &\in K_2, \qquad \text{for some } 1 \leq i(1) \leq n(1);\text{ so,} \\
4x - 2x_{i(1)} &\in 2K_2, \\
4x - 2x_{i(1)} - x_{i(2)} &\in K_3, \qquad \text{for some } n(1)+1 \leq i(2) \leq n(2);\text{ so,} \\
8x - 4x_{i(1)} - 2x_{i(2)} &\in 2K_3, \\
8x - 4x_{i(1)} - 2x_{i(2)} - x_{i(3)} &\in K_4, \qquad \text{for some } n(2)+1 \leq i(3) \leq n(3);\text{ so,}
\end{aligned}$$

etc. Alternatively,

$$\begin{aligned}
x - \frac{x_{i(1)}}{2} &\in \tfrac{1}{2}K_2, \\
x - \frac{x_{i(1)}}{2} - \frac{x_{i(2)}}{4} &\in \tfrac{1}{4}K_3, \\
x - \frac{x_{i(1)}}{2} - \frac{x_{i(2)}}{4} - \frac{x_{i(3)}}{8} &\in \tfrac{1}{8}K_4, \ldots.
\end{aligned}$$

It follows that

$$x = \lim_n \sum_{k=1}^{n} \frac{x_{i(k)}}{2^k}$$

and $x \in \overline{co}(0, x_{i(1)}, x_{i(2)}, \cdots) \subseteq \overline{co}(0, x_1, x_2, \cdots)$. □

Exercises

1. *A theorem of Mazur.* The closed convex hull of a norm-compact subset of a Banach space is norm compact.

2. *Distinguishing between finite-dimensional Banach spaces of the same dimension.* Let n be a positive integer. Denote by l_1^n, l_2^n, and l_∞^n the n-dimensional real

Banach spaces determined by the norms $\| \|_1$, $\| \|_2$, and $\| \|_\infty$, respectively,

$$\|(a_1, a_2, \ldots, a_n)\|_1 = |a_1| + |a_2| + \cdots + |a_n|,$$
$$\|(a_1, a_2, \ldots, a_n)\|_2 = \left(|a_1|^2 + |a_2|^2 + \cdots + |a_n|^2\right)^{1/2},$$
$$\|(a_1, a_2, \ldots, a_n)\|_\infty = \max\{|a_1|, |a_2|, \ldots, |a_n|\}.$$

(i) No pair of the spaces l_1^n, l_2^n, and l_∞^n are mutually isometric.

(ii) If T is a linear isomorphism between l_1^n and l_2^n or between l_∞^n and l_2^n, then the product of the operator norm of T and the operator norm of T^{-1} always exceeds $\sqrt{n}$.

If T is a linear isomorphism between l_1^n and l_∞^n, then $\|T\| \|T^{-1}\| \geq n$.

3. *Limitations in Riesz's lemma.*

(i) Let X be the closed linear subspace of $C[0,1]$ consisting of those $x \in C[0,1]$ that vanish at 0. Let $Y \subseteq X$ be the closed linear subspace of x in X for which $\int_0^1 x(t)\, dt = 0$. Prove that there is no $x \in S_X$ such that distance $(x, Y) \geq 1$.

(ii) If X is a Hilbert space and Y is a proper closed linear subspace of X, then there is an $x \in S_X$ so that distance $(x, S_Y) = \sqrt{2}$.

(iii) If Y is a proper closed linear subspace of l_p $(1 < p < \infty)$, then there is an $x \in S_X$ so that distance $(x, Y) \geq 1$.

4. *Compact operators between Banach spaces.* A linear operator $T: X \to Y$ between the Banach spaces X and Y is called *compact* if TB_X is relatively compact.

(i) Compact linear operators are bounded. Compact isomorphic embeddings and compact quotients (between Banach spaces) have finite-dimensional range.

(ii) The sum of two compact operators is compact, and any product of a compact operator and a bounded operator is compact.

(iii) A subset K of a Banach space X is relatively compact if and only if for every $\varepsilon > 0$ there is a relatively compact set K_ε in X such that

$$K \subseteq \varepsilon B_X + K_\varepsilon.$$

Consequently, the compact operators from X to Y form a closed (linear) subspace of the space of all bounded linear operators.

(iv) Let $T: X \to Y$ be a bounded linear operator, and suppose that for each $\varepsilon > 0$ there is a Banach space X_ε and a compact linear operator $T_\varepsilon: X \to X_\varepsilon$ for which

$$\|Tx\| \leq \|T_\varepsilon x\| + \varepsilon.$$

for all $x \in B_X$. Show that T is itself compact.

(v) Let $T: X \to Y$ be a compact linear operator and suppose $S: Z \to Y$ is a bounded linear operator with $SZ \subseteq TX$. Show that S is a compact operator.

5. *Compact subsets of C(K) spaces for compact metric K.* Let (K, d) be any compact metric space, denote by $C(K)$ the Banach space of continuous scalar-valued

functions on K.

(i) A totally bounded subset $\mathscr{K}$ of $C(K)$ is equicontinuous, i.e., given $\varepsilon > 0$ there is a $\delta > 0$; so $d(k, k') \leq \delta$ implies that $|f(k) - f(k')| \leq \varepsilon$ for all $f \in \mathscr{K}$.

(ii) If $\mathscr{K}$ is a bounded subset of $C(K)$ and D is any countable (dense) subset of K, then each sequence of members of $\mathscr{K}$ has a subsequence converging pointwise on D.

(iii) Any equicontinuous sequence that converges pointwise on the set $S \subseteq K$ converges uniformly on $\bar{S}$.

Recalling that a compact metric space is separable, we conclude to the Ascoli-Arzelà theorem.

Ascoli-Arzelà theorem. *A bounded subset $\mathscr{K}$ of $C(K)$ is relatively compact if and only if $\mathscr{K}$ is equicontinuous.*

6. *Relative compactness in l_p $(1 \leq p < \infty)$.* For any p, $1 \leq p < \infty$, a bounded subset K of l_p is relatively compact if and only if

$$\lim_n \sum_{i=n}^{\infty} |k_i|^p = 0$$

uniformly for $k \in K$.

Notes and Remarks

Theorem 1 was certainly known to Polish analysts in the twenties, though a precise reference seems to be elusive. In any case, A. Tychonoff (of product theorem fame) proved that all finite-dimensional Hausdorff linear topological spaces of the same dimension are linearly homeomorphic.

As we indicate all too briefly in the exercises, the isometric structures of finite-dimensional Banach spaces can be quite different. This is as it should be! In fact, much of the most important current research concerns precise estimates regarding the relative isometric structures of finite-dimensional Banach spaces.

Riesz's lemma was established by F. Riesz (1918); it was he who first noted Theorem 4 as well. As the exercises may well indicate, strengthening Riesz's lemma is a delicate matter. R. C. James (1964) proved that a Banach space X is reflexive if and only if each x^* in X^* achieves its norm on B_X. Using this, one can establish the following: *For a Banach space X to have the property that given a proper closed linear subspace Y of X there exists an x of norm-one such that $d(x, Y) \geq 1$ it is necessary and sufficient that X be reflexive.*

There is another proof of Theorem 4 that deserves mention. It is due to G. Choquet and goes like this: Suppose the Heine-Borel theorem holds in

X; so closed bounded subsets of the Banach space X are compact. Then the closed unit ball B_X is compact. Therefore, there are points $x_1, \ldots, x_n \in B_X$ such that $B_X \subseteq \bigcup_{i=1}^n (x_i + \frac{1}{2} B_X)$. Let Y be the linear span of $\{x_1, x_2, \ldots, x_n\}$; Y is closed. Look at the Banach space X/Y; let $\varphi: X \to X/Y$ be the canonical map. Notice that $\varphi(B_X) \subseteq \varphi(B_X)/2$! Therefore, $\varphi(B_X) = \{0\}$ and X/Y is zero dimensional. $Y = X$.

Theorem 5 is due to A. Grothendieck who used it to prove that every compact linear operator between two Banach spaces factors through a subspace of c_0; look at the exercises following Chapter II. Grothendieck used this factorization result in his investigations into the approximation property for Banach spaces.

An Afterthought to Riesz's Theorem

(This could have been done by Banach!)

Thanks to Cliff Kottman a substantial improvement of the Riesz lemma can be stated and proved. In fact, *if X is an infinite-dimensional normed linear space, then there exists a sequence (x_n) of norm-one elements of X for which $\|x_m - x_n\| > 1$ whenever $m \neq n$.*

Kottman's original argument depends on combinatorial features that live today in any improvements of the cited result. In Chapter XIV we shall see how this is so; for now, we give a noncombinatorial proof of Kottman's result. We were shown this proof by Bob Huff who blames Tom Starbird for its simplicity. Only the Hahn-Banach theorem is needed.

We proceed by induction. Choose $x_1 \in X$ with $\|x_1\| = 1$ and take $x_1^* \in X^*$ such that $\|x_1^*\| = 1 = x_1^* x_1$.

Suppose $x_1^*, \ldots, x_k^*$ (linearly independent, norm-one elements of X^*) and $x_1, \ldots, x_k$ (norm-one elements) have been chosen. Choose $y \in X$ so that $x_1^* y, \ldots, x_k^* y < 0$ and take any nonzero vector x common to $\cap_{i=1}^k \ker x_i^*$. Choose K so that

$$\|y\| < \|y + Kx\|.$$

Then for any nontrivial linear combination $\sum_{i=1}^k \alpha_i x_i^*$ of the x_i^* we know that

$$\left|\sum_{i=1}^k \alpha_i x_i^*(y + Kx)\right| = \left|\sum_{i=1}^k \alpha_i x_i^*(y)\right|$$
$$\leq \left\|\sum_{i=1}^k \alpha_i x_i^*\right\| \|y\| < \left\|\sum_{i=1}^k \alpha_i x_i^*\right\| \|y + Kx\|.$$

Let $x_{k+1} = (y + Kx)\|y + Kx\|^{-1}$ and choose x_{k+1}^* to be a norm-one functional satisfying $x_{k+1}^* x_{k+1} = 1$. Since $|\sum_{i=1}^k \alpha_i x_i^*(y + Kx)| < \|\sum_{i=1}^k \alpha_i x_i^*\| \, \|y +$

$kx\|$, x_{k+1}^* is not a linear combination of $x_1^*, \ldots, x_k^*$. Also, if $1 \le i \le k$, then

$$\begin{aligned}\|x_{k+1} - x_i\| &\ge |x_i^*(x_{k+1} - x_i)| \\ &= |x_i^* x_{k+1} - x_i^* x_i| > 1\end{aligned}$$

since $x_i^* x_i = 1$ and $x_i^* x_{k+1} < 0$.

This proof is complete.

Bibliography

Choquet, G. 1969. *Lectures on Analysis, Vol. I: Integration and Topological Vector Spaces*, J. Marsden, T. Lance, and S. Gelbart (eds.). New York–Amsterdam: W. A. Benjamin.

James, R. C. 1964. Weakly compact sets. *Trans. Amer. Math. Soc.*, **113**, 129–140.

Kottman, C. A. 1975. Subsets of the unit ball that are separated by more than one. *Studia Math.*, **53**, 15–27.

Grothendieck, A. 1955. *Produits tensorials topologiques et espaces nucléaires. Memoirs Amer. Math. Soc.*, **16**.

Riesz, F. 1918. Über lineare Funktionalgleichungen. *Acta Math.*, **41**, 71–98.

Tychonoff, A. 1935. Ein Fixpunktsatz. *Math. Ann.* **111**, 767–776.

CHAPTER II

The Weak and Weak* Topologies: An Introduction

As we saw in our brief study of compactness in normed linear spaces, the norm topology is too strong to allow any widely applicable subsequential extraction principles. Indeed, in order that each bounded sequence in X have a norm convergent subsequence, it is necessary and sufficient that X be finite dimensional. This fact leads us to consider other, weaker topologies on normed linear spaces which are related to the linear structure of the spaces and to search for subsequential extraction principles therein. As so often happens in such ventures, the roles of these topologies are not restricted to the situations initially responsible for their introduction. Rather, they play center court in many aspects of Banach space theory.

The two weaker-than-norm topologies of greatest importance in Banach space theory are the weak topology and the weak-star (or weak*) topology. The first (the weak topology) is present in every normed linear space, and in order to get any results regarding the existence of convergent or even Cauchy subsequences of an arbitrary bounded sequence in this topology, one must assume additional structural properties of the Banach space. The second (the weak* topology) is present only in dual spaces; this is not a real defect since it is counterbalanced by the fact that the dual unit ball will always be weak* compact. *Beware*: This compactness need not of itself ensure good *subsequential* extraction principles, but it does get one's foot in the door.

The Weak Topology

Let X be a normed linear space. We describe the weak topology of X by indicating how a net in X converges weakly to a member of X. Take the net (x_d); we say that (x_d) converges weakly to x_0 if for each $x^* \in X^*$,

$$x^*x_0 = \lim_d x^*x_d.$$

Whatever the weak topology may be, it is linear (addition and scalar multiplication are continuous) and Hausdorff (weak limits are unique).

Alternatively, we can describe a basis for the weak topology. Since the weak topology is patently linear, we need only specify the neighborhoods of 0; translation will carry these neighborhoods throughout X. A typical basic neighborhood of 0 is generated by an $\varepsilon > 0$ and finitely many members $x_1^*, \ldots, x_n^*$ of X^*. Its form is

$$W(0; x_1^*, \ldots, x_n^*, \varepsilon) = \{x \in X : |x_1^* x|, \ldots, |x_n^* x| < \varepsilon\}.$$

Weak neighborhoods of 0 can be quite large. In fact, each basic neighborhood $W(0; x_1^*, \ldots, x_n^*, \varepsilon)$ of 0 contains the intersection $\cap_{i=1}^n \ker x_i^*$ of the null spaces $\ker x_i^*$ of the x_i^*, a linear subspace of finite codimension. In case X is infinite dimensional, weak neighborhoods of 0 are big!

Though the weak topology is smaller than the norm topology, it produces the same continuous linear functionals. In fact, if f is a weakly continuous linear functional on the normed linear space X, then $U = \{x : |f(x)| < 1\}$ is a weak neighborhood of 0. As such, U contains a $W(0; x_1^*, \ldots, x_n^*, \varepsilon)$. Since f is linear and $W(0, x_1^*, \ldots, x_n^*, \varepsilon)$ contains the linear space $\cap_{i=1}^n \ker x_i^*$, it follows that $\ker f$ contains $\cap_{i=1}^n \ker x_i^*$ as well. But here's the catch: if the kernel of f contains $\cap_{i=1}^n \ker x_i^*$, then f must be a linear combination $x_1^*, \ldots, x_n^*$, and so $f \in X^*$. This follows from the following fact from linear algebra.

Lemma. *Let E be a linear space and $f, g_1, \ldots, g_n$ be linear functionals on E such that* $\ker f \supseteq \cap_{i=1}^n \ker g_i$. *Then f is a linear combination of the g_i's.*

PROOF. Proceed by induction on n. For $n = 1$ the lemma clearly holds. Let us assume it has been established for $k \leq n$. Then, for given $\ker f \supseteq \cap_{i=1}^{n+1} \ker g_i$, the inductive hypothesis applies to

$$f|_{\ker g_{n+1}}, g_1|_{\ker g_{n+1}}, \ldots, g_n|_{\ker g_{n+1}}.$$

It follows that, on $\ker g_{n+1}$, f is a linear combination $\Sigma_{i=1}^n a_i g_i$ of $g_1, \ldots, g_n$; $f - \Sigma_{i=1}^n a_i g_i$ vanishes on $\ker g_{n+1}$. Now apply what we know about the case $n = 1$ to conclude that $f - \Sigma_{i=1}^n a_i g_i$ is a scalar multiple of g_{n+1}. □

It is important to realize that the weak topology is really of quite a different character than is the norm topology (at least in the case of *infinite-dimensional* normed spaces). For example, *if the weak topology of a normed linear space X is metrizable, then X is finite dimensional.* Why is this so? Well, metrizable topologies satisfy the first axiom of countability. So if the weak topology of X is metrizable, there exists a sequence (x_n^*) in X^* such that given any weak neighborhood U of 0, we can find a rational $\varepsilon > 0$ and an $n(U)$ such that U contains $W(0; x_1^*, \ldots, x_{n(U)}^*, \varepsilon)$. Each $x^* \in X^*$ generates the weak neighborhood $W(0; x^*, 1)$ of 0 which in turn contains one of the sets $W(0; x_1^*, \ldots, x_{n(W(0;x^*,1))}^*, \varepsilon)$. However, we have seen that this entails x^* being a linear combination of $x_1^*, \ldots, x_{n(W)}^*$. If we let F_m be the linear span of $x_1^*, \ldots, x_m^*$, then each F_m is a finite-dimensional linear subspace of X^* which is a fortiori closed; moreover, we have just seen that $X^* = \cup_m F_m$. The Baire category theorem now alerts us to the fact that one of the F_m has nonempty interior, a fact which tells us that the F_m has to be all of X^*. X^* (and hence X) must be finite dimensional.

It can also be shown that *in case X is an infinite-dimensional normed linear space, then the weak topology of X is not complete.* Despite its contrary nature, the weak topology provides a useful vehicle for carrying on analysis in infinite-dimensional spaces.

Theorem 1. *If K is a convex subset of the normed linear space X, then the closure of K in the norm topology coincides with the weak closure of K.*

PROOF. There are no more open sets in the weak topology than there are in the norm topology; consequently, the norm closure is harder to get into than the weak closure. In other words $\overline{A}^{\|\cdot\|} \subseteq \overline{A}^{\text{weak}}$.

If K is a *convex* set and if there were a point $x_0 \in \overline{K}^{\text{weak}} \setminus \overline{K}^{\|\cdot\|}$, then there would be an $x_0^* \in X^*$ such that

$$\sup x_0^* \overline{K}^{\|\cdot\|} \leq \alpha < \beta \leq x_0^*(x_0)$$

for some α, β. This follows from the separation theorem and the convexity of $\overline{K}^{\|\cdot\|}$. However, $x_0 \in \overline{K}^{\text{weak}}$ implies there is a net (x_d) in K such that

$$x_0 = \text{weak} \lim_d x_d.$$

It follows that

$$x_0^* x_0 = \lim_d x_0^* x_d,$$

an obvious contradiction to the fact that $x_0^* x_0$ is separated from all the $x_0^* x_d$ by the gulf between α and β. □

A few consequences follow.

Corollary 2. *If (x_n) is a sequence in the normed linear space for which weak* $\lim_n x_n = 0$, *then there is a sequence (σ_n) of convex combinations of the x_n such that* $\lim_n \|x_n\| = 0$.

A natural hope in light of Corollary 2 would be that given a weakly null sequence (x_n) in the normed linear space X, one might be able (through very judicious pruning) to extract a subsequence (y_n) of (x_n) whose arithmetic means $n^{-1}\Sigma_{k=1}^n y_k$ tend to zero in norm. Sometimes this is possible and sometimes it is not; discussions of this phenomenon will appear throughout this text.

Corollary 3. *If Y is a linear subspace of the normed linear space X, then* $\overline{Y}^{\text{weak}} = \overline{Y}^{\|\cdot\|}$.

Corollary 4. *If K is a convex set in the normed linear space X, then K is norm closed if and only if K is weakly closed.*

The weak topology is defined in a projective manner: it is the weakest topology on X that makes each member of X^* continuous. As a consequence of this and the usual generalities about projective topologies, *if Ω is a topological space and $f: \Omega \to X$ is a function, then f is weakly continuous if and only if x^*f is continuous for each $x^* \in X^*$.*

Let $T: X \to Y$ be a linear map between the normed linear spaces X and Y. Then T is weak-to-weak continuous if and only if for each $y^* \in Y^*$, y^*T is a weakly continuous linear functional on X; this, in turn, occurs if and only if y^*T is a norm continuous linear functional on X for each $y^* \in Y^*$.

Now if $T: X \to Y$ is a norm-to-norm continuous linear map, it obviously satisfies the last condition enunciated in the preceding paragraph. On the other hand, if T is not norm-to-norm continuous, then TB_X is not a bounded subset of Y. Therefore, the Banach-Steinhaus theorem directs us to a $y^* \in Y^*$ such that y^*TB_X is not bounded; y^*T is not a bounded linear functional. Summarizing we get the following theorem.

Theorem 5. *A linear map $T: X \to Y$ between the normed linear spaces X and Y is norm-to-norm continuous if and only if T is weak-to-weak continuous.*

The Weak* Topology

Let X be a normed linear space. We describe the weak* topology of X^* by indicating how a net (x_d^*) in X^* converges weak* to a member x_0^* of X^*. We say that (x_d^*) *converges weak* to* $x_0^* \in X^*$ if for each $x \in X$,

$$x_0^*x = \lim_d x_d^*x.$$

As with the weak topology, we can give a description of a typical basic weak* neighborhood of 0 in X^*; this time such a neighborhood is generated by an $\varepsilon > 0$ and a finite collection of elements in X, say $x_1, \ldots, x_n$. The form is

$$W^*(0; x_1, \ldots, x_n, \varepsilon) = \left\{ x^* \in X^* : |x^*x_1|, \ldots, |x^*x_n| \leq \varepsilon \right\}.$$

The weak* topology is a linear topology; so it is enough to describe the neighborhoods of 0, and neighborhoods of other points in X^* can be obtained by translation. Notice that weak* basic neighborhoods of 0 are also weak neighborhoods of 0; in fact, they are just the basic neighborhoods generated by those members of X^{**} that are actually in X. Of course, any x^{**} that are left over in X^{**} after taking away X give weak neighborhoods of 0 in X^* that are not weak* neighborhoods. A conclusion to be drawn is this: the weak* topology is no bigger than the weak topology. Like the weak topology, excepting finite-dimensional spaces, duals are never weak* metrizable or weak* complete; also, proceeding as we did with the weak topology, it's easy to show that the weak* dual of X^* is X. An important consequence of this is the following theorem.

Goldstine's Theorem. *For any normed linear space X, B_X is weak* dense in $B_{X^{**}}$, and so X is weak* dense in X^{**}.*

Proof. The second assertion follows easily from the first; so we concentrate our attentions on proving B_X is always weak* dense in $B_{X^{**}}$. Let $x^{**} \in X^{**}$ be any point not in $\overline{B_X}^{\text{weak}^*}$. Since $\overline{B_X}^{\text{weak}^*}$ is a weak* closed convex set and $x^{**} \notin \overline{B_X}^{\text{weak}^*}$, there is an $x^* \in X^{**}$'s weak* dual X^* such that

$$\sup\left\{ x^*y^{**} : y^{**} \in \overline{B_X}^{\text{weak}^*} \right\} < x^{**}x^*.$$

Of course we can assume $\|x^*\| = 1$; but now the quantity on the left is at least $\|x^*\| = 1$, and so $\|x^{**}\| > 1$. It follows that every member of $B_{X^{**}}$ falls inside $\overline{B_X}^{\text{weak}^*}$. □

As important and useful a fact as Goldstine's theorem is, the most important feature of the weak* topology is contained in the following compactness result.

Alaoglu's Theorem. *For any normed linear space X, B_{X^*} is weak* compact. Consequently, weak* closed bounded subsets of X^* are weak* compact.*

Proof. If $x^* \in B_{X^*}$, then for each $x \in B_X$, $|x^*x| \le 1$. Consequently, each $x^* \in B_{X^*}$ maps B_X into the set D of scalars of modulus ≤ 1. We can therefore identify each member of B_{X^*} with a point in the product space D^{B_X}. Tychonoff's theorem tells us this latter space is compact. On the other hand, the weak* topology is defined to be that of pointwise convergence on B_X, and so this identification of B_{X^*} with a subset of D^{B_X} leaves the weak* topology unscathed; it need only be established that B_{X^*} is closed in D^{B_X} to complete the proof.

Let (x_d^*) be a net in B_{X^*} converging pointwise on B_X to $f \in D^{B_X}$. Then it is easy to see that f is "linear" on B_X: in fact, if $x_1, x_2 \in B_X$ and a_1, a_2 are scalars such that $a_1x_1 + a_2x_2 \in B_X$, then

$$\begin{aligned} f(a_1x_1 + a_2x_2) &= \lim_d x_d^*(a_1x_1 + a_2x_2) \\ &= \lim_d a_1x_d^*(x_1) + a_2x_d^*(x_2) \\ &= \lim_d a_1x_d^*(x_1) + \lim_d a_2x_d^*(x_2) \\ &= a_1f(x_1) + a_2f(x_2). \end{aligned}$$

It follows that f is indeed the restriction to B_X of a linear functional x' on X; moreover, since $f(x)$ has modulus ≤ 1 for $x \in B_X$, this x' is even in B_{X^*}. This completes the proof. □

A few further remarks on the weak* topology are in order.

First, it is a locally convex Hausdorff linear topology, and so the separation theorem applies. In this case it allows us to separate points (even weak* compact convex sets) from weak* closed convex sets by means of the weak* continuous linear functionals on X^*, i.e., members of X.

Second, though it is easy to see that the weak* and weak topologies are not the same (unless $X = X^{**}$), it is conceivable that weak* convergent *sequences* are weakly convergent. Sometimes this does occur, and we will, in fact, run across cases of this in the future. Because the phenomenon of weak* convergent sequences being weakly convergent automatically brings one in contact with checking pointwise convergence on $B_{X^{**}}$, it is not too surprising that this phenomenon is still something of a mystery.

Exercises

1. *The weak topology need not be sequential.* Let $A \subseteq l_2$ be the set $\{e_m + me_n : 1 \leq m < n < \infty\}$. Then $0 \in \overline{A}^{\text{weak}}$, yet no sequence in A is weakly null.

2. *Helly's theorem.*

   (i) Given $x_1^*, \ldots, x_n^* \in X^*$, scalars $\alpha_1, \ldots, \alpha_n$, and $\varepsilon > 0$, there exists an $x_\varepsilon \in X$ for which $\|x\| \leq \gamma + \varepsilon$ and such that $x_1^* x = \alpha_1, \ldots, x_n^* x = \alpha_n$ if and only if for any scalars $\beta_1, \ldots, \beta_n$

   $$\left| \sum_{i=1}^{n} \beta_1 \alpha_1 \right| \leq \gamma \left\| \sum_{i=1}^{n} \alpha_i x_i^* \right\|.$$

   (ii) Let $x^{**} \in X^{**}$, $\varepsilon > 0$ and $x_1^*, \ldots, x_n^* \in X^*$. Then there exists $x \in X$ such that $\|x\| \leq \|x^{**}\| + \varepsilon$ and $x_1^*(x) = x^{**}(x_1^*), \ldots, x_n^*(x) = x^{**}(x_n^*)$.

3. *An infinite-dimensional normed linear space is never weakly complete.*

   (i) A normed linear space X is finite dimensional if and only if every linear functional on X is continuous.

   (ii) An infinite-dimensional normed linear space is never weakly complete. *Hint:* Apply (i) to get a discontinuous linear functional φ on X^*; then using (i), the Hahn-Banach theorem, and Helly's theorem, build a weakly Cauchy net in X indexed by the finite-dimensional subspaces of X^* with φ the only possible weak limit point.

4. *Schauder's theorem.*

   (i) If $T: X \to Y$ is a bounded linear operator between the Banach spaces X and Y, then for any $y^* \in Y^*, y^*T \in X^*$, the operator $T^*: Y^* \to X^*$ that takes a $y^* \in Y^*$ to $y^*T \in X^*$ is a bounded linear operator, called T^*, for which $\|T\| = \|T^*\|$.

(ii) A bounded linear operator $T: X \to Y$ between Banach spaces is compact if and only if its adjoint $T^*: Y^* \to X^*$ is.

(iii) An operator $T: X \to Y$ whose adjoint is weak*-norm continuous is compact. However, not every compact operator has a weak*-norm continuous adjoint.

(iv) An operator $T: X \to Y$ is compact if and only if its adjoint is weak*-norm continuous on weak* compact subsets of Y^*.

5. *Dual spaces.* Let X be a Banach space and $E \subseteq X^*$. Suppose E separates the points of X and B_X is compact in the topology of pointwise convergence on E. Then X is a dual space whose predual is the closed linear span of E in X^*.

6. *Factoring compact operators through subspaces of c_0.*

(i) A subset $\mathscr{K}$ of c_0 is relatively compact if and only if there is an $x \in c_0$ such that

$$|k_n| \le |x_n|$$

holds for all $k \in \mathscr{K}$ and all $n \ge 1$.

(ii) A bounded linear operator $T: X \to Y$ between two Banach spaces is compact if and only if there is a norm-null sequence (x_n^*) in X^* for which

$$\|Tx\| \le \sup_n |x_n^* x|$$

for all x. Consequently, T is compact if and only if there is a $\lambda \in c_0$ and a bounded sequence (y_n^*) in X^* such that

$$\|Tx\| \le \sup_n |\lambda_n|^2 |y_n^* x|$$

for all x.

(iii) Every compact linear operator between Banach spaces factors compactly through some subspace of c_0; that is, if $T: X \to Y$ is a compact linear operator between the Banach spaces X and Y, then there if a closed linear subspace Z of c_0 and compact linear operators $A: X \to Z$ and $B: Z \to Y$ such that $T = BA$.

Notes and Remarks

The notion of a weakly convergent sequence in $L_2[0,1]$ was used by Hilbert and, in $L_p[0,1]$, by F. Riesz, but the first one to recognize that the weak topology was just that, a topology, was von Neumann. Exercise 1 is due to von Neumann and clearly indicates the highly nonmetrizable character of the weak topology in an infinite-dimensional Banach space. The nonmetrizability of the weak topology of an infinite-dimensional normed space was discussed by Wehausen.

Theorem 1 and the consequences drawn from it here (Corollaries 2 to 4) are due to Mazur (1933). Earlier, Zalcwasser (1930) and, independently, Gillespie and Hurwitz (1930) had proved that any weakly null sequence in

$C[0,1]$ admits of a sequence of convex combinations that converge uniformly to zero. The fact that weakly closed linear subspaces of a normed linear space are norm closed appears already in Banach's "Operationes Lineaires."

The weak continuity of a bounded linear operator was first noticed by Banach in his masterpiece; the converse of Theorem 5 was proved by Dunford. Generalizations to locally convex spaces were uncovered by Dieudonné and can be found in most texts on topological vector spaces.

As one ought to suspect, Goldstine's theorem and Alaoglu's theorem are named after their discoverers. Our proof of Goldstine's theorem is far from the original, being closer in spirit to proofs due to Dieudonné and Kakutani; for a discussion of Goldstine's original proof, as well as an application of its main theme, the reader is advised to look to the Notes and Remarks section of Chapter IX. Helly's theorem (Exercise 2) is closely related to Goldstine's and often can be used in its place. In the form presented here, Helly's theorem is due to Banach; of course, like the Hahn-Banach theorem, Helly's theorem is a descendant of Helly's selection principle.

The fact that infinite-dimensional Banach spaces are never weakly complete seems to be due to Kaplan; our exercise was suggested to us by W. J. Davis.

Alaoglu's theorem was discovered by Banach in the case of a separable Banach space; many refer to the result as the Banach-Alaoglu theorem. Alaoglu (1940) proved the version contained here for the expressed purpose of differentiating certain vector-valued measures.

Bibliography

Alaoglu, L. 1940. Weak topologies of normed linear spaces. *Ann. Math.*, **41** (2), 252–267.

Gillespie, D. C. and Hurwitz, W. A. 1930. On sequences of continuous functions having continuous limits. *Trans. Amer. Math. Soc.*, **32**, 527–543.

Kaplan, S. 1952. Cartesian products of reals. *Amer. J. Math.*, **74**, 936–954.

Mazur, S. 1933. Über konvexe Mengen in linearen normierte Räumen. *Studia Math.*, **4**, 70–84.

von Neumann, J. 1929–1930. Zur Algebra der Funktionaloperationen und Theorie der Normalen Operatoren. *Math. Ann.*, **102**, 370–427.

Wehausen, J. 1938. Transformations in linear topological spaces. *Duke Math. J.*, **4**, 157–169.

Zalcwasser, Z. 1930. Sur une propriété der champ des fonctions continues. *Studia Math.*, **2**, 63–67.

CHAPTER III

The Eberlein-Šmulian Theorem

We saw in the previous chapter that regardless of the normed linear space X, weak* closed, bounded sets in X^* are weak* compact. How does a subset K of a Banach space X get to be weakly compact? The two are related. Before investigating their relationship, we look at a couple of necessary ingredients for weak compactness and take a close look at two illustrative nonweakly compact sets.

Let K be a weakly compact set in the normed linear space X. If $x^* \in X^*$, then x^* is weakly continuous; therefore, x^*K is a compact set of scalars. It follows that x^*K is bounded for each $x^* \in X^*$, and so K is bounded. Further, K is weakly compact, hence weakly closed, and so norm closed. Conclusion: *Weakly compact sets are norm closed and norm bounded.*

Fortunately, closed bounded sets need not be weakly compact.

Consider B_{c_0}. Were B_{c_0} weakly compact, each sequence in B_{c_0} would have a weak cluster point in B_{c_0}. Consider the sequence σ_n defined by $\sigma_n = e_1 + \cdots + e_n$, where e_k is the kth unit vector in c_0. The sup norm of c_0 is rigged so that $\|\sigma_n\| = 1$ for all n. What are the possible weak cluster points of the sequence (σ_n)? Take a $\lambda \in B_{c_0}$ that is a weak cluster point of (σ_n). For each $x^* \in c_0^*$, $(x^*\sigma_n)$ has $x^*\lambda$ for a cluster point; i.e., the values of $x^*\sigma_n$ get as close as you please to $x^*\lambda$ infinitely often. Now evaluation of a sequence in c_0 at its kth coordinate is a continuous linear functional; call it e_k^*. Note that $e_k^*(\sigma_n) = 1$ for all $n \geq k$. Therefore, $e_k^*\lambda = 1$. This holds true for all k. Hence, $\lambda = (1, 1, \ldots, 1, \ldots) \notin c_0$. *$B_{c_0}$ is not weakly compact.*

Another example: *B_{l_1} is not weakly compact.* Since $l_1 = c_0^*$ (isometrically), were B_{l_1} weakly compact, the weak and weak* topologies on B_{l_1} would have to coincide (comparable compact Hausdorff topologies coincide). However, consider the sequence (e_n) of unit vectors in l_1. If $\lambda \in c_0$, then $e_n\lambda = \lambda_n \to 0$ as $n \to \infty$. So (e_n) is weak* null. If we suppose B_{l_1} weakly compact, then (e_n) is weakly null, but then there ought to be a sequence (γ_n) of convex combinations of the e_n such that $\|\gamma_n\|_1 \to 0$. Here's the catch: Take a convex combination of e_n's—the resulting vector's l_1 norm is 1. The supposition that B_{l_1} is weakly compact is erroneous.

There is, of course, a common thread running through both of the above examples. In the first, the natural weak cluster point fails to be in c_0; not all is lost though, because it is in B_{l_∞}. Were $B_{c_0} = B_{l_\infty}$, this would have been enough to ensure B_{c_0}'s weak compactness. In the second case, the weak compactness of B_{l_1} was denied because of the fact that the weak* and weak topologies on B_{l_1} were not the same; in other words, there were more x^{**}'s than there were x's to check against for convergence. Briefly, B_{c_0} is smaller than B_{l_∞}.

Suppose $B_X = B_{X^{**}}$. Naturally, this occurs when and only when $X = X^{**}$; such X are called *reflexive*. Then the natural embedding of X into X^{**} is a weak-to-weak* homeomorphism of X onto X^{**} that carries B_X exactly onto $B_{X^{**}}$. It follows that B_X is weakly compact.

On the other hand, should B_X be weakly compact, then any $x^{**} \in X^{**}$ not in B_X can be separated from the weak* compact convex set B_X by an element of the weak* dual of X^{**}; i.e., there is an $x^* \in B_{X^*}$ such that

$$\sup_{\|x\| \leq 1} x^*x(= \|x^*\| = 1) < x^{**}x^*.$$

It follows that $\|x^{**}\| > 1$ and so $B_X = B_{X^{**}}$.

Summarizing: *B_X is weakly compact if and only if X is reflexive.*

Let's carry the above approach one step further. Take a bounded set A in the Banach space X. Suppose we want to show that A is relatively weakly compact. If we take $\overline{A}^{\text{weak}}$ and the resulting set is weakly compact, then we are done. How do we find $\overline{A}^{\text{weak}}$ though? Well, we have a helping hand in Alaoglu's theorem: start with A, look at $\overline{A}^{\text{weak}^*}$ up in X^{**}, and see what elements of X^{**} find themselves in $\overline{A}^{\text{weak}^*}$. We know that $\overline{A}^{\text{weak}^*}$ is weak* compact. Should each element in $\overline{A}^{\text{weak}^*}$ actually be in X, then $\overline{A}^{\text{weak}^*}$ is just $\overline{A}^{\text{weak}}$; what's more, the weak* and weak topologies are the same, and so $\overline{A}^{\text{weak}}$ is weakly compact.

So, to show a bounded set A is relatively weakly compact, the strategy is to look at $\overline{A}^{\text{weak}^*}$ and see that each of its members is a point of X. We employ this strategy in the proof of the main result of this chapter.

Theorem (Eberlein-Šmulian). *A subset of a Banach space is relatively weakly compact if and only if it is relatively weakly sequentially compact.*

In particular, a subset of a Banach space is weakly compact if and only if it is weakly sequentially compact.

PROOF. To start, we will show that a relatively weakly compact subset of a Banach space is relatively weakly sequentially compact. This will be accomplished in two easy steps.

Step 1. If K is a (relatively) weakly compact set in a Banach space X and X^* contains a countable total set, then $\overline{K}^{\text{weak}}$ is metrizable. Recall that a set $F \subseteq X^*$ is called *total* if $f(x) = 0$ for each $f \in F$ implies $x = 0$.

Suppose that K is weakly compact and $\{x_n^*\}$ is a countable total subset of nonzero members of X^*. The function $d: X \times X \to R$ defined by

$$d(x, x') = \sum_n |x_n^*(x - x')| \|x_n^*\|^{-1} 2^{-n}$$

is a metric on X. The formal identity map is weakly-to-d continuous on the bounded set K. Since a continuous one-to-one map from a compact space to a Hausdorff space is a homeomorphism, we conclude that d restricted to $K \times K$ is a metric that generates the weak topology of K.

Step 2. Suppose A is a relatively weakly compact subset of the Banach space X and let (a_n) be a sequence of members of A. Look at the closed linear span $[a_n]$ of the a_n; $[a_n]$ is weakly closed in X. Therefore, $A \cap [a_n]$ is relatively weakly compact in the separable Banach space $[a_n]$. Now the dual of a *separable* Banach space contains a countable total set: if $\{d_n\}$ is a countable dense set in the unit sphere of the separable space and $\{d_n^*\}$ is chosen in the dual to satisfy $d_n^* d_n = 1$, it is easy to verify that $\{d_n^*\}$ is total. From our first step we know that $\overline{A \cap [a_n]}^{\text{weak}}$ is metrizable in the weak topology of $[a_n]$. Since compactness and sequential compactness are equivalent in metric spaces, $\overline{A \cap [a_n]}^{\text{weak}}$ is a weakly sequentially compact subset of $[a_n]$. In particular, if a is any weak limit point of (a_n), then there is a subsequence (a_n') of (a_n) that converges weakly to a in $[a_n]$. It is plain that (a_n') also converges weakly to a in X.

We now turn to the converse. We start with an observation: if E is a finite-dimensional subspace of X^{**}, then there is a finite set E' of S_{X^*} such that for any x^{**} in E

$$\frac{\|x^{**}\|}{2} \le \max\{|x^{**}x^*| : x^* \in E'\}.$$

In fact, S_E is norm compact. Therefore, there is a finite $\frac{1}{4}$ net $F = \{x_1^{**}, \dots, x_n^{**}\}$ for S_E. Pick $x_1^*, \dots, x_n^* \in S_{X^*}$ so that

$$x_k^{**} x_k^* > \tfrac{3}{4}.$$

Then whenever $x^{**} \in S_E$, we have

$$\begin{aligned} x^{**}x_k^* &= x_k^{**}x_k^* + \left(x^{**}x_k^* - x_k^{**}x_k^*\right) \\ &\ge \tfrac{3}{4} - \tfrac{1}{4} = \tfrac{1}{2} \end{aligned}$$

for a suitable choice of k.

This observation is the basis of our proof.

Let A be a relatively weakly sequentially compact subset of X; each infinite subset of A has a weak cluster point in X since A is also relatively weakly countably compact. Consider $\overline{A}^{\text{weak}^*}$. $\overline{A}^{\text{weak}^*}$ is weak* compact since A, and therefore $\overline{A}^{\text{weak}^*}$, is bounded due to the relative weak sequential compactness of A. We use the strategy espoused at the start of this section to show A is relatively weakly compact; that is, we show $\overline{A}^{\text{weak}^*}$ actually lies in X.

Take $x^{**} \in \bar{A}^{\text{weak}^*}$, and let $x_1^* \in S_{X^*}$. Since $x^{**} \in \bar{A}^{\text{weak}^*}$ each weak* neighborhood of x^{**} contains a member of A. In particular, the weak* neighborhood generated by $\varepsilon = 1$ and x_1^*, $\{y^{**} \in X^{**} : |(y^{**} - x^{**})(x_1^*)| < 1\}$, contains a member a_1 of A. From this we get

$$|(x^{**} - a_1)(x_1^*)| < 1.$$

Consider the linear span $[x^{**}, x^{**} - a_1]$ of x^{**} and $x^{**} - a_1$; this is a finite-dimensional subspace of X^{**}. Our observation deals us $x_2^*, \ldots, x_{n(2)}^* \in S_{X^*}$ such that for any y^{**} in $[x^{**}, x^{**} - a_1]$,

$$\frac{\|y^{**}\|}{2} \leq \max\{|y^{**}(x_k^*)| : 1 \leq k \leq n(2)\}.$$

x^{**} is not going anywhere, i.e., it is still in $\bar{A}^{\text{weak}^*}$; so each weak* neighborhood of x^{**} intersects A. In particular, the weak* neighborhood about x^{**} generated by $\frac{1}{2}$ and $x_1^*, x_2^*, \ldots, x_{n(2)}^*$ intersects A to give us an a_2 in A such that

$$|(x^{**} - a_2)(x_1^*)|, |(x^{**} - a_2)(x_2^*)|, \ldots, |(x^{**} - a_2)(x_{n(2)}^*)| < \tfrac{1}{2}.$$

Now look at the linear span $[x^{**}, x^{**} - a_1, x^{**} - a_2]$ of $x^{**}, x^{**} - a_1$, and $x^{**} - a_2$. As a finite-dimensional subspace, $[x^{**}, x^{**} - a_1, x^{**} - a_2]$ provides us with $x_{n(2)+1}^*, \ldots, x_{n(3)}^*$ in S_{X^*} such that

$$\frac{\|y^{**}\|}{2} \leq \max\{|y^{**}(x_k^*)| : 1 \leq k \leq n(3)\}$$

for any $y^{**} \in [x^{**}, x^{**} - a_1, x^{**} - a_2]$.

Once more, quickly. Choose a_3 in A such that $x^{**} - a_3$ charges against $x_1^*, \ldots, x_{n(3)}^*$ for no more than $\frac{1}{3}$ value. Observe that the finite-dimensional linear space $[x^{**}, x^{**} - a_1, x^{**} - a_2, x^{**} - a_3]$ provides us with a finite subset $x_{n(3)+1}^*, \ldots, x_{n(4)}^*$ in S_{X^*} such that

$$\frac{\|y^{**}\|}{2} \leq \max\{|y^{**}(x_k^*)| : 1 \leq k \leq n(4)\}$$

for any $y^{**} \in [x^{**}, x^{**} - a_1, x^{**} - a_2, x^{**} - a_3]$.

Where does all this lead us? Our hypothesis on A (being relatively weakly sequentially compact) allows us to find an $x \in X$ that is a weak cluster point of the constructed sequence $(a_n) \subseteq A$. Since the closed linear span $[a_n]$ of the a_n is weakly closed, $x \in [a_n]$. It follows that $x^{**} - x$ is in the weak* closed linear span of $\{x^{**}, x^{**} - a_1, x^{**} - a_2, \ldots\}$. Our construction of the x_i^* and the a_i assures us that

$$\frac{\|y^{**}\|}{2} \leq \sup_m |y^{**} x_m^*| \tag{1}$$

holds for any y^{**} in the linear span of $x^{**}, x^{**} - a_1, x^{**} - a_2, \ldots$. An easy continuity argument shows that (1) applies as well to any y^{**} in the weak* closed linear span of $x^{**}, x^{**}-a_1, x^{**}-a_2, \ldots$. In particular, we can apply (1) to $x^{**}-x$. However,

$$|(x^{**}-x)(x_m^*)| \leq |(x^{**}-a_k)(x_m^*)| + |x_m^*(a_k) - x_m^*(x)|$$
$$\leq \frac{1}{p} + \text{as little as you please}$$

if $m \leq n(p)$, $p \leq k$ and you take advantage of the fact that x is a weak cluster point of (a_n). So $x^{**}-x=0$, and this ensures that $x^{**}=x$ is in X. □

Exercises

1. *The failure of the Eberlein-Šmulian theorem in the weak* topology.* Let Γ be any set and denote by $l_1(\Gamma)$ the set of all functions $x: \Gamma \to$ scalars for which

$$\|x\|_1 = \sum_{\gamma \in \Gamma} |x(\gamma)| < \infty.$$

$l_1(\Gamma)$ is a Banach space whose dual space in $l_\infty(\Gamma)$, the space of bounded scalar-valued functions on Γ normed by the sup norm; the action of $\varphi \in l_\infty(\Gamma) = l_1(\Gamma)^*$ on $x \in l_1(\Gamma)$ is given by

$$\varphi(x) = \sum_{\gamma \in \Gamma} \varphi(\gamma) x(\gamma)$$

   (i) If Γ is an uncountable set, then $B_{l_\infty(\Gamma)}$ is weak* compact but not weak* sequentially compact.

   (ii) If Γ is infinite, then $B_{l_\infty(\Gamma)^*}$ contains a weak* compact set that has no nontrivial weak* convergent sequences.

2. *Weakly compact subsets of l_∞ are norm separable.*

   (i) Weak* compact subsets of X^* are metrizable in their weak* topology whenever X is separable.

   (ii) Weakly compact subsets of l_∞ are norm separable.

3. *Gantmacher's theorem.* A bounded linear operator $T: X \to Y$ between the Banach spaces X and Y is *weakly compact* if $\overline{TB_X}$ is weakly compact in Y.

   (i) A bounded linear operator $T: X \to Y$ is weakly compact if and only if $T^{**}(X^{**}) \subseteq Y$.

   (ii) A bounded linear operator $T: X \to Y$ is weakly compact if and only if T^* is weak*-weak continuous from Y^* to X^*.

   (iii) A bounded linear operator $T: X \to Y$ is weakly compact if and only if T^* is.

   (iv) A Banach space X is reflexive if and only if its dual X^* is.

Notes and Remarks

Šmulian (1940) showed that weakly compact subsets of Banach spaces are weakly sequentially compact. He also made several interesting passes at the converse as did Phillips (1943). The proof of the converse was to wait for Eberlein (1947). Soon after Eberlein's proof, Grothendieck (1952) provided a considerable generalization by showing that relatively weakly sequentially compact sets are relatively weakly compact in any locally convex space that is quasi-complete in its Mackey topology; in so doing, Grothendieck noted that Eberlein's proof (on which Grothendieck closely modeled his) required no tools that were not available to Banach himself, making Eberlein's achievement all the more impressive.

As one might expect of a theorem of the quality of the Eberlein-Šmulian theorem, there are many generalizations and refinements.

The most common proof of the Eberlein-Šmulian theorem, found, for instance, in Dunford and Schwartz, is due to Brace (1955). Those who have used Brace's proof will naturally see much that is used in the proof presented here. We do not follow Brace, however, since Whitley (1967) has given a proof (the one we do follow) that offers little room for conceptual improvement. Incidentally, Pelczynski (1964) followed a slightly different path to offer a proof of his own that uses basic sequences; we discuss Pelczynski's proof in Chapter V.

Weakly compact sets in Banach spaces are plainly different from general compact Hausdorff spaces. Weakly compact sets have a distinctive character: they are sequentially compact, and each subset of a weakly compact set has a closure that is sequentially determined. There is more to weakly compact sets than just these consequences of the Eberlein-Smulian theorem, and a good place to start learning much of what there is is Lindenstrauss's survey paper on the subject (1972). Floret's monograph also provides a readable, informative introduction to the subject.

Bibliography

Bourgin, D. G. 1942. Some properties of Banach spaces. *Amer. J. Math.*, **64**, 597–612.

Brace, J. W. 1955. Compactness in the weak topology. *Math. Mag.*, **28**, 125–134.

Eberlein, W. F. 1947. Weak compactness in Banach spaces, I. *Proc. Nat. Acad. Sci. USA*, **33**, 51–53.

Grothendieck, A. 1952. Critères de compacité dans les espaces fonctionnels généraux. *Amer. J. Math.*, **74**, 168–186.

Floret, K. 1980. *Weakly Compact Sets*. Springer Lecture Notes in Mathematics, Volume 801. New York: Springer-Verlag.

Lindenstrauss, J. 1972. Weakly compact sets—their topological properties and the Banach spaces they generate. Proceedings of the Symposium on Infinite Dimensional Topology, Annals of Math. Studies, no. 69, 235–263.

Pelczynski, A. 1964. A proof of Eberlein-Smulian theorem by an application of basic sequences. *Bull. Acad. Polon. Sci.*, **12**, 543–548.
Phillips, R. S. 1943. On weakly compact subsets of a Banach space. *Amer. J. Math.*, **65**, 108–136.
Smulian, V. L. 1940. Über lineare topologische Räume. *Mat. Sbornik N.S.*, **7**(49), 425–448.
Whitley, R. J. 1967. An elementary proof of the Eberlein-Smulian theorem. *Math. Ann.*, **172**, 116–118.

CHAPTER IV
The Orlicz-Pettis Theorem

In this chapter we prove the following theorem.

The Orlicz-Pettis Theorem. *Let $\Sigma_n x_n$ be a series whose terms belong to the Banach space X. Suppose that for each increasing sequence (k_n) of positive integers*

$$\text{weak}\lim_n \sum_{j=1}^{n} x_{k_j}$$

exists. Then for each increasing sequence (k_n) of positive integers

$$\text{norm}\lim_n \sum_{j=1}^{n} x_{k_j}$$

exists.

Put succinctly, the Orlicz-Pettis theorem says that weak subseries convergence implies subseries convergence in Banach spaces.

Our proof relies on the theory of the Bochner integral, and its success derives from the marvelous measurability theorem of Pettis. It is the exposition of the theory of the Bochner integral that occupies most of our time in this chapter; however, with the payoff including the Orlicz-Pettis theorem, our work will be highly rewarded.

Start by letting (Ω, Σ, μ) be a probability space and X be a Banach space. We first establish the ground rules for measurability.

$f: \Omega \to X$ is called *simple* if there are disjoint members $E_1, \dots, E_n$ of Σ and vectors $x_1, \dots, x_n \in X$ for which $f(\omega) = \Sigma_{i=1}^{n} \chi_{E_i}(\omega) x_i$ holds for all $\omega \in \Omega$, where χ_E denotes the indicator function of the set $E \subseteq \Omega$. Obviously such functions should be deemed measurable. Next, any function $f: \Omega \to X$ which is the μ-almost everywhere limit of a sequence of simple functions is μ-measurable. The usual facts regarding the stability of measurable functions under sums, scalar multiples, and pointwise almost everywhere convergence are quickly seen to apply. Egoroff's theorem on almost uniform convergence generalizes directly to the vector-valued case—one need only replace absolute values with norms at the appropriate places in the standard proof.

A function $f: \Omega \to X$ is *scalarly μ-measurable* if x^*f is μ-measurable for each $x^* \in X^*$. A crucial step in this proof of the Orlicz-Pettis theorem will have been taken once we demonstrate the following theorem.

Pettis Measurability Theorem. *A function $f: \Omega \to X$ is μ-measurable if and only if f is scalarly μ-measurable and there exists an $E \in \Sigma$ with $\mu(E) = 0$ such that $f(\Omega \setminus E)$ is a norm-separable subset of X.*

PROOF. It is plain to see that a μ-measurable function $f: \Omega \to X$ is scalarly μ-measurable and μ-essentially separably valued. We concentrate on the converse. Suppose $f: \Omega \to X$ is scalarly μ-measurable and $E \in \Sigma$ can be found for which $\mu(E) = 0$ and $f(\Omega \setminus E)$ is a separable subset of X. Let $\{x_n : n \geq 1\}$ be a countable dense subset of $f(\Omega \setminus E)$. Choose $\{x_n^* : n \geq 1\} \subseteq S_{X^*}$ in such a way that $x_n^* x_n = \|x_n\|$. Given $\omega \in \Omega \setminus E$ it is plain that $\|f(\omega)\| = \sup_n |x_n^*(f(\omega))|$. It follows that $\|f(\cdot)\|$ is μ-measurable. Similarly for each n, $\|f(\cdot) - x_n\|$ is μ-measurable.

Let $\varepsilon > 0$ be given. Look at $[\|f(\omega) - x_n\| < \varepsilon] = E_n$ (we prefer to use the probabilists' notation here; so $[\|f(\omega) - x_n\| < \varepsilon]$ is $\{\omega \in \Omega : \|f(\omega) - x_n\| < \varepsilon\}$). Each E_n is almost in Σ (and, if μ is complete, actually does belong to Σ), and so for each n there is a $B_n \in \Sigma$ such that $\mu(E_n \Delta B_n) = 0$. Define $g: \Omega \to X$ by

$$g(\omega) = \begin{cases} x_n & \text{if } \omega \in B_n \setminus (B_1 \cup \cdots \cup B_{n-1}), \\ 0 & \text{if } \omega \notin \bigcup_n B_n. \end{cases}$$

It is clear that $\|g(\omega) - f(\omega)\| < \varepsilon$ for any ω outside of both E and $\cup_n (E_n \Delta B_n)$.

We have shown that given $\varepsilon > 0$ there is a countably valued function g and a μ-null set $N_\varepsilon \in \Sigma$ such that g assumes distinct values on disjoint members of Σ and such that f and g are uniformly within ε of each other on $\Omega \setminus N_\varepsilon$. Giving a little (of Ω) to get a little (and make g simple) quickly produces a sequence of simple functions converging μ-almost everywhere to f, which completes the proof. □

Now for the Bochner integral.

If $f: \Omega \to X$ is simple, say $f(\omega) = \sum_{i=1}^n \chi_{E_i}(\omega) x_i$, then for any $E \in \Sigma$

$$\int_E f \, d\mu = \sum_{i=1}^n \mu(E \cap E_i) x_i.$$

A μ-measurable function $f: \Omega \to X$ is called *Bochner integrable* if there exists a sequence of simple functions (f_n) such that

$$\lim_n \int_\Omega \|f_n(\omega) - f(\omega)\| d\mu(\omega) = 0.$$

In this case $\int_E f\, d\mu$ is defined for each $E \in \Sigma$ by

$$\int_E f\, d\mu = \lim_n \int_E f_n\, d\mu.$$

Our first result regarding the Bochner integral is due to Bochner himself and is in a sense the root of all that is "trivial" about the Bochner integral.

Bochner's Characterization of Integrable Functions. *A μ-measurable function $f: \Omega \to X$ is Bochner integrable if and only if $\int_\Omega \|f\|\, d\mu < \infty$.*

PROOF. If f is Bochner integrable, then there's a simple function g such that $\int_\Omega \|f - g\|\, d\mu < 7$; it follows that

$$\int \|f\|\, d\mu \le \int \|f - g\|\, d\mu + \int \|g\|\, d\mu < \infty.$$

Conversely, suppose f(and so $\|f\|$) is μ-measurable with $\int \|f\|\, d\mu < \infty$. Choose a sequence of countably valued measurable functions (f_n) such that $\|f - f_n\| \le 1/n$, μ-almost everywhere. Here a peek at the proof of the Pettis measurability theorem is acceptable. Since $\|f_n(\cdot)\| \le \|f(\cdot)\| + 1/n$ almost all the time, we see that $\int \|f_n\|\, d\mu < \infty$. For each n write f_n in its native form

$$f_n(\omega) = \sum_{m=1}^{\infty} \chi_{E_{n,m}}(\omega) x_{n,m},$$

where $E_{n,i} \cap E_{n,j} = \varnothing$ whenever $i \ne j$, all $E_{n,m}$ belong to Σ, and all the $x_{n,m}$ belong to X. For each n pick p_n so large that

$$\int_{\bigcup_{m=p_n+1}^{\infty} E_{n,m}} \|f_n\|\, d\mu < \frac{1}{n}.$$

What is left of f_n is $\sum_{m=1}^{p_n} \chi_{E_{n,m}} x_{n,m} = g_n$, a simple function for which

$$\int \|f - g_n\|\, d\mu \le \frac{2}{n}.$$

f is Bochner integrable, and this proof is complete. □

In a very real sense Bochner's characterization of Bochner-integrable functions trivializes the Bochner integral, reducing as it does much of the development to the Lebesgue integral. This reduction has as a by-product the resultant elegance and power of the Bochner integral. We'll say a bit more about this elsewhere and restrict our attentions herein to a few more-or-less obvious consequences of the work done to this point.

Corollary

1. *(Dominated Convergence Theorem). If (f_n) are Bochner-integrable X-valued functions on Ω, $f: \Omega \to X$ is the almost everywhere limit of (f_n) and $\|f_n(\cdot)\| \leq g(\cdot)$ almost all the time and for all n, where $g \in L_1(\mu)$, then f is Bochner integrable and $\int_\Omega \|f - f_n\| \, d\mu \to 0$ and $\int_E f_n \, d\mu \to \int_E f \, d\mu$ for each $E \in \Sigma$.*
2. *If f is Bochner integrable, then $\|\int_E f \, d\mu\| \leq \int_E \|f\| \, d\mu$ holds for all $E \in \Sigma$. Consequently, $\int_E f \, d\mu$ is a countably additive μ-continuous X-valued set function on Σ.*

PROOF. Part 1 follows from Bochner's characterization and the scalar dominated convergence theorem: $\|f_n(\cdot) - f(\cdot)\| \leq 2g(\cdot)$ almost all the time. Part 2 is obvious if f is simple and simple for other f. □

One noteworthy conclusion to be drawn from 2 above is the fact that *if f is Bochner integrable, then $\{\int_E f \, d\mu : E \in \Sigma\}$ is a relatively compact subset of X.* In case f is a simple function, this follows from the estimate $\|\int_E f \, d\mu\| \leq \int_\Omega \|f\| \, d\mu < \infty$ and the resulting boundedness of $\{\int_E f \, d\mu : E \in \Sigma\}$ in the finite-dimensional linear span of the range of f. For arbitrary Bochner-integrable $f: \Omega \to X$ one need only pick a simple $g: \Omega \to X$ for which $\int_\Omega \|f - g\| \, d\mu$ is very small to see that $\{\int_E f \, d\mu : E \in \Sigma\}$ is closely approximable by $\{\int_E g \, d\mu : E \in \Sigma\}$, a totally bounded subset of X. Of course this says that given $\varepsilon > 0$ each vector in $\{\int_E f \, d\mu : E \in \Sigma\}$ can be approximated within $\varepsilon/2$ by a vector in the totally bounded set $\{\int_E g \, d\mu : E \in \Sigma\}$, so $\{\int_E f \, d\mu : E \in \Sigma\}$ is itself totally bounded.

Now for the proof of the Orlicz-Pettis theorem.

Let's imagine what could go wrong with the theorem. If $\Sigma_n x_n$ is weakly subseries convergent (i.e., satisfies the hypotheses of the Orlicz-Pettis theorem) yet fails to be norm subseries convergent, it's because there's an increasing sequence (k_n) of positive integers for which $(\Sigma_{j=1}^n x_{k_j})$ is not a Cauchy sequence in X. This can only happen if there is an $\varepsilon > 0$ and an intertwining pair of increasing sequences (j_n) and (l_n) of positive integers for which $j_1 < l_1 < j_2 < l_2 < \cdots$ satisfying $\|\Sigma_{i=j_n}^{l_n} x_{k_i}\| > \varepsilon$ for all n. The series $\Sigma_n y_n$ formed by letting $y_n = \Sigma_{i=j_n}^{l_n} x_{k_i}$ is a subseries of $\Sigma_n x_n$ and so is weakly summable in X; in particular, (y_n) is weakly null. On the other hand, $\|y_n\| > \varepsilon$ for all n. In short, if the Orlicz-Pettis theorem fails at all, it is possible to find a weakly subseries convergent series $\Sigma_n y_n$ for which $\|y_n\| > \varepsilon$ holds for all n. Preparations are now complete; it's time for the main course.

Let Ω be the compact metric space $\{-1,1\}^{\mathbb{N}}$ of all sequences (ε_n) of signs $\varepsilon_n = \pm 1$. Let Σ denote the σ-field of Borel subsets of Ω. Let μ be the product measure on $\{-1,1\}^{\mathbb{N}}$ resulting from the identical coordinate measures on $\{-1,1\}$ that assign to each elementary event $\{-1\}$ and $\{1\}$ the probability $\frac{1}{2}$. The reader might recognize (Ω, Σ, μ) as the Cantor group with its resident Haar measure. No matter—we have a probability measure space and a natural function $f\colon \Omega \to X$, namely, if (ε_n) is a sequence of signs, $\varepsilon_n = \pm 1$, then

$$f((\varepsilon_n)) = \text{weak} \lim_n \sum_{k=1}^{n} \varepsilon_k y_k.$$

Of course the weak subseries convergence of $\Sigma_n y_n$ is just what is needed to make sense of f's definition for any $(\varepsilon_n) \in \{-1,1\}^{\mathbf{N}}$. Each coordinate function is continuous on Δ so that f is scalarly μ-measurable on Δ to X. Moreover, the range of f is contained in the (weakly) closed linear span of the vectors y_n; so $f(\Omega)$ is separable. Pettis's measurability theorem applies to f; f is μ-measurable. Finally, the range of f is contained in the weak closure of $\{\Sigma_{k \in \Delta} \varepsilon_k y_k : \Delta \text{ is a finite set of positive integers, } \varepsilon_k = \pm 1 \text{ for } k \in \Delta\}$, a set easily seen to be weakly bounded; f is itself weakly bounded, hence bounded. Bochner's characterization theorem applies to show f is Bochner integrable with respect to μ.

Let's compute. Let E_n be the set of all sequences ε of ± 1's, whose nth coordinate ε_n is 1; $E_n \in \Sigma$ and $\int_{E_n} f \, d\mu = y_n/2$. The sequence (y_n) is weakly null and sits inside the relatively norm compact set $\{2\int_E f \, d\mu : E \in \Sigma\}$. It follows that each subsequence of (y_n) has a norm convergent subsequence whose only possible limit is 0 since (y_n) is weakly null. In other words, (y_n) is norm null! This is a very difficult thing for (y_n) to endure: $\|y_n\| > \varepsilon > 0$ for all n and $\lim_n \|y_n\| = 0$, a contradiction.

Exercises

1. *Weakly countably additive vector measures are countably additive.* Let Σ be a σ-field of subsets of the set Ω and X be a Banach space. Show that any weakly countably additive measure $F\colon \Sigma \to X$ is countably additive in the norm topology of X.

   By means of a counterexample, show that the aforementioned result fails if Σ is but a field of sets.

2. *The Pettis integral.* Let (Ω, Σ, μ) be a probability measure space and X be a Banach space. A function $f\colon \Omega \to X$ is called *scalarly measurable* if x^*f is measurable for each $x^* \in X^*$; f is called *scalarly integrable* if $x^*f \in L_1(\mu)$ for each $x^* \in X^*$.

   (i) If $f\colon \Omega \to X$ is scalarly integrable, then for each $E \in \Sigma$ there is an $x_E^{**} \in X^{**}$ such that

   $$x_E^{**}x^* = \int_E x^*f(\omega) \, d\mu(\omega)$$

   holds for each $x^* \in X^*$.

   (ii) If $f\colon \Omega \to X$ is bounded and scalarly measurable, then f is scalarly integrable and each of the x_E^{**} from (i) is weak* sequentially continuous on X^*.

   We say that f is *Pettis integrable* if each x_E^{**} is actually in X, in which case we denote x_E by Pettis $\int_E f \, d\mu$.

(iii) If f is Pettis integrable, then the map taking $E \in \Sigma$ into Pettis $\int_E f\,d\mu$ is countably additive. Bochner-integrable functions are Pettis integrable.

A Banach space X is said to have *Mazur's property* if weak* sequentially continuous functionals on X^* are actually weak* continuous, i.e., belong to X.

(iv) If X has Mazur's property, then bounded scalarly measurable X-valued functions are Pettis integrable.

(v) Separable Banach spaces enjoy Mazur's property, as do reflexive spaces.

Let Γ be a set and denote by $c_0(\Gamma)$ the Banach space of all scalar-valued functions x on Γ for which given $\varepsilon > 0$ the set

$$\{\gamma \in \Gamma : |x(\gamma)| > \varepsilon\}$$

is finite; $x \in c_0(\Gamma)$ has norm $\sup_{\gamma \in \Gamma}|x(\gamma)|$; so $c_0(\Gamma)^* = l_1(\Gamma)$.

(vi) $c_0(\Gamma)$ has Mazur's property.

(vii) l_∞ does not have Mazur's property.

3. *A theorem of Krein and Šmulian.* The object of this exercise is to prove the following:

Theorem (Krein-Šmulian). *The closed convex hull of a weakly compact subset of a Banach space is weakly compact.*

Let K be a weakly compact set sitting inside the Banach space X.

(i) X may be assumed to be separable. Do so!

(ii) The function $\varphi : K \to X$ defined by

$$\varphi(k) = k$$

is Bochner integrable with respect to every regular Borel measure defined on (K, weak).

(iii) The operator $I_\varphi : C(K, \text{weak})^* \to X$ defined by

$$I_\varphi(\mu) = \text{Bochner} \int \varphi(k)\, d\mu(k)$$

is weak*-weak continuous.

(iv) The closed convex hull of K lies inside of $I_\varphi(B_{C(K,\text{weak})^*})$.

4. *The bounded multiplier test.* A series $\Sigma_n x_n$ in a Banach space X is unconditionally convergent if and only if for any $(t_n) \in l_\infty$ the series $\Sigma_n t_n x_n$ converges.

Notes and Remarks

The story of the Orlicz-Pettis theorem is a curious one. Proved by Orlicz in the late twenties, it was lost to much of its mathematical public for most of a decade because of a fluke. In the (original) 1929 Polish edition of Banach's

"Operationes Lineaires," note was made of Orlicz's theorem; on translation into French the note on Orlicz's theorem was not amended either to indicate that with the passage of time the proof had already appeared or to include exact bibliographic data. As a result, when Pettis was writing his thesis, he found himself in need of a proof of the Orlicz-Pettis theorem; in addition to providing said proof, Pettis gave several basic applications of the result. These applications are the bulk of Exercises 1 and 2.

Our proof is due to Kwapien (1974). It was shown to us by Iwo Labuda and Jerry Uhl. Somehow it is appropriate that there be a proof of the Orlicz-Pettis theorem that depends ultimately on Pettis's measurability theorem, since so much of Pettis's mathematical work was concerned with the subtle interplay between the weak and norm topologies in separable Banach spaces.

That the Krein-Šmulian theorem (Exercise 3) can be derived from the theory of the Bochner integral seems to be due to Dunford and Schwartz. The reader will no doubt realize that Mazur's theorem (to the effect that the closed convex hull of a norm-compact set is norm compact) can also be derived in this fashion.

There are other proofs of the Orlicz-Pettis theorem, and we will present two of them in later chapters.

It is noteworthy that Grothendieck (1953) and McArthur (1967) have proved the Orlicz-Pettis theorem in locally convex spaces.

We mention in passing that the failure of Pettis's "weak measures are measures" theorem for algebras of sets (indicated in Exercise 1) has been investigated by Schachermayer, who has discovered a number of non-σ-complete Boolean algebras where Pettis's theorem holds. Schachermayer goes on to give several interesting characterizations of this phenomena and pose a number of problems related to it.

Finally, we must mention that Kalton (1971, 1980) has underlined the separable character of the Orlicz-Pettis theorem by proving a version of the theorem in topological groups. Picking up on Kalton's lead, Anderson and Christenson (1973) have established a permanent link between subseries convergence in a space and the measure-theoretic structure of the space.

For an informative, lively discussion of the Orlicz-Pettis theorem we recommend both Kalton's lecture and Uhl's lecture as reported in the proceedings of the Pettis Memorial Conference.

Bibliography

Anderson, N. J. M. and Christenson, J. P. R. 1973. Some results on Borel structures with applications to subseries convergence in Abelian topological groups. *Israel J. Math.*, **15**, 414–420.

Grothendieck, A. 1953. Sur les applications lineaires faiblement compactes d'espaces du type $C(K)$. *Can. J. Math.*, **5**, 129–173.

Kalton, N. J. 1971. Subseries convergence in topological groups and vector spaces. *Israel J. Math.*, **10**, 402–412.

Kalton, N. J. 1980. The Orlicz-Pettis theorem. In Proceedings of the Conference on Integration, Topology and Geometry in Linear Spaces, Contemporary Math 2.

Kwapien, S. 1974. On Banach spaces containing c_0. *Studia Math.*, **52**, 187–188.

McArthur, C. W. 1967. On a theorem of Orlicz and Pettis. *Pacific J. Math.*, **22**, 297–302.

Orlicz, W. 1929. Beiträge zu Theorie der Orthogonalentwicklungen, II. *Studia Math.*, **1**, 241–255.

Pettis, B. J. 1938. On integration in vector spaces. *Trans. Amer. Math. Soc.*, **44**, 277–304.

Schachermayer, W. 1982. On some classical measure-theoretic theorems for non-sigma-complete Boolean algebras. Dissertations in Math, Volume 214.

Uhl, J. J. 1980. Pettis's Measurability theorem. In Proceedings of the Conference on Integration, Topology and Geometry in Linear Spaces, Contemporary Math 2.

CHAPTER V
Basic Sequences

In any earnest treatment of sequences and series in Banach spaces a featured role must be reserved for basic sequences. Our initial discussion of this important notion will occupy this whole chapter. A foundation will be laid on which we will build several of the more interesting constructs in the theory of sequences and series in Banach spaces.

Let's give a brief hint of what's planned. After introductory remarks about bases and basic sequences, we show how Mazur proved the existence of basic sequences in any infinite-dimensional Banach space and take immediate advantage of those ideas to present Pelczynski's proof of the Eberlein-Smulian theorem. The Bessaga-Pelczynski selection principle will then be derived and, after a brief discussion of weakly unconditionally Cauchy series, this principle will be applied to characterize spaces containing isomorphs of c_0. Here we must mention that the Orlicz-Pettis theorem is rederived along with an improvement thereof in spaces without c_0 subspaces. Finally, we see that c_0's appearance or absence in a dual coincides with l_∞'s and use this to describe still another sharpening of the Orlicz-Pettis theorem, this time in *duals* without c_0 subspaces. It's a full program; so it's best that we get on with it.

A sequence (x_n) in a Banach space X is called a *Schauder basis* (or *basis*) for X if for each $x \in X$ there exists a unique sequence (α_n) of scalars such that

$$x = \lim_n \sum_{k=1}^{n} \alpha_k x_k .$$

It is easy to see that a Schauder basis consists of independent vectors. Of great importance to our goals is the notion of *basic sequence*: a basic sequence in a Banach space X is a sequence (x_n) that is a basis for its closed linear span $[x_n]$.

Of some note is the fact that *if* (x_n) *is a Schauder basis for the Banach space* X, *then each of the coefficient functionals* $x_k^*: \sum_n \alpha_n x_n \to \alpha_k$, that go hand in hand with the x_n, *is continuous on* X. Indeed, let S denote (for the moment) the linear space of all scalar sequences (s_n) for which $\lim_n \Sigma_{k=1}^n s_k x_k$ exists in X. We define $|||(s_n)|||$ to be $\sup_n \|\Sigma_{k=1}^n s_k x_k\|$. Using the uniqueness of expansions with respect to the system (x_n), one sees that the operator $B:(S, |||\cdot|||) \to (X, \|\cdot\|)$ given by $B(s_n) = \lim_n \Sigma_{k=1}^n s_k x_k$ is a norm-decreasing, one-to-one, linear operator from S onto X. B is in fact an isomorphism. To see why this is so, we need only show that $(S, |||\cdot|||)$ is a Banach space and appeal to the open mapping theorem. Now $(S, |||\cdot|||)$ is quickly seen to be a normed linear space; so completeness is the issue at hand. Let $(y_p) = ((s_{pi}))$ be a $|||\cdot|||$-Cauchy sequence in S. Since

$$\begin{aligned} |s_{pi} - s_{qi}|\,\|x_i\| &\leq 2 \sup_n \left\| \sum_{i=1}^n (s_{pi} - s_{qi}) x_i \right\| \\ &= 2|||y_p - y_q|||, \end{aligned}$$

$(s_{pi})_p$ converges for each i. Let (s_i) be the sequence of scalars obtained by letting $p \to \infty$: $s_i = \lim_p s_{pi}$. Let r be an index so chosen that for $p \geq r$, $|||y_p - y_r||| < \varepsilon$, ε a preassigned positive number. In light of the definition of S's norm, we see that whenever $p \geq r$, $\|\Sigma_{i=1}^n (s_{pi} - s_{ri}) x_i\| \leq \varepsilon$ for all n. Since $y_r = (s_{ri}) \in S$, there is a cutoff n_ε such that whenever $m, n \geq n_\varepsilon$ with $m \geq n$ say, $\|\Sigma_{i=n}^m s_{ri} x_i\| \leq \varepsilon$. It is now easy to see (after letting $p \to \infty$) that for $m, n \geq n_\varepsilon$ we get, for $m \geq n \geq n_\varepsilon$,

$$\left\| \sum_{i=n}^m s_i x_i \right\| \leq 3\varepsilon,$$

and so $s = (s_i) \in S$, too, and is in fact the limit of the sequence $(y_i) = ((s_{pi})_{p \geq 1})$ from S. Now that B's isomorphic nature has been established, it is clear that, for any $k \geq 1$, the coefficient functional x_k^* is continuous as

$$|\alpha_k|\,\|x_k\| \leq 2\|B^{-1}\| \left\| \sum_n \alpha_n x_n \right\|.$$

A space with a basis is always separable, and it is indeed the case that most of the natural separable Banach spaces have bases. It ought to be, in fact, it *must* be pointed out that finding a basis for a well-known space is not always an easy task. A few examples will be cited; proofs will not be presented.

In the case of the classical separable sequence spaces c_0 and l_p (for $1 \leq p < \infty$), the sequence (e_n) of unit coordinate vectors

$$e_n = \Big(0, 0, \ldots, 0, \underset{n\text{th place}}{1,} 0, 0, \ldots\Big)$$

is a basis. This *is* easy to show. In the case of c, the space of convergent sequences, we must supplement the sequence (e_n) with the constant sequence $1 = (1, 1, \ldots, 1, \ldots)$; the sequence $(1, e_1, e_2, \ldots)$ is a basis for c.

What about function spaces? Here life becomes more complicated. In the case of $C[0,1]$, *J*. Schauder showed that *the* Schauder basis is a basis, where the terms of the basis are given as follows:

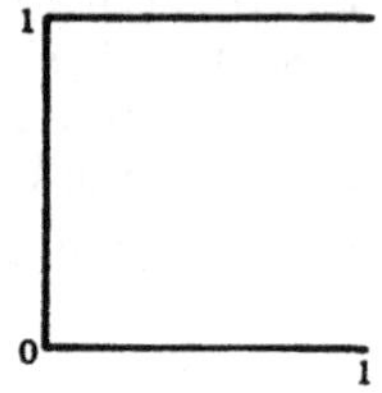

$$f_1(t) = 1 \quad \text{for all } t \in [0,1].$$

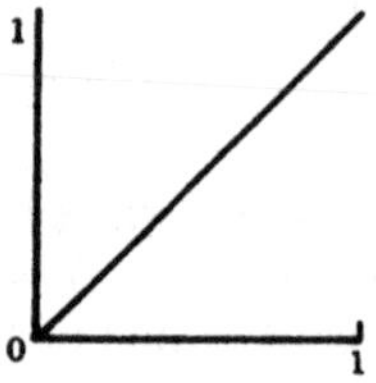

$$f_2(t) = t \quad \text{for each } t \in [0,1].$$

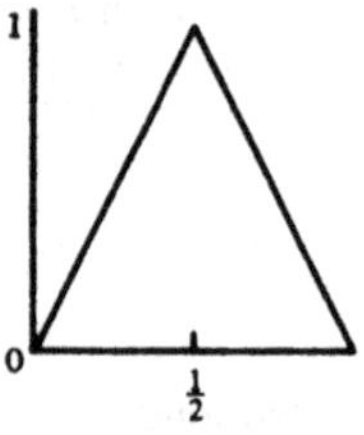

$$f_3(t) = \begin{cases} 2t & \text{for each } t \in \left[0, \frac{1}{2}\right], \\ 2 - 2t & \text{for each } t \in \left[\frac{1}{2}, 1\right]. \end{cases}$$

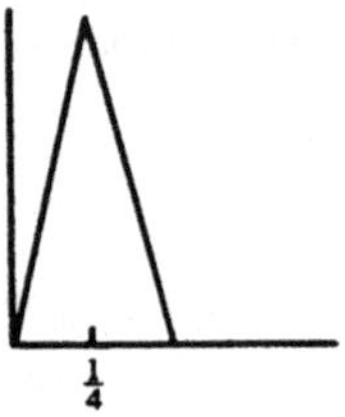

$$f_4(t) = \begin{cases} 4t & \text{for each } t \in \left[0, \frac{1}{4}\right], \\ 2 - 4t & \text{for each } t \in \left[\frac{1}{4}, \frac{1}{2}\right], \\ 0 & \text{for } t \geq \frac{1}{2}. \end{cases}$$

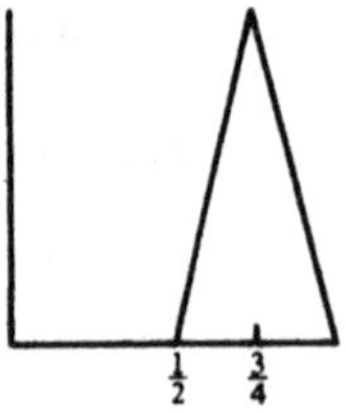

$$f_5(t) = \begin{cases} 0 & \text{for } t \leq \frac{1}{2}, \\ 4t - 2 & \text{for each } t \in \left[\frac{1}{2}, \frac{3}{4}\right], \\ 4 - 4t & \text{for each } t \in \left[\frac{3}{4}, 1\right]. \end{cases}$$

Generally, if $n \geq 1$ and $1 \leq i \leq 2^n$, then we can define f_{2^n+i+1} as follows:

$$f_{2^n+i+1}(t) = f_3(2^n t + 1 - i) \quad \text{whenever } 2^n t + 1 - i \in [0,1].$$

In the case of $L_p[0,1]$, where $1 \le p < \infty$, the *Haar basis* is given by

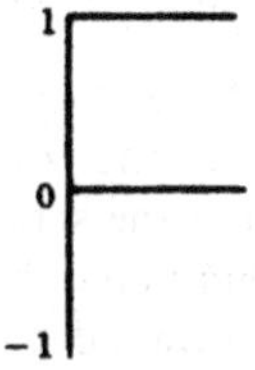

$$f_1(t) = 1 \quad \text{for all } t \in [0,1].$$

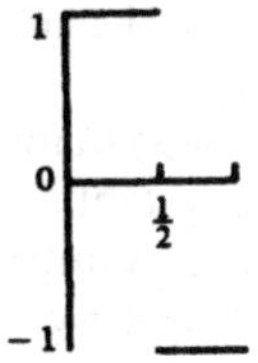

$$f_2(t) = c_{[0,\frac{1}{2}]}(t) - c_{(\frac{1}{2},1]}(t).$$

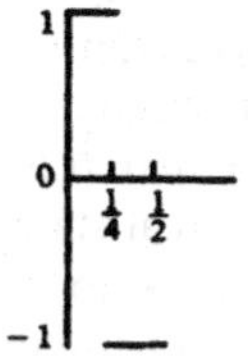

$$f_3(t) = c_{[0,\frac{1}{4}]}(t) - c_{(\frac{1}{4},\frac{1}{2}]}(t).$$

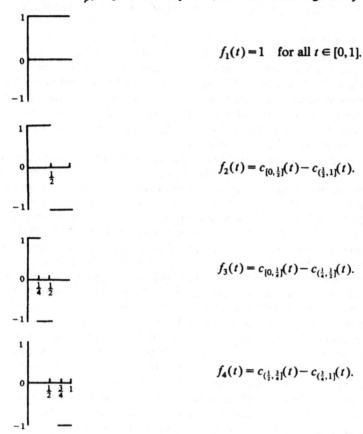

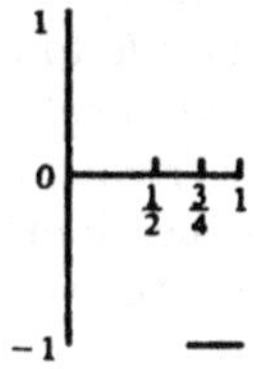

$$f_4(t) = c_{(\frac{1}{2},\frac{3}{4}]}(t) - c_{(\frac{3}{4},1]}(t).$$

Generally, if $n \ge 1$ and $1 \le i \le 2^n$, then f_{2^n+i} is given by

$$f_{2^n+i}(t) = c_{[(2i-2)/2^{n+1},(2i-1)/2^{n+1}]}(t) - c_{((2i-1)/2^{n+1},2i/2^{n+1}]}(t).$$

It is now well known that there *are* separable Banach spaces without bases. Per Enflo (1973), the first to find such a space, looked inside c_0 and was duly rewarded.

Therefore, the fact that a separable Banach space has a basis does provide some structural information about the space. Unfortunately, unless the space and/or the basis packs extra punch, little can be derived from this minimal, yet hard-to-achieve, bit of information.

$C[0,1]$ has a basis. This is of interest—*not* because it registers $C[0,1]$ as a member of the "basis club," but because $C[0,1]$ plays a central role in the theory of Banach spaces, and so the fact that it has a basis can on occasion be exploited. One special property of $C[0,1]$ that indicates the kind of exploitation possible is its universality among separable Banach spaces:

every separable Banach space is isometrically isomorphic to a closed linear subspace of $C[0,1]$. $C[0,1]$*'s universality, in tandem with the fact that* $C[0,1]$ *has a basis, pays off.*

The Haar system is a basis for all the L_p, $1 \leq p < \infty$. For $1 < p < \infty$, it is more: it is an *unconditional basis*; i.e., not only does each member of the space have a unique series expansion in terms of the basis, but the series is unconditionally convergent. The spaces spanned by unconditional bases enjoy finer structural properties than spaces without unconditional bases; the exercises hint at a few of the added pleasures. Incidentally, the Haar system is *not* an unconditional basis for $L_1[0,1]$; in fact, $L_1[0,1]$ does not have an unconditional basis of any kind.

It is worth remarking that showing the Schauder and Haar systems are bases for the spaces indicated above is not difficult; to establish the unconditionality of the Haar system (in case $1 < p < \infty$) is highly nontrivial.

Oftentimes, whether a space has a basis is in itself difficult to answer, and even on responding to this question, the possibility of the existence of an unconditional basis looms large. For instance, it was not until 1974 that Botschkariev showed that the disk algebra has a basis: the Franklin system (i.e., the Gram-Schmidt orthogonalization of the Schauder system in the Hilbert space $L_2[0,1]$); soon thereafter, Pelczynski showed that the disk algebra does *not* have an unconditional basis. Each proof has real claims to depth. Again, the Franklin system was shown by Wojtaszczyk to be an unconditional basis for the classical Hardy space $H^1(D)$ of functions analytic inside the disk and with integrable boundary values; it is an absolute must to point out that earlier, Maurey in a real tour de force of analytical know-how had shown that $H^1(D)$ *has* an unconditional basis *without* explicitly citing one. After Carleson had had some clarifying effect on the question, Wojtaszczyk got into the act. None of these developments has the faintest resemblance to "easy" mathematics, not the work of Wojtaszczyk, or Carleson, or Maurey, especially not Maurey!

Bases are important; bases with added features, even more so. Basic sequences are likewise important, especially for general structure-theoretic studies. Since our purpose is, to some extent, the study of convergence of sequences and series and the effect thereof on the structure of a Banach space, it is not too unbelievable that basic sequences will occupy some of our attention. How does one recognize a basic sequence?

The basic test is provided by our first real result.

Theorem 1. *Let* (x_n) *be a sequence of nonzero vectors in the Banach space* X. *Then in order that* (x_n) *be a basic sequence, it is both necessary and sufficient that there be a finite constant* $K > 0$ *so that for any choice of scalars* $(a_n)_{n \geq 1}$ *and any integers* $m < n$ *we have*

$$\left\| \sum_{i=1}^{m} a_i x_i \right\| \leq K \left\| \sum_{i=1}^{n} a_i x_i \right\|.$$

The proof is easy but well worth the time to be carefully studied. We present it in all its important (and perhaps in a few of its other) details.

PROOF. Suppose (x_n) is a basis for its closed linear span $[x_n]$ and define $P_k:[x_n]\to[x_n]$ by

$$P_k\left(\sum_n a_n x_n\right) = \sum_{n=1}^{k} a_n x_n.$$

Each P_k is a bounded linear operator [since each of the coordinate functionals x_j^* $(1 \le j \le k)$ is continuous], and for any $x \in [x_n]$, we have $x = \lim_{k\to\infty} P_k x$. It follows from the Banach-Steinhaus theorem that $\sup_n \|P_n\| < \infty$. Thus, should $m < n$ and $\sum_k a_k x_k \in X$, then

$$\begin{aligned}\left\|\sum_{k=1}^{m} a_k x_k\right\| &= \left\|P_m \sum_k a_k x_k\right\| \\ &= \left\|P_m P_n \sum_k a_k x_k\right\| \\ &= \left\|P_m \sum_{k=1}^{n} a_k x_k\right\| \\ &\le \|P_m\| \left\|\sum_{k=1}^{n} a_k x_k\right\| \\ &\le \sup_n \|P_n\| \cdot \left\|\sum_{k=1}^{n} a_k x_k\right\|.\end{aligned}$$

Let $K = \sup_n \|P_n\|$.

Now suppose (x_n) is a sequence of nonzero vectors for which there is a $K > 0$ such that whenever $m < n$,

$$\left\|\sum_{i=1}^{m} a_i x_i\right\| \le K \left\|\sum_{i=1}^{n} a_i x_i\right\|$$

holds. Plainly, if a vector x has a representation in the form $\sum_n a_n x_n = \lim_n \sum_{i=1}^{n} a_i x_i$, that representation is unique; this follows, for instance, from the fact that for any $j, k \ge 1$,

$$|a_j|\|x_j\| = \|a_j x_j\| \le K \left\|\sum_{i=j}^{j+k} a_i x_i\right\|,$$

so that

$$|a_j| \le \frac{K}{\|x_j\|} \left\|\sum_{i \ge j} a_i x_i\right\|.$$

Regarding representable elements, we notice that each vector in the linear span of the x_n is clearly representable, by a finite sum in fact. The condition that whenever $m < n$,

$$\left\|\sum_{i=1}^{m} a_i x_i\right\| \leq K \left\|\sum_{i=1}^{n} a_i x_i\right\|$$

ensures that the operators P_m, from the linear span of $\{x_n\}$ to itself, given by $P_m(\Sigma a_i x_i) = \Sigma_{k=1}^{m} a_k x_k$, are bounded linear operators each of whose operator norms are $\leq K$; it follows that each P_m has a bounded linear extension, still called P_m, projecting $[x_n : n \geq 1]$ onto $[x_n : 1 \leq n \leq m] = \text{lin}\{x_1, \ldots, x_m\}$. A noteworthy effect of this is the continuity of the "coordinate functionals" x_k^* defined on the span of $\{x_n\}$ by $x_k^*(\Sigma_i a_i x_i) = a_k$; the x_k^* have unique extensions to all of $[x_n : n \geq 1]$, too, given by $x_k^*(x)x_k = P_k(x) - P_{k-1}(x)$. Now we're ready for some action. We claim that every element of $[x_n]$ has a representation (necessarily unique, as we have seen) in the form $\lim_n \Sigma_{k=1}^{n} a_k x_k = \Sigma_n a_n x_n$. Let $x \in [x_n]$ and $\varepsilon > 0$ be given. Then there is a $\sigma \in \text{lin}\{x_1, \ldots, x_{n(\varepsilon)}\}$, for some $n(\varepsilon)$, such that $\|x - \sigma\| < \varepsilon$. But now if $n \geq n_\varepsilon$, then

$$\begin{aligned} \|x - P_n x\| &\leq \|x - \sigma\| + \|\sigma - P_n \sigma\| + \|P_n \sigma - P_n x\| \\ &= \|x - \sigma\| + \|\sigma - \sigma\| + \|P_n(\sigma - x)\| \\ &\leq \varepsilon + \|P_n\| \varepsilon \leq (1 + K)\varepsilon. \end{aligned}$$

It follows that $x = \lim_n P_n x = \lim_n \Sigma_{k=1}^{n} x_k^*(x) x_k$. □

As an application of Theorem 1 we prove that every infinite-dimensional Banach space contains a subspace with a basis. We follow S. Mazur's lead.

Lemma 2. *Let F be a finite-dimensional linear subspace of the infinite-dimensional Banach space X, and let $\varepsilon > 0$. Then there is an $x \in X$ such that $\|x\| = 1$ and*

$$\|y\| \leq (1 + \varepsilon)\|y + \lambda x\| \tag{1}$$

for all $y \in F$ and all scalars λ.

Proof. Assuming (as we may) that $\varepsilon < 1$, pick a finite $\varepsilon/2$ net $\{y_1, \ldots, y_k\}$ for S_F and select $y_1^*, \ldots, y_k^*$ in S_{X^*} so that $y_i^* y_i = 1$ for $i = 1, 2, \ldots, k$. Take any x in S_X for which $y_1^* x = y_2^* x = \cdots = y_k^* x = 0$. This x will do. In fact, if $y \in S_F$, then there is a y_i within $\varepsilon/2$ of y; find that y_i. Take any scalar λ and compute

$$\begin{aligned} \|y + \lambda x\| &\geq \|y_i + \lambda x\| - \|y - y_i\| \geq \|y_i + \lambda x\| - \frac{\varepsilon}{2} \\ &\geq y_i^*(y_i + \lambda x) - \frac{\varepsilon}{2} = 1 - \frac{\varepsilon}{2} \geq \frac{1}{1 + \varepsilon}. \end{aligned}$$

This shows (1) in case $\|y\| = 1$; homogeneity takes care of the rest of F. □

Corollary 3. *Every infinite-dimensional Banach space contains an infinite-dimensional closed linear subspace with a basis.*

PROOF. Let X be the ambient space and $\varepsilon > 0$. Choose a sequence (ε_n) of positive numbers such that $\prod_{n=1}^{\infty}(1+\varepsilon_n) \leq 1+\varepsilon$. Take $x_1 \in S_X$ and pick $x_2 \in S_X$ such that

$$\|x\| \leq (1+\varepsilon_1)\|x + \lambda x_2\|$$

for every scalar multiple x of x_1; a look at the preparatory lemma will tell you where to look for x_2. Let F be the linear span of x_1 and x_2. Pick $x_3 \in S_X$ such that

$$\|x\| \leq (1+\varepsilon_2)\|x + \lambda x_3\|$$

for every x in F; again, a look at the preparatory lemma should help in the selection of x_3. Continue. The sequence (x_n) so generated is basic with basis constant $\leq 1+\varepsilon$. What's more, if P_n is the nth projection operator, then $\|P_n\| \leq \prod_{i=n}^{\infty}(1+\varepsilon_i)$. □

A short detour seems well advised at this juncture. This detour is suggested by A. Pelczynski's proof of the Eberlein-Smulian theorem via basic sequences. This proof, of which Whitley's is a sympathetic cousin, builds on a modification of Mazur's construction of basic sequences.

Lemma 4. *Let B be a bounded subset of the Banach space X and $x_0^{**} \in X^{**}$ be a point of $\overline{B}^{weak^*}$ in X^{**} such that $\|x_0^{**} - b\| \geq \delta > 0$ for all $b \in B$. Then there exist a sequence (x_n) in B and an $x_0^* \in X^*$ such that*

1. $\lim_n x_0^* x_n = x_0^{**} x_0^* \geq \|x_0^{**}\|/2$.
2. $(x_n - x_0^{**})$ *is a basic sequence in* X^{**}.
3. *Should* $x_0^{**} \neq 0$, *then* $x_0^{**} \notin [x_n - x_0^{**}]$, *the closed linear span of* $(x_n - x_0^{**})$.

PROOF. Choose $(c_n)_{n\geq 0}$ so that $0 < c_n < 1$ for all $n \geq 0$ and so that whenever $1 \leq p < q < \infty$, $\prod_{i=p}^{q-1}(1-c_i) > 1 - c_0$.

Take any $x_0^* \in X^*$ such that $x_0^{**}x_0^* \geq \|x_0^{**}\|/2$. By hypothesis, there is an $x_1 \in B$ such that

$$|x_0^* x_1 - x_0^{**}x_0^*| < 1.$$

Let E_1 denote the 1-dimensional subspace of X^{**} spanned by $x_1 - x_0^{**}$; S_{E_1} is compact, and so we can pick a $c_1/3$ net $e_1, \ldots, e_{N(1)}$ for S_{E_1}. Let $x_1^*, \ldots, x_{N(1)}^*$ be chosen from S_{X^*} in such a way that

$$|x_i^* e_i| > 1 - \frac{c_1}{3}.$$

By hypothesis, there is an $x_2 \in B$ so that

$$|x_0^* x_2 - x_0^{**}x_0^*| < \tfrac{1}{2}, \qquad |x_1^* x_2 - x_0^{**}x_1^*|, \ldots, |x_{N(1)}^* x_2 - x_0^{**}x_{N(1)}^*| < \frac{\delta c_1}{6}$$

all hold. *Notice* that for any $e \in E_1$ and any scalar t we get

$$\left\|e + t\left(x_2 - x_0^{**}\right)\right\| \geq (1 - c_1)\|e\|. \tag{2}$$

Homogeneity of the norm allows us to prove (2) for $e \in S_{E_1}$ and conclude to its validity for all members of E_1. Two possibilities come to mind: $|t| \leq 2/\delta$ and $|t| > 2/\delta$. First, $|t| \leq 2/\delta$: pick e_i so that $\|e - e_i\| < c_1/3$ and look at what happens.

$$\begin{aligned}\left\|e + t\left(x_2 - x_0^{**}\right)\right\| &\geq \left|\left(e + t\left(x_2 - x_0^{**}\right)\right)\left(x_i^*\right)\right| \\ &\geq |x_i^* e_i| - \left|t\left(x_2 - x_0^{**}\right)\left(x_i^*\right)\right| - \|x_i^*\|\,\|e - e_i\| \\ &\geq \left(1 - \frac{c_1}{3}\right) - \frac{2}{\delta} \cdot \frac{\delta c_1}{6} - \frac{c_1}{3} \\ &= 1 - c_1 = (1 - c_1)\|e\|.\end{aligned}$$

The second possibility, $|t| > 2/\delta$, is easy, too:

$$\begin{aligned}\left\|e + t\left(x_2 - x_0^{**}\right)\right\| &\geq \frac{2}{\delta}\|x_2 - x_0^{**}\| - \|e\| \\ &\geq \frac{2}{\delta}\delta - \|e\| \geq 2 - 1 = 1 \geq (1 - c_1)\|e\|.\end{aligned}$$

Let's check up on a linear combination of $x_1 - x_0^{**}$ and $x_2 - x_0^{**}$, say $t_1(x_1 - x_0^{**}) + t_2(x_0, x_0^{**})$. Letting e in (2) be $t_1(x_1 - x_0^{**})$, we get

$$\left\|t_1\left(x_1 - x_0^{**}\right) + t_2\left(x_2 - x_0^{**}\right)\right\| \geq (1 - c_1)\left\|t_1\left(x_1 - x_0^{**}\right)\right\|.$$

Suppose we repeat the above procedure.

Let E_2 denote the 2-dimensional subspace of X^{**} spanned by $x_1 - x_0^{**}$ and $x_2 - x_0^{**}$. There are elements $e_1, \ldots, e_{N(2)} \in S_{E_2}$ (not necessarily related to the $c_1/3$ net) which form a $c_2/3$ net for S_{E_2}. Pick $x_1^*, \ldots, x_{N(2)}^* \in S_{X^*}$ so that

$$|x_i^* e_i| > 1 - \frac{c_2}{3}.$$

By hypothesis, there is an $x_3 \in B$ such that

$$|x_0^* x_3 - x_0^{**} x_0^*| < \tfrac{1}{3}, \qquad |x_1^* x_3 - x_0^{**} x_1^*|, \ldots, |x_{N(2)}^* x_3 - x_0^{**} x_{N(2)}^*| < \frac{\delta c_2}{6}$$

all hold. Notice that for any $e \in E_2$ and any scalar t we have

$$\left\|e + t\left(x_3 - x_0^{**}\right)\right\| \geq (1 - c_2)\|e\|. \tag{3}$$

We leave the verification to the reader; actually two possibilities ought to come to mind (on reducing the problem to $\|e\| = 1$), and each is handled precisely as before with only the names being changed. If a linear combination $t_1(x_1 - x_0^{**}) + t_2(x_2 - x_0^{**}) + t_3(x_3 - x_0^{**})$ is under consideration, then

(3) tells us [on letting $e = t_1(x_1 - x_0^{**}) + t_2(x_2 - x_0^{**})$, naturally] that

$$\left\| \sum_{i=1}^{3} t_i(x_i - x_0^{**}) \right\| \geq (1 - c_2) \left\| \sum_{i=1}^{2} t_i(x_i - x_0^{**}) \right\|.$$

Proceeding thusly, we find a sequence (x_n) in B such that for all $n \geq 1$,

$$|x_0^* x_n - x_0^{**} x_0^*| < \frac{1}{n}$$

and for which given $1 \leq p < q < \infty$ and scalars $t_1, \ldots, t_q$,

$$\left\| \sum_{i=1}^{p} t_i(x_i - x_0^{**}) \right\| \leq \prod_{i=p}^{q-1} \left(\frac{1}{1 - c_i} \right) \left\| \sum_{i=1}^{q} t_i(x_i - x_0^{**}) \right\|.$$

It is now plain that we can find (x_n) and x_0^* to satisfy 1 and 2. To see that should $x_0^{**} \neq 0$, we could achieve 3 as well, we must notice that

$$\bigcap_{k=1}^{\infty} \text{closed linear span } \{x_k - x_0^{**}, x_{k+1} - x_0^{**}, \ldots\} = 0;$$

so eventually the subspaces $[x_n - x_0^{**}]_{n \geq k}$ expel x_0^{**} from their premises. If done at $k = k_0$, just look at the sequence $(x_{n+k_0})_{n \geq 1} \subseteq B$; it achieves 1, 2, *and* 3. □

Now we are ready for the Eberlein-Šmulian theorem.

The Eberlein-Šmulian Theorem (Pelczynski Style). *Let B be a bounded subset of the Banach space X. Then the following statements about B are equivalent*:

1. *The weak closure of B is not weakly compact.*
2. *B contains a countable set C with no weak limit point in X.*
3. *There's a basic sequence (x_n) in B such that for some $x_0^* \in X^*$, $\lim_n x_0^* x_n > 0$.*
4. *B is not weakly sequentially compact in X.*

PROOF. Statement 1 implies 3. By statement 1 there must exist an $x_0^{**} \in X^{**} \setminus X$ in the weak* closure of B up in X^{**}. Notice that $d(x_0^{**}, B) \geq d(x_0^{**}, X) > 0$. Applying Lemma 4, we find a sequence (x_n) in B and an $x_0^* \in X^*$ such that

(i) $\lim_n x_0^* x_n = x_0^{**} x_0^* \geq \|x_0^{**}\|/2$.
(ii) $(x_n - x_0^{**})$ is a basic sequence in X^{**}.
(iii) $x_0^{**} \notin [x_n - x_0^{**}] =$ closed linear span of $\{x_n - x_0^{**}\}_{n \geq 1}$.

Let $Z = [x_0^{**}, \{x_n\}_{n \geq 1}]$. Since x_0^{**} is not in $[x_n]$, nor is it in $[x_n - x_0^{**}]$, each of these subspaces is of codimension 1 in Z. Therefore, there are bounded linear projections $A, P: Z \to Z$ such that $PZ = [x_n]$ and $AZ = [x_n - x_0^{**}]$, where $Ax_0^{**} = 0 = Px_0^{**}$. Obviously, if $z^{**} \in Z$, then there's a scalar $t_{z^{**}} = t$ such that $z^{**} - Pz^{**} = tx_0^{**}$; therefore, if $z^{**} \in [x_n - x_0^{**}]$, $z^{**} = Az^{**} = APz^{**}$. By symmetry, $PAx = x$ for any $x \in [x_n]$. It follows that P maps $[x_n - x_0^{**}]$ onto $[x_n]$ in an isomorphic manner. Since $P(x_n - x_0^{**}) = x_n$ for all n, (x_n) is a basic sequence which satisfies $\underline{\lim}_n x_0^* x_n \geq \|x_0^*\|/2 > 0$, thanks to (i).

Statement 3 implies 2. Let $C = \{x_n\}$, where (x_n) is the basic sequence alluded to in 3. The inequality $\underline{\lim}_n x_0^* x_n > 0$ eliminates the origin as a potential weak limit point of C, yet the origin serves as the only possible weak limit point of any basic sequence. The verdict: C has no weak limit points.

That 2 implies 1 and 4 is plain; therefore, we concentrate on showing that 4 in the absence of 2 is contradictory. The assumption of statement 4 leads to a sequence (y_n) of points of B, none of whose subsequences are weakly convergent to a member of X. Since no subsequence of (y_n) is norm convergent, we can pass to a subsequence and assume that $\{y_n\}$ is norm discrete; $\{y_n\}$ has a weak limit point x_0 in X—after all, we are denying 2. x_0 is not a norm limit point of $\{y_n\}$; so, with the exclusion of but a few y_n, we can assume $d(x_0, \{y_n\}) > 0$. We can apply Lemma 4 again to extract a subsequence (x_n) from (y_n) so that $(x_n - x_0)$ is a basic sequence. Remember we're denying 2; so $\{x_n\}$ has a weak limit point, *but* x_0 is the only candidate for the position since $(x_n - x_0)$ is basic! (x_n) converges weakly to x_0, which is a contradiction to 4. □

More mimicry of Mazur's technique provides us with a utility-grade version of a principle for selecting basic sequences due to Bessaga and Pelczynski (1958). Though we will soon be presenting the complete unexpurgated story of the Bessaga-Pelczynski selection principle, the following milder form is worth pursuing at this imprecise moment.

Bessaga-Pelczynski Selection Principle (Utility-Grade). *Let (x_n) be a weakly null, normalized sequence in the Banach space X. Then (x_n) admits of a basic sequence.*

Proof. Let $(\varepsilon_n)_{n \geq 0}$ be a sequence of positive numbers each less than 1 for which $\prod_{n=1}^{\infty}(1 - \varepsilon_n) > 1 - \varepsilon_0$.

Suppose that in our quest for a basic subsequence we have fought our way through to choosing $x_{n_1}, x_{n_2}, \ldots, x_{n_k}$ with $n_1 < n_2 < \cdots < n_k$, of course. Let $Y(k)$ be the linear span of $x_1, \ldots, x_{n_k}$.

Pick $z_1, \ldots, z_m$ in $S_{Y(k)}$ so that each $y \in S_{Y(k)}$ lies within $\varepsilon_k/4$ of a z. Correspondingly, there are $z_1^*, \ldots, z_m^*$ in S_{X^*} so that $z_i^* z_i > 1 - \varepsilon_k/4$ for each $i = 1, 2, \ldots, m$. Eventually we run across an $x_{n_{k+1}}$, where $n_{k+1} > n_k$, for which $|z_1^* x_{n_{k+1}}|, |z_2^* x_{n_{k+1}}|, \ldots, |z_m^* x_{n_{k+1}}|$ are all less than $\varepsilon_k/4$. We claim that for any $y \in S_{Y(k)}$ and any scalar α,

$$\|y + \alpha x_{n_{k+1}}\| \geq (1 - \varepsilon_k)\|y\|. \tag{4}$$

Sound familiar? It should. A quick peek at what we did in Corollary 3 or in

Lemma 4 will tell the story: (x_{n_k}) is a basic subsequence of (x_n). Let's verify (4) (again). Two possibilities come to mind: $|\alpha| < 2$ and $|\alpha| \geq 2$. If $|\alpha| < 2$, then on picking z_i $(1 \leq i \leq m)$ so that $\|y - z_i\| < \varepsilon_k/4$, we see that

$$\begin{aligned}\|y + \alpha x_{n_{k+1}}\| &\geq \left|z_i^*(y + \alpha x_{n_{k+1}})\right| \\ &\geq |z_i^* z_i| - \left|z_i^*(y - z_i)\right| - \left|z_i^*(\alpha x_{n_{k+1}})\right| \\ &\geq \left(1 - \frac{\varepsilon_k}{4}\right) - \|y - z_i\| - 2|z_i^* x_{n_{k+1}}| \\ &\geq \left(1 - \frac{\varepsilon_k}{4}\right) - \frac{\varepsilon_k}{4} - \frac{2\varepsilon_k}{4} = 1 - \varepsilon_k = (1 - \varepsilon_k)\|y\|.\end{aligned}$$

If $|\alpha| \geq 2$, then

$$\|y + \alpha x_{n_{k+1}}\| \geq |\alpha|\,\|x_{n_{k+1}}\| - \|y\| = 2 - 1 \geq (1 - \varepsilon_k)\|y\|. \qquad \square$$

The natural bases for classical spaces play a central role in the study of Banach spaces, and the ability to recognize their présence (as a basic sequence) in different circumstances is worth developing. For this reason we introduce the notion of equivalent bases.

Let (x_n) be a basis for X and (y_n) be a basis for Y. We say that (x_n) and (y_n) are *equivalent* if the convergence of $\sum_n a_n x_n$ is equivalent to that of $\sum_n a_n y_n$.

Theorem 5. *The bases (x_n) and (y_n) are equivalent if and only if there is an isomorphism between X and Y that carries each x_n to y_n.*

PROOF. Recall that in our earlier comments about bases we renormed X by taking any $x = \sum_n s_n x_n$ and defining $|||x|||$ by

$$|||x||| = \sup_n \left\| \sum_{k=1}^{n} s_k x_k \right\|.$$

Result: An isomorph of X in which (x_k) is still a basis but is now a "monotone" basis, i.e., $|||\sum_{k=1}^{m} s_k x_k||| \leq |||\sum_{k=1}^{m+n} s_k x_k|||$ for any $m, n \geq 1$. Notice that if (x_n) and (y_n) are equivalent, then they are equivalent regardless of which equivalent norm is put on their spans. So we might as well assume each is monotone to begin with; we do so and now look at the operator $T: X \to Y$ that takes $\sum_n a_n x_n$ to $\sum_n a_n y_n$ (what other operator could there be?); T is one to one and onto. T also has a closed graph; this is easy to see from the monotonicity of each basis. T is an isomorphism and takes x_n to y_n. Enough said about the necessity of the condition; sufficiency requires but a moment of reflection, and we recommend such to the reader. $\square$

Equivalence of bases is a finer gradation than the isomorphic nature of their spans. Indeed, Pelczynski and Singer showed that anv infinite-dimensional Banach space admitting a basis has uncountably many nonequivalent bases! What's the situation with natural bases for special spaces? How can we recognize them? For some bases, satisfactory answers are known. One such case is the unit vector basis of c_0. Corollary 7 below characterizes c_0's unit vector basis, and Theorem 8 gives an elegant application to the theory of series in Banach spaces.

A series $\Sigma_n x_n$ is said to be *weakly unconditionally Cauchy* (wuC) if, given any permutation π of the natural numbers, $(\Sigma_{k=1}^n x_{\pi(k)})$ is a weakly Cauchy sequence; alternatively, $\Sigma_n x_n$ is wuC if and only if for each $x^* \in X^*$, $\Sigma_n |x^* x_n| < \infty$.

Theorem 6. *The following statements regarding a formal series $\Sigma_n x_n$ in a Banach space are equivalent:*

1. $\Sigma_n x_n$ *is* wuC.
2. *There is a $C > 0$ such that for any $(t_n) \in l_\infty$*
$$\sup_n \left\| \sum_{k=1}^n t_k x_k \right\| \leq C \sup_n |t_n|.$$
3. *For any $(t_n) \in c_0$, $\Sigma_n t_n x_n$ converges.*
4. *There is a $C > 0$ such that for any finite subset Δ of N and any signs $\pm$ we have $\|\Sigma_{n \in \Delta} \pm x_n\| \leq C$.*

Proof. Suppose 1 holds and define $T: X^* \to l_1$ by
$$Tx^* = (x^* x_n).$$
T is a well-defined linear map with a closed graph; therefore, T is bounded. From this we see that for any $(t_n) \in B_{l_\infty}$ and any $x^* \in B_{X^*}$,
$$\left| x^* \sum_{k=1}^n t_k x_k \right| = |(t_1, \ldots, t_n, 0, 0, \ldots)(Tx^*)|$$
$$\leq \|T\|.$$
Part 2 follows from this.

If we suppose 2 holds and let $(t_n) \in c_0$, then keeping $m < n$ and letting both go off to ∞, we have
$$\left\| \sum_{k=m}^n t_k x_k \right\| \leq C \sup_{m \leq k < n} |t_k| \to 0$$
from which 3 follows easily.

If 3 holds, then the operator $T: c_0 \to X$ defined by
$$T(t_n) = \sum_n t_n x_n$$
cannot be far behind; part 3 assures us that T is well-defined. T is plainly linear and has a closed graph. T is bounded. The values of T on B_{c_0} are

bounded. In particular, vectors of the form $\Sigma_{n \in \Delta} \pm x_n$, where Δ ranges over the finite subsets Δ of N and we allow all the $\pm$'s available, are among the values of T on B_{c_0}, and that is statement 4.

Finally, if 4 is in effect, then for any $x^* \in B_{X^*}$ we have

$$x^* \sum_{n \in \Delta} \pm x_n = \sum_{n \in \Delta} \pm x^* x_n$$
$$\leq \left\| \sum_{n \in \Delta} \pm x_n \right\| \leq C$$

for any finite subset Δ of N and any choice of signs $\pm$. That $\Sigma_n |x^* x_n| < \infty$ follows directly from this, and along with it we get part 1. □

Corollary 7. *A basic sequence for which* $\inf_n \|x_n\| > 0$ *and* $\Sigma_n x_n$ *is* wuC *is equivalent to the unit vector basis of* c_0.

PROOF. If (x_n) is a basic sequence and $\Sigma_n t_n x_n$ is convergent, then $(\Sigma_{k=1}^n t_k x_k)$ is a Cauchy sequence. Therefore, letting n tend to infinity, the sequence

$$|t_n| \|x_n\| = \left\| \sum_{k=1}^{n} t_k x_k - \sum_{k=1}^{n-1} t_k x_k \right\|$$

tends to 0; from this and the restraint $\inf_n \|x_n\| > 0$, it follows that $(t_n) \in c_0$.

On the other hand, if (x_n) is a basic sequence and $\Sigma_n x_n$ is wuC, then $\Sigma_n t_n x_n$ converges for each $(t_n) \in c_0$, thanks to Theorem 6, part 3.

Consequently, a basic sequence (x_n) with $\inf_n \|x_n\| > 0$ and for which $\Sigma_n x_n$ is wuC is equivalent to the unit vector basis of c_0. □

Theorem 8. *Let X be a Banach space. Then, in order that each series* $\Sigma_n x_n$ *in* X *with* $\Sigma_n |x^* x_n| < \infty$ *for each* $x^* \in X^*$ *be unconditionally convergent, it is both necessary and sufficient that X contains no copy of* c_0.

PROOF. If X contains a copy of c_0, then the series corresponding to $\Sigma_n e_n$, where e_n is the nth unit coordinate vector, is wuC but not unconditionally convergent.

On the other hand, if X admits a series $\Sigma_n x_n$ which is not unconditionally convergent yet satisfies $\Sigma_n |x^* x_n| < \infty$ for each $x^* \in X^*$, then for some sequences $(p_n), (q_n)$ of positive integers with $p_1 < q_1 < p_2 < q_2 < \cdots$, we have $\inf_n \|\Sigma_{k=p_n}^{q_n} x_k\| > 0$. Letting $y_n = \Sigma_{k=p_n}^{q_n} x_k$, we see that (y_n) is weakly null and $\inf_n \|y_n\| > 0$. Normalizing (y_n), we keep the weakly null feature and can utilize the Bessaga-Pelczynski selection principle and Corollary 7 to find a basic subsequence of (y_n) equivalent to c_0's unit vector basis. Theorem 5 takes over: a copy of c_0 is contained in X. □

The above results of Bessaga and Pelczynski can be used to give another proof of the Orlicz-Pettis theorem. Indeed, a weakly subseries convergent series $\Sigma_n x_n$ in a Banach space is wuC in that space. Should $\Sigma_n x_n$ not be subseries convergent, three increasing sequences (p_n), (q_n), and (r_n) of positive integers could be found with $p_1 < q_1 < p_2 < q_2 < \cdots$, such that the sequence (y_n) given by

$$y_n = \sum_{i=p_n}^{q_n} x_{r_i}$$

satisfies $\|y_n\| \geq \varepsilon > 0$ for some judiciously chosen ε. Now $\Sigma_n y_n$ is a subseries of $\Sigma_n x_n$ and so is weakly subseries convergent too. In particular, (y_n) is weakly null and $\inf_n \|y_n\| > 0$; there is a subsequence (z_n) of (y_n) that is basic. A look at Corollary 7 will tell you that (z_n) is equivalent to the unit vector basis (e_n) of c_0, yet a further look will convince you that $\Sigma_n e_n$ is *not* weakly convergent. This flaw proves the theorem.

The study of a sequential problem ofttimes reduces to analysis inside some space with a basis, and approximation in terms of expansions with respect to this basis plays an important role in the study under way. Frequently useful in such ventures is the notion of a *block basic sequence*: Let (x_n) be a basis for a Banach space, (p_n) and (q_n) be intertwining sequences of positive integers (i.e., $p_1 < q_1 < p_2 < q_2 < \cdots$), and $y_n = \Sigma_{i=p_n}^{q_n} a_i x_i$ be nontrivial linear combinations of the x_i; we call the sequence (y_n) a block basic sequence taken with respect to (x_n), or simply a block basic sequence. It is easy (and safe) to believe that (y_n) is basic (just look at Theorem 1). The following results of Bessaga and Pelczynski establishes the fundamental criterion for locating block basic sequences.

Bessaga-Pelczynski Selection Principle. *Let (x_n) be a basis for X and suppose (x_n^*) is the sequence of coefficient functionals. If (y_m) is a sequence in X for which*

$$\varliminf_m \|y_m\| > 0$$

and

$$\lim_m x_n^* y_m = 0 \quad \textit{for each } n,$$

then (y_n) has a subsequence that is equivalent to a block basic sequence taken with respect to (x_n).

PROOF. First, we find a way of ensuring that a constructed basic sequence is equivalent to an existent one. We prove a stability result of enough interest by itself that we call it Theorem 9.

Theorem 9. *Let (z_n) be a basic sequence in the Banach space X, and suppose (z_n^*) is the sequence of coefficient functionals (extended to all of X in a Hahn-Banach fashion). Suppose (y_n) is a sequence in X for which $\Sigma_n \|z_n^*\| \, \|z_n - y_n\| < 1$. Then (y_n) is a basic sequence equivalent to (z_n).*

In fact, if we define $T\colon X \to X$ by

$$Tx = \sum_n z_n^*(x)(z_n - y_n),$$

then $\|T\| \leq \Sigma_n \|z_n^*\| \|z_n - y_n\| < 1$. It follows that $(I + T + T^2 + \cdots + T^n)$ converges in operator norm to $(I - T)^{-1}$; $I - T$ is a bounded linear operator from X onto X with a bounded inverse. Of course, $(I - T)(z_n) = y_n$.

Back to the Bessaga-Pelczynski selection principle, let $K > 0$ be chosen so that for any $m, n \geq 1$

$$\left\| \sum_{k=1}^{m} a_k x_k \right\| \leq K \left\| \sum_{k=1}^{m+n} a_k x_k \right\|.$$

By passing to a subsequence, we might as well assume that $\|y_m\| \geq \varepsilon > 0$ for all n. With but a slight loss of generality (none of any essential value), we can assume that $\|y_m\| = 1$ for all m. Now we get on with the proof.

Since (x_n) is a basis for X, y_1 admits an expansion,

$$y_1 = \sum_n x_n^*(y_1) x_n.$$

Hence there is a q_1 such that

$$\left\| \sum_{k=q_1+1}^{\infty} x_k^*(y_1) x_k \right\| < \frac{1}{4K2^3}.$$

Since $\lim_m x_n^*(y_m) = 0$ for each n, there is a $p_2 > 1 = p_1$ such that

$$\left\| \sum_{k=1}^{q_1} x_k^*(y_{p_2}) x_k \right\| < \frac{1}{4K2^4}.$$

Again, (x_n) is a basis for X; so y_{p_2} admits a representation,

$$y_{p_2} = \sum_n x_n^*(y_{p_2}) x_n.$$

Hence there is a $q_2 > q_1$ such that

$$\left\| \sum_{k=q_2+1}^{\infty} x_k^*(y_{p_2}) x_k \right\| < \frac{1}{4K2^4}.$$

Once more, appeal to the assumption that $\lim_m x_n^* y_m = 0$ for each n to pick a $p_3 > p_2$ such that

$$\left\| \sum_{k=1}^{q_2} x_k^*(y_{p_3}) x_k \right\| < \frac{1}{4K2^5}.$$

Got your p's and q's straight? Let

$$z_n = \sum_{k=q_n+1}^{q_{n+1}} x_k^*(y_{p_{n+1}}) x_k.$$

Note that

$$1=\|y_{p_{n+1}}\|=\left\|\left(\sum_{k=1}^{q_n}+\sum_{k=q_n+1}^{q_{n+1}}+\sum_{k=q_{n+1}+1}^{\infty}\right)x_k^*(y_{p_{n+1}})x_k\right\|$$
$$\leq\left\|\left(\sum_{k=1}^{q_n}+\sum_{k=q_{n+1}+1}^{\infty}\right)x_k^*(y_{p_{n+1}})x_k\right\|+\|z_n\|$$
$$\leq\frac{1}{4K2^{n+2}}+\frac{1}{4K2^{n+2}}+\|z_n\|.$$

It follows that $\|z_n\|\geq\frac{1}{2}$ for all n. (z_n) is a block basic sequence taken with respect to (x_n) and has the same expansion constant K as does (x_n); i.e., whenever $k\leq j$, we have.

$$\left\|\sum_{i=1}^{k}\alpha_i z_i\right\|\leq K\left\|\sum_{i=1}^{j}\alpha_i z_i\right\|.$$

From this and the fact that $\|z_n\|\geq\frac{1}{2}$ we see that the coefficient functionals of (z_n) satisfy $\|z_n^*\|\leq 4K$. Now we look to Theorem 9:

$$\sum_n\|z_n^*\|\,\|z_n-y_{p_{n+1}}\|\leq 4K\sum_n\|z_n-y_{p_{n+1}}\|$$
$$\leq 4K\sum_n\left\|\left(\sum_{k=1}^{q_n}+\sum_{k=q_{n+1}+1}^{\infty}\right)x_k^*(y_{p_{n+1}})x_k\right\|$$
$$\leq 4K\sum_n\left(\frac{1}{4K2^{n+1}}+\frac{1}{4K2^{n+2}}\right)$$
$$=\sum_n\left(\frac{1}{2^{n+2}}+\frac{1}{2^{n+2}}\right)=\frac{1}{2}.\qquad\square$$

For a quick application of the selection principle we present the following theorem of Bessaga and Pelczynski.

Theorem 10. *The following are equivalent*:

1. X^* *contains a copy of* c_0.
2. X *contains a complemented copy of* l_1.
3. X^* *contains a copy* Z *of* l_∞ *for which*
   a. Z *is isomorphic to* l_∞ *when* Z *is given the relative weak* topology of* X^* *and* l_∞ *has its usual weak* topology as* l_1*'s dual.*
   b. *There is a projection* $P: X^*\to X^*$ *which is weak*-weak* continuous and for which* $PX^*=Z$.

Proof. Only the derivation of 2 from 1 needs proof.

Our derivation of 2 will turn on the following property of l_1: *if* l_1 *is a quotient of the Banach space* X, *then* l_1 *is isomorphic to a complemented subspace of X*. The easy proof of this can be found in Chapter VII, but insofar as it is key to the present situation, a word or two is appropriate. Let $Q: X \to l_1$ be a bounded linear operator of X onto l_1; by the open mapping theorem there is a bounded sequence (b_n) in X for which $Qb_n = e_n$. The sequence (b_n) is equivalent to the unit vector basis of l_1; furthermore, if $R: l_1 \to X$ is defined by $Re_n = b_n$, then QR is the identity operator. From this it follows that $RQ: X \to X$ is a bounded linear projection from X onto $[b_n]$, a space isomorphic to l_1.

Let $T: c_0 \to X^*$ be an isomorphism and denote, as usual, by (e_n) the unit vector basis of c_0. Look at $T^*: X^{**} \to l_1$ and let $S = T^*|_X$; for any $x \in X$, $Sx = (Te_1(x), Te_2(x), \ldots)$. Since T is an isomorphism, T^* is a quotient map. B_X is weak* dense in $B_{X^{**}}$ thanks to Goldstine's theorem; therefore, $S(B_X)$ is weak* dense in $T^*B_{X^{**}}$, a neighborhood of the origin in l_1. It follows that for some sequence (λ_n) of scalars bounded away from 0 and some sequence (x_n) in B_X, the Sx_n are weak* close to the $\lambda_n e_n^*$, where e_n^* is the nth unit vector in l_1. How close? Well, close enough to ensure that $\lim_n (Sx_n)(e_k) = 0$ for each k and that the $(Sx_n)(e_n)$ are bounded away from zero. The norm of Sx_n is kept away from zero by its value on e_n; also the values of e_k on the Sx_n tend to zero as n goes off toward infinity. By the Bessaga-Pelczynski selection principle, (Sx_n) must have a subsequence (Sx_{k_n}) that is equivalent to a block basic sequence taken with respect to the unit vector basis of l_1. But it is easy to see that block bases built out of l_1's unit vectors are equivalent to the original unit vector basis of l_1 and, in fact, span a subspace of l_1 complemented in l_1 and, of course, isomorphic to l_1.

Therefore, S followed by a suitable isomorphism produces an operator from X onto a space isomorphic to l_1. X admits l_1 as a quotient. l_1 is isomorphic to a complemented subspace of X. □

Now to return to series in Banach spaces we note the following:

Corollary 11. *In order that each series $\Sigma_n x_n^*$ in the dual X^* of a Banach space X for which $\Sigma_n |x_n^* x| < \infty$ for each $x \in X$ be unconditionally convergent, it is both necessary and sufficient that X^* contain no isomorphic copy of l_∞.*

PROOF. If X^* contains an isomorphic copy of l_∞, then it contains a weak* isomorphic copy Z of l_∞ as described in part 3 of Theorem 10. Looking at the unit vectors of c_0 as they appear in Z, they look just as they do in $l_\infty : \Sigma_n e_n$ is weak* unconditionally convergent in l_∞ to 1 but certainly not norm convergent to anything; the same can be done in Z.

On the other hand, if $\Sigma_n |x_n^* x| < \infty$ for each $x \in X$, then $(\Sigma_{k=1}^n x_k^*)$ is a weak* Cauchy sequence in X^*, and so

$$\text{weak}^* \lim_n \sum_{k=1}^n x_k^*$$

exists by Alaoglu's theorem. Furthermore, if $(t_n) \in c_0$, then $\Sigma_n |t_n x_n^*(x)| < \infty$

for each $x \in X$ and

$$\text{weak* } \lim_n \sum_{k=1}^{n} t_k x_k^*$$

exists as well. An operator is "born"; define $T: c_0 \to X^*$ by

$$T((t_n)) = \text{weak* } \lim_n \sum_{k=1}^{n} t_k x_k^*.$$

T is linear and has closed graph; hence, T is a bounded linear operator. Regardless of the finite set Δ of positive integers considered or of the choices of signs $\pm$ made,

$$\left\| \sum_{n \in \Delta} \pm x_n^* \right\| = \left\| T\left(\sum_{n \in \Delta} \pm e_n \right) \right\| \leq \|T\|.$$

$\Sigma_n x_n^*$ is wuC. If X^* does not contain l_∞, it cannot contain c_0 by Theorem 10; in such a case, $\Sigma_n x_n^*$ is unconditionally convergent by Theorem 8. □

Just as Theorem 9 ensures that sequences close to basic sequences are themselves basic, our next result tells us that if a basic sequence spans a complemented subspace and if you nudge the sequence with delicate enough stroke, then the resulting sequence is basic and spans a complemented subspace.

Theorem 12. *Let (z_n) be a basic sequence in the Banach space X with coefficient functionals (z_n^*). Suppose that there is a bounded linear projection $P: X \to X$ onto the closed linear span $[z_n]$ of the z_n. If (y_n) is any sequence in X for which*

$$\sum_n \|P\| \|z_n^*\| \|z_n - y_n\| < 1,$$

then (y_n) is a basic sequence equivalent to (z_n) and the closed linear span $[y_n]$ of the y_n is also complemented in X.

PROOF. Since P is a linear projection with nontrivial range, $\|P\| \geq 1$. It follows then from Theorem 9 that (y_n) is a basic sequence equivalent to (z_n). The condition set forth in the hypotheses is easily seen to be just what is needed to prove that the operator $A: X \to X$ defined by

$$Ax = x - Px + \sum_n z_n^*(Px) y_n$$

satisfies $\|A - I\| < 1$. Therefore, A is an isomorphism of X onto itself. It is easy to see that $Az_n = y_n$. Finally, if we look at $Q = APA^{-1}$, then we should see that $Q^2 = APA^{-1}APA^{-1} = APPA^{-1} = APA^{-1} = Q$; since the range of Q is $[y_n]$, the proof is complete. □

We remark that Theorem 12 finds frequent use in the study of the structure of Banach spaces; in fact, we will have an opportunity to apply it in a somewhat typical situation in Chapter VII.

There is a more-or-less natural sequence of events that precedes the application of Theorem 12 in special spaces. Suppose, for instance, you're working in the space l_p (some finite $p \geq 1$). One way to produce a complemented subspace of l_p is to build vectors in the following fashion: Take sequences (m_k) and (n_k) of positive integers with

$$1 \leq m_1 \leq n_1 < m_2 \leq n_2 < \cdots < m_k \leq n_k < m_{k+1} \cdots,$$

and build nonzero blocks

$$b_k = \sum_{j=m_k}^{n_k} a_j e_j.$$

Then the closed linear span of the b_k is isomorphic to l_p (this is not hard to see), the sequence $(b_k / \|b_k\|)$ is a basic sequence equivalent to the unit vector basis (e_k) of l_p, and the closed linear span of the b_k is complemented in l_p.

Indeed, only the last of these statements needs any real demonstration. The basic sequence $(b_k / \|b_k\|)$ has a companion sequence (β_n^*) of coefficient functionals defined on all of l_p (after suitably extending via the Hahn-Banach theorem). If $x \in l_p$, then $Px = \sum_k \beta_k^*(\sum_{m_k}^{n_k} x_j e_j) b_k / \|b_k\|$ defines a bounded linear projection $P: l_p \to l_p$ whose range is the closed linear span of the b_k.

It is one of the more pleasant facts of life that many of the situations in which one wants to find a complemented copy of l_p somewhere, there is a sequence like b_k near by, close enough in fact to apply Theorem 12.

Exercises

1. *Renorming spaces to improve basis constants.* Let (x_n) be a basis for the Banach space X.

(i) Show that X can be given an equivalent norm $|||\cdot|||$ such that for any scalars $a_1, a_2, \ldots, a_m, a_{m+1}, \ldots, a_n$, we have

$$\left|\left|\left| \sum_{i=1}^{m} a_i x_i \right|\right|\right| \leq \left|\left|\left| \sum_{i=1}^{n} a_i x_i \right|\right|\right|.$$

(ii) Suppose (x_n) is an unconditional basis for X. Show that there is a constant $K > 0$ so that given any permutation π of the natural numbers and any $x = \sum_n x_n^*(x) x_n \in X$, we have

$$\left\| \sum_n x_{\pi(n)}^*(x) x_{\pi(n)} \right\| \leq K \left\| \sum_n x_n^*(x) x_n \right\|.$$

(iii) Show that if (x_n) is an unconditional basis for X, then X can be renormed so that, whenever π is a permutation of the natural numbers and $x =$

$\Sigma_n x_n^*(x)x_n \in X$, we have

$$\left|\left|\left|\sum_n x_{\pi(n)}^*(x)x_{\pi(n)}\right|\right|\right| \leq \left|\left|\left|\sum_n x_n^*(x)x_n\right|\right|\right|.$$

(iv) If (x_n) is an unconditional basis for X, $x \in X$, and $t = (t_n) \in l_\infty$, then $\Sigma_n t_n x_n^*(x)x_n \in X$. Show that there exists a constant $K > 0$ such that for any $x = \Sigma_n x_n^*(x)x_n \in X$ and any $(t_n) \in B_{l_\infty}$, we have

$$\left\|\sum_n t_n x_n^*(x)x_n\right\| \leq K\left\|\sum_n x_n^*(x)x_n\right\|.$$

(v) If (x_n) is an unconditional basis for X, then X can be renormed so that for any $x \in X$ and any $(t_n) \in B_{l_\infty}$, we have

$$\left|\left|\left|\sum_n t_n x_n^*(x)x_n\right|\right|\right| \leq \left|\left|\left|\sum_n x_n^*(x)x_n\right|\right|\right|.$$

2. *The unit vector bases of c_0 and l_1.*

(i) A normalized basic sequence (x_n) is equivalent to the unit vector basis of c_0 if and only if there is a constant $K > 0$ such that

$$\left\|\sum_{i=1}^n c_i x_i\right\| \leq K \sup_{1 \leq i \leq n} |c_i|$$

holds for any n and any scalars $c_1, c_2, \ldots, c_n$.

(ii) A normalized basic sequence (x_n) is equivalent to the unit vector basis of l_1 if and only if there is a constant $K > 0$ such that

$$\sum_{i=1}^n |c_i| \leq K\left\|\sum_{i=1}^n c_i x_i\right\|$$

holds for any n and any scalars $c_1, c_2, \ldots, c_n$.

(iii) Any time there is an $x^* \in S_{X^*}$ such that $x^* x_n \geq \theta > 0$ for some fixed θ and all terms x_n of a normalized basic sequence (x_n), then (x_n) is equivalent to the unit vector basis of l_1.

3. *Shrinking bases and boundedly complete bases.* Let (x_n) be a basis for X and (x_n^*) be the coefficient functionals.

(i) (x_n^*) is always a basis for its closed linear span in X^*; further, (x_n^*) is a "weak* basis" for X^*, i.e., each $x^* \in X^*$ has a unique representation in the form $x^* = \text{weak}^* \lim_n \Sigma_{k=1}^n a_k x_k^*$.

(ii) Each of the following is necessary and sufficient for (x_n^*) to be a basis for X^*:

(a) The closed linear span of $\{x_n^* : n \geq 1\}$ is X^*.

(b) $\lim_n \|x^*\|_n = 0$ for each $x^* \in X^*$, where $\|x^*\|_n$ is the norm of x^* when x^* is restricted to the linear span of $\{x_{n+1}, x_{n+2}, \ldots\}$.

A basis having the properties enunciated in (ii) is called *shrinking*. A companion notion to that of a shrinking basis is that of a *boundedly complete* basis; the basis (x_n) is called *boundedly complete* whenever given a sequence (a_n) of scalars for which $\{\Sigma_{k=1}^n a_k x_k : n \geq 1\}$ is bounded, then $\lim_n \Sigma_{k=1}^n a_k x_k$ exists.

(iii) If (x_n) is a shrinking basis for X, then (x_n^*) is a boundedly complete basis for X^*.

4. *Boundedly complete bases span duals.* Let (x_n) be a boundedly complete basis for X, let (x_n^*) denote the sequence of coefficient functionals, and let $[x_n^*]$ denote the norm-closed linear span of the x_n^* in X^*.

(i) Show that for each $x^{**} \in X^{**}$ the series

$$\sum_n x^{**}(x_n^*) x_n$$

converges to an element of X. [*Hint*: A diagonal argument can be used to find a sequence (y_n) in B_X such that $\lim_n x_k^* y_n = x^{**} x_k^*$ holds for $k = 1, 2, \ldots$. This lets one realize vectors of the form $\Sigma_{i=1}^m x^{**}(x_i^*) x_i$ as limits of vectors that look like $\Sigma_{i=1}^m x_i^*(y_n) x_i$; these vectors—and hence their limits—all lie inside a fixed ball of X.]

(ii) The map P that takes an x^{**} in X^{**} to the vector $\Sigma_n x^{**}(x_n^*) x_n$ in X is a bounded linear projection on X^{**} that has for a kernel $\{x^{**} \in X^{**} : x^{**} x^* = 0 \text{ for all } x^* \in [x_n^*]\}$.

(iii) X is isomorphic to $[x_n^*]^*$.

(iv) (x_n^*) is a shrinking basis for $[x_n^*]$.

NB One can conclude from this exercise and its predecessor that a basis (y_n) for a space Y is shrinking if and only if the sequence (y_n^*) of coefficient functionals is a boundedly complete basis for Y^*.

5. *Bases spanning reflexive spaces.* Let X be a Banach space with basis (x_n) whose coefficient functionals will be denoted by (x_n^*). X is reflexive if and only if (x_n) is shrinking and boundedly complete.

6. *Unconditional bases.* Let X be a Banach space with an unconditional basis (x_n).

(i) If (x_n) is not boundedly complete, then X contains an isomorphic copy of c_0.

(ii) If (x_n) is not shrinking, then X contains an isomorphic copy of l_1.

[*Hints*: The renorming of Exercise 1(v) helps matters in each case. Similarly, it helps to know what to look for if you are looking for c_0's unit vector basis or l_1's unit vector basis; a peek at Exercise 2 may be worth your while.]

(iii) A Banach space with an unconditional basis is reflexive if and only if the space contains no copy of c_0 or l_1.

7. *Weak* basic sequences.* Let X be a separable Banach space. A sequence (y_n^*) in X^* is called *weak* basic* provided that there is a sequence (y_n) in X so that (y_n, y_n^*) is a biorthogonal sequence ($y_m^* y_n = \delta_{mn}$) and for each y^* in the weak*-closed linear span of the y_n^* we have $y^* = \text{weak}^* \lim_n \Sigma_{i=1}^n y^*(y_i) y_i^*$.

If (x_n^*) is a weak*-null normalized sequence in X^*, then (x_n^*) admits a subsequence (y_n^*) that is weak* basic.
[*Hint*: Pick $\varepsilon_n > 0$, $\varepsilon_n < 1$ so that $\Sigma_n \varepsilon_n$ and $\Pi_n(1-\varepsilon_n)^{-1} < \infty$. Using Helly's theorem and X's separability, extract a subsequence (y_n^*) of (x_n^*) and locate an increasing sequence (F_n) of finite subsets of S_X so that the linear span of $\cup_n F_n$ is dense in X in such a way as to simultaneously achieve (a) given $\varphi \in \text{lin}\{y_1^*, \ldots, y_n^*\}$, $\|\varphi\| = 1$, there is $x \in F_n$ so that $x - \varphi$ has functional norm $< \varepsilon_n/3$ on $\text{lin}\{y_1^*, \ldots, y_n^*\}$ and (b) $|y_{n+1}(x)| < \varepsilon_n/3$, $x \in F_n$.]

8. *Unconditionally converging operators.* Let X and Y be Banach spaces. A bounded linear operator $T: X \to Y$ is said to be *unconditionally converging* if $\Sigma_n Tx_n$ is unconditionally convergent whenever $\Sigma_n x_n$ is weakly unconditionally Cauchy; T is called *completely continuous* if T maps weakly convergent sequences into norm convergent sequences; T is called *weakly completely continuous* if T maps weakly Cauchy sequences into weakly convergent sequences.

(i) A bounded linear operator $T: X \to Y$ fails to be unconditionally converging if and only if there is a subspace S of X isomorphic to c_0 such that the restriction $T|_S$ of T to S is an isomorphism.

(ii) Weakly compact operators and completely continuous operators are weakly completely continuous; in turn, weakly completely continuous operators are unconditionally converging.

9. *Auerbach bases.* If X is an n-dimensional Banach space, then there exist $x_1, \ldots, x_n \in S_X$ and $x_1^*, \ldots, x_n^* \in S_{X^*}$ satisfying $x_i^* x_j = \delta_{ij}$. [*Hint*: On choosing $x_1, \ldots, x_n \in S_X$ so as to maximize the determinant $D(x_1, \ldots, x_n)$, with respect to some designated coordinate system, think of Cramer's rule.]

10. *A Banach space is reflexive if each subspace with a basis is.* It is an easy consequence of the Eberlein-Smulian theorem that a Banach space is reflexive if and only if each of its separable closed linear subspaces is. In this exercise we outline a proof that leads to the claim of the exercise.

(i) A set G in the dual Y^* of a Banach space Y is called *norming* if for each $y \in Y$, $\|y\| = \sup\{|g(y)|: g \in G, \|g\| = 1\}$. If G is a norming set in Y^* and (y_n) is a normalized sequence in Y for which $\lim_n g(y_n) = 0$ for each $g \in G$, then (y_n) has a basic subsequence, with first term y_1 if you please.

(ii) If X is a (separable) Banach space containing a weakly Cauchy sequence that isn't weakly convergent, then X contains a subspace with a nonshrinking basis.

[*Hint*: Let x_1^{**} be the weak* $\lim_n x_n$, where (x_n) is weakly Cauchy but not weakly convergent, and set $x_n^{**} = x_1^{**} - x_{n-1}$ for $n \geq 2$. Applying (i) to $(y_n) = (x_n^{**})$, $Y = X^{**}$, and $G = X^*$, obtain a basic subsequence $(x_{n_k}^{**})$ of (x_n^{**}) with $x_{n_1}^{**} = x_1^{**}$. Let $Z_1 = [x_{n_k}]$, $Z_2 = [x_{n_k}^{**}]$, and $Z_3 = [x_{n_{k+1}}^{**}]$, all taken up in X^{**}. Then $Z_1 \subseteq X$, $Z_3 \subseteq Z_2$, $\dim(Z_2/Z_3) = 1$, $Z_1 \subseteq Z_2$, and $\dim(Z_2/Z_1) = 1$. Show Z_1 and Z_3 are isomorphic. Now using the fact that (x_{n_k}) has no weak limit in X, show that $(x_{n_k}^{**})$ and $(x_{n_{k+1}}^{**})$ are not shrinking bases.]

(iii) If X is a (separable) Banach space containing a sequence (x_n) in B_X having no weak Cauchy subsequence, then X contains a subspace with a nonshrinking basis.

[*Hint*: Pick a countable norming set G in S_{X^*}, using the attainable assumption of X's separability, diagonalize, and use (i) on an appropriate sequence of differences of the distinguished sequence (x_n).]

11. *Subspaces of l_p $(1 \leq p < \infty)$ or c_0.* If $X = l_p$ $(1 \leq p < \infty)$ or $X = c_0$, then every infinite-dimensional closed linear subspace Y of X contains a subspace Z isomorphic to X and complemented in X.

Notes and Remarks

Schauder bases were introduced by J. Schauder who, in addition to noting that the unit coordinate vectors form a basis for the spaces c_0 and l_p (if $1 \leq p < \infty$), constructed the Schauder basis for the space $C[0,1]$. Schauder is also responsible for the proof that the Haar system forms a basis for $L_p[0,1]$ if $1 \leq p < \infty$.

The automatic continuity of coefficient functionals was first noted by Banach whose method of proof has been the model for all further improvements. It's plain from the proof where the ideas behind Exercise 1 were born. Theorem 1 was known to Banach, as was Corollary 3. On the one hand, the proof of Theorem 1 appears in Banach's "Operationes Lineairès," whereas only the statement of Corollary 3 is to be found there. Indeed, it was not until 1958 before any claim to a proof of Corollary 3 was made, at which time *three* proofs appeared! M. M. Day (1962), B. Gelbaum (1958), and C. Bessaga and A. Pelczynski (1958) each gave correct proofs of Corollary 3. Interestingly enough it is probable that none of these proofs was the one known to Banach; it seems likely that Banach knew of Mazur's technique for producing basic sequences, and it is that technique that we follow here. The first exposition of Mazur's technique for the general mathematical public is found in a 1962 note of A. Pelczynski. In any case, this technique has found numerous applications since, with the exercise on weak* basic sequences being typical; the result expressed in Exercise 7 is due to W. B. Johnson and H. P. Rosenthal.

From Theorem 6 on, the results of this chapter are right out of the Bessaga-Pelczynski classic, "Bases and unconditional convergence in Banach spaces." The influence that paper has on this chapter is, or ought to be, plain.

It is an arguable choice to include as exercises, rather than as part of the text, the results of R. C. James (1950, 1951, 1982). In any case, it is certain that this material is now accessible to the hard-working student, and so, with a few hints provided, we have chosen to reward that student with Exercises 3 to 6. It is a fact that the material of these exercises is fundamental Banach space theory and the stymied student would do well to take an occasional peek at the originator's words on these topics, particularly his wonderful exposition in the *American Mathematical Monthly*,

(1982). Actually, regarding Exercise 4, the fact that boundedly complete bases span duals was first noted by L. Alaoglu (1940).

Exercise 9 is due to Auerbach and, as yet, has no perfect infinite-dimensional analogue. On the one hand, not all separable Banach spaces even have a basis, whereas, on the other hand, those that do, need not have a basis where both the basis members and the coefficient functionals have norm one; each of these facts were first found to be so by Enflo (1973). However, there is another notion that offers a viable alternative for generalization, the notion of a Markushevich basis. A biorthogonal system $(x_i, x_i^*)_{i\in I}$ is called a *Markushevich basis* for the Banach space X if the span of the x_i is dense in X and the span of the x_i^* is weak* dense in X^*. Separable Banach spaces have long been known to have (countable) Markushevich bases; whether one can choose the sequence $(x_n, x_n^*)_{n\geq 1}$ so that $\|x_n\| = 1 = \|x_n^*\|$ as well is still unknown. The best attempt has been by R. Ovsepian and A. Pelczynski (1975), modified by Pelczynski, to prove that *if X is a separable Banach space and $\varepsilon > 0$, then there exists a (countable) Markushevich basis $(x_n, x_n^*)_{n\geq 1}$ for X for which $\|x_n\|\,\|x_n^*\| \leq 1+\varepsilon$ for all n.*

Exercise 10 outlines the proof of a theorem of Pelczynski, following his footsteps quite closely. The use of bases to characterize reflexivity has been one of the more fruitful pastimes of general basis theory. In addition to James's results (outlined in these exercises) and Pelczynski's, we cite the beautiful (and useful) result of M. Zippin (1968): *If X is a separable Banach space with a basis, then X is reflexive if and only if each basis of X is shrinking if and only if each basis of X is boundedly complete.*

Bibliography

Alaoglu, L. 1940. Weak topologies in normed linear spaces. *Ann. Math.* **41**, 252–267.

Bessaga, C. and Pelczynski, A. 1958. On bases and unconditional convergence of series in Banach spaces. *Studia Math.*, **17**, 151–164.

Botschkariev, S. V. 1974. Existence of a basis in the space of analytic functions, and some properties of the Franklin system. *Mat. Sb.*, **24**, 1–16.

Carleson, L. An explicit unconditional basis in H^1.

Day, M. M. 1962. On the basis problem in normed spaces. *Proc. Amer. Math. Soc.*, **13**, 655–658.

Enflo, P. 1973. A counterexample to the approximation property in Banach spaces. *Acta Math.*, **130**, 309–317.

Enflo, P. Banach space with basis without a normalized monotone basis. *Ark. Math.*

Gelbaum, B. R. 1958. Banach spaces and bases. *An. Acad. Brasil. Ci.*, **30**, 29–36.

James, R. C. 1950. Bases and reflexivity of Banach spaces. *Ann. of Math.*, **52**, 518–527.

James, R. C. 1951. A non-reflexive Banach space isometric to its second conjugate space. *Proc. Nat. Acad. Sci. (USA)*, **37**, 174–177.

James, R. C. 1982. Bases in Banach spaces. *Amer. Math. Mon.*, **89**, 625–640.
Johnson, W. B. and Rosenthal, H. P. 1972. On w^* basic sequences and their applications to the study of Banach spaces. *Studia Math.*, **43**, 77–92.
Kadec, M. I. and Pelczynski, A. 1965. Basic sequences, biorthogonal systems and norming sets in Banach and Frechet spaces. *Studia Math.*, **25**, 297–323.
Marcinkiewicz, J. 1937. Quelques théormès sur les séries orthogonales. *Ann. Soc. Polon. Math.*, **16**, 84–96.
Markusevich, A. I. 1943. On a basis in the wide sense for linear spaces. *Dokl. Akad. Nauk. SSSR*, **41**, 241–244.
Maurey, B. 1981. Isomorphismes entre espaces H_1. *Acta Math.*, **145**, 79–120.
Ovsepian, R. I. and Pelczynski, A. 1975. The existence in every separable Banach space of a fundamental total and bounded biorthogonal sequence and related constructions of uniformly bounded orthonormal systems in L^2. *Studia Math.*, **54**, 149–159.
Pelczynski, A. 1962. A note on the paper of I. Singer "Basic sequences and reflexivity of Banach spaces." *Studia Math.*, **21**, 371–374.
Pelczynski, A. 1976. All separable Banach spaces admit for every $\varepsilon > 0$ fundamental and total biorthogonal sequences bounded by $1 + \varepsilon$. *Studia Math.*, **55**, 295–304.
Schauder, J. 1927. Zur Theorie stetiger Abbildungen in Funktionalräumen. *Math. Zeit.*, **26**, 47–65.
Schauder, J. 1928. Eine Eigenschaft des Haarschen Orthogonalsystems. *Math. Z.*, **28**, 317–320.
Wojtasczyck, P. The Franklin system is an unconditional basis in H^1.
Zippin, M. 1968. A remark on bases and reflexivity in Banach spaces. *Israel J. Math.*, **6**, 74–79.

CHAPTER VI
The Dvoretsky-Rogers Theorem

Recall that a normed linear space X is a Banach space if and only if given any absolutely summable series $\Sigma_n x_n$ in X, $\lim_n \Sigma_{k=1}^n x_k$ exists. Of course, *in case X is a Banach space*, this gives the following implication for a series $\Sigma_n x_n$: *if $\Sigma_n \|x_n\| < \infty$, then $\Sigma_n x_n$ is unconditionally convergent*; that is, $\Sigma_n x_{\pi(n)}$ converges for each permutation π of the natural numbers.

What of the converse? Our memories of calculus jar the mind to recall that *for a series of scalars to be absolutely convergent, it is both necessary and sufficient that the series be unconditionally convergent.* This fact, in tandem with the equivalence of coordinatewise convergence with norm convergence in any finite-dimensional Banach space, bootstraps to prove that *in any finite-dimensional Banach space, unconditionally convergent series are absolutely-convergent.*

In infinite-dimensional Banach spaces the situation is readily seen to be quite different. For instance, in c_0, if we look at $x_n = e_n/n$, where e_n is the nth unit vector, then $\Sigma_n x_n$ converges unconditionally to the member $(1/n)$ of c_0; of course, $\|x_n\| = 1/n$, and so $\Sigma_n x_n$ is not absolutely convergent. Similar examples can be constructed in any of the classical Banach spaces. (An aside: The aforementioned examples are *not* always trivially discovered; a particularly trying case is l_1.) The Polish founders of Banach space theory were led to conjecture that in every infinite-dimensional Banach space there is an unconditionally convergent series $\Sigma_n x_n$ for which $\Sigma_n \|x_n\| = \infty$.

In 1950, A. Dvoretsky and C. A. Rogers established this conjecture's validity. Within a very short while, A. Grothendieck (1956) was able to give a substantially different proof of the Dvoretsky-Rogers theorem; in fact, Grothendieck went so far as to classify those Frechet spaces (i.e., complete metric locally convex spaces) for which unconditionally convergent series are absolutely convergent. The proof we give below is modeled on ideas of Grothendieck but follows a bit more direct path to the Dvoretsky-Rogers theorem. The ideas used will appear again later. Presently, we are concerned with the proof of the following.

Dvoretsky-Rogers Theorem. *If every unconditionally convergent series in the Banach space X is absolutely convergent, then X is finite dimensional.*

Let $1 \le p < \infty$ and X, Y be Banach spaces.

We say that the bounded linear operator $T: X \to Y$ is absolutely p-summing [denoted by $T \in \Pi_p(X; Y)$] if given any sequence (x_n) from X for which $\Sigma_n |x^* x_n|^p < \infty$, for each $x^* \in X^*$, we have $\Sigma_n \|Tx_n\|^p < \infty$.

A number of remarks about the notation of an absolutely p-summing operator are in order.

Suppose (x_n) is a sequence in X for which $\Sigma_n |x^* x_n|^p < \infty$ for each $x^* \in X^*$. Then the mapping from X^* to l_p that takes an x^* to the sequence $(x^* x_n)$ is well-defined, linear, and, having a closed graph, continuous. Consequently, there is a $C > 0$ such that

$$\sup_{\|x^*\| \le 1} \left\{ \sum_n |x^* x_n|^p \right\}^{1/p} \le C. \tag{1}$$

Now, a straightforward argument shows that if we consider the linear space of sequences (x_n) in X for which $\Sigma_n |x^* x_n|^p < \infty$, for each $x^* \in X^*$, then the resulting space, called here $l_p^{\text{weak}}(X)$, is a Banach space with the norm

$$\|(x_n)\|_{l_p^{\text{weak}}(X)} = \inf\{ C > 0 : (1) \text{ holds} \}.$$

Next, we have the space $l_p^{\text{strong}}(Y)$ of all sequences in Y for which $\Sigma_n \|y_n\|^p < \infty$. $l_p^{\text{strong}}(Y)$ is a Banach space with the norm

$$\|(y_n)\|_{l_p^{\text{strong}}(Y)} = \left(\sum_n \|y_n\|^p \right)^{1/p}$$

An operator $T: X \to Y$ is absolutely p-summing if and only if $(Tx_n) \in l_p^{strong}(Y)$ whenever $(x_n) \in l_p^{weak}(X)$. This is trivial. Not much harder is the fact that if $T: X \to Y$ is absolutely p-summing, then the linear operation that takes an (x_n) in $l_p^{\text{weak}}(X)$ to (Tx_n) in $l_p^{\text{strong}}(Y)$ has a closed graph and is, therefore, a bounded linear operator—call it $\hat{T}$. We define the absolutely p-summing norm $\pi_p(T)$ of T to be the operator norm of $\hat{T}$ viewed as an operator from $l_p^{\text{weak}}(X)$ to $l_p^{\text{strong}}(Y)$. A bit of care reveals that the collection of $\hat{T}$ is a closed linear subspace of the Banach space of all bounded linear operators from $l_p^{\text{weak}}(X)$ to $l_p^{\text{strong}}(Y)$. From this it follows that $\Pi_p(X; Y)$ is a Banach space in the norm π_p. Further, it is easy to see that if T is absolutely p-summing, then

$$\pi_p(T) = \inf\{ \rho > 0 : \text{inequality (2) holds for any } x_1, x_2, \ldots, x_n \in X \},$$

$$\left(\sum_{i=1}^n \|Tx_i\|^p \right)^{1/p} \le \rho \sup_{B_{X^*}} \left(\sum_{i=1}^n |x^* x_i|^p \right)^{1/p} \tag{2}$$

A fundamental result linking measure theory to the theory of absolutely p-summing operators is the following.

Grothendieck-Pietsch Domination Theorem. *Suppose $T: X \to Y$ is an absolutely p-summing operator. Then there exists a regular Borel probability measure μ defined on B_{X^*} (in its weak* topology) for which*

$$\|Tx\|^p \le \pi_p^p(T) \int_{B_{X^*}} |x^*x|^p \, d\mu(x^*)$$

holds for each $x \in X$.

PROOF. Suppose $x_1, \dots, x_n \in X$. Define the function

$$f_{x_1,\dots,x_n}: B_{X^*} \to \mathbb{R}$$

by

$$f_{x_1,\dots,x_n}(x^*) = \pi_p^p(T) \sum_{k=1}^n |x^*x_k|^p - \sum_{k=1}^n \|Tx_k\|^p.$$

Each $f_{x_1,\dots,x_n}$ is weak* continuous on B_{X^*}, and the collection $C = \{ f_{x_1,\dots,x_n} \in C(B_{X^*}, \text{weak}^*) : x_1, \dots, x_n \in X \}$ is a convex cone in $C(B_{X^*}, \text{weak}^*)$, each of whose members is somewhere nonnegative—this last fact being due to the absolutely p-summing nature of T. Now C is disjoint from the convex cone $\mathfrak{N} = \{ f \in C(B_{X^*}, \text{weak}^*) : f(x^*) < 0 \text{ for each } x^* \in B_{X^*} \}$, and this latter cone has an interior. Therefore, there is a nonzero continuous linear functional $\mu \in C(B_{X^*}, \text{weak}^*)^*$ (i.e., regular Borel measures on B_{X^*} in its weak* topology) such that

$$\int f \, d\mu = \mu(f) \le 0 \le \mu(g) = \int g \, d\mu,$$

for $f \in \mathfrak{N}$, $g \in C$. The measure μ has the distinction of being nonpositive on strictly negative functions; therefore, it is nonnegative on strictly positive functions, and it follows that μ is a nonnegative measure. Normalizing μ gives a probability measure. Also, μ is nonnegative on C; so $\int f_x \, d\mu \ge 0$ for each $x \in X$. But this just says that

$$\|Tx\|^p \le \pi_p^p(T) \int_{B_{X^*}} |x^*x|^p \, d\mu(x^*),$$

which is what was wanted. □

Let's look at the above inequality a bit closer.

Let $T: X \to Y$ be absolutely p-summing. As a bounded linear operator, T satisfies the inequality

$$\|Tx\| \le \|T\| \|x(\cdot)\|_\infty$$

for each $x \in X$, where we may interpret each $x \in X$ as acting (continuously) on (B_{X^*}, weak^*). However, in light of the Grothendieck-Pietsch domination theorem there is a regular Borel probability measure μ on (B_{X^*}, weak^*) for which

$$\|Tx\| \le \pi_p(T) \|x(\cdot)\|_{L_p(\mu)} \tag{3}$$

holds for each $x \in X$. Inequality (3) tells us that T acts in a continuous linear fashion from X to Y even when X is viewed as sitting in $L_p(\mu)$. If we let X_p denote the closure of X in $L_p(\mu)$, then we can find a unique continuous linear extension $P: X_p \to Y$ of T to all of X_p. Let $G: X \to X_p$ be the natural inclusion mapping of X in its original norm into X_p, the $L_p(\mu)$-completion of X. G is a continuous linear operator too. One more thing: $T = PG$. Pictorially, the diagram

$$\begin{array}{ccc} X & \xrightarrow{T} & Y \\ & G\searrow \quad \nearrow P & \\ & X_p & \end{array}$$

commutes.

There are two things about G that must be mentioned.

First, G is a weakly compact operator; that is, G takes B_X into a weakly compact set in X_p. If $p > 1$, then this follows from the reflexivity of X_p. If $p = 1$, then one need only notice that G is the restriction to X of the inclusion operator taking $C(B_{X^*}, \text{weak}^*)$ into $L_1(\mu)$; on its way from $C(B_{X^*}, \text{weak}^*)$ into $L_1(\mu)$, the inclusion operator passes through $L_2(\mu)$—making it, and G, weakly compact.

Next, G is completely continuous; that is, G takes weakly convergent sequences to norm convergent sequences. In fact, if (x_n) is a weakly convergent sequence in X and $x_0 = \text{weak}\lim_n x_n$, then there is an $M > 0$ such that $\|x_n\| \le M$ for all n and $x^*x_0 = \lim_n x^*x_n$ for each $x^* \in X^*$ as well. Viewing X as acting on B_{X^*}, we get $\lim_n x_n(x^*) = x_0(x^*)$ for each $x^* \in B_{X^*}$ and $|x_n(x^*)| \le M$ holding for each $x^* \in B_{X^*}$. By Lebesgue's bounded convergence theorem, this gives us

$$\lim_n \|Gx_n - Gx_0\|_{L_p(\mu)} = \lim_n \|x_n(\cdot) - x_0(\cdot)\|_{L_p(\mu)} = 0.$$

Reflect for a moment on these developments. Since the operator $P: X_p \to Y$ is weakly continuous as well as continuous, the above properties of G are passed along to T. T is weakly compact and completely continuous.

What if $T: X \to Y$ is absolutely p-summing and $S: Y \to Z$ is absolutely r-summing? Each is weakly compact and completely continuous. It follows that for any bounded sequence (x_n) in X, (Tx_n) admits of a weakly convergent subsequence; so (STx_n) admits of a norm convergent subsequence. $ST(B_X)$ is relatively norm compact. Consequently, we have the following theorem.

Theorem. *If $1 \le p < \infty$ and X is infinite dimensional, then the identity operator on X is not absolutely p-summing.*

Alternatively, if $1 \le p < \infty$ and $\Sigma_n \|x_n\|^p < \infty$ holds whenever $\Sigma_n |x^*x_n|^p < \infty$ for each $x^* \in X^*$, then X is finite dimensional.

The Dvoretsky-Rogers theorem follows easily from this. How? Well consider any Banach space X in which the unconditional convergence of a series implies its absolute convergence. X cannot contain any isomorph of c_0 since we saw earlier that c_0 admits of non-absolutely convergent unconditionally convergent series. It follows from the Bessaga-Pelczynski c_0 theorem that $\Sigma_n x_n$ is unconditionally convergent whenever $\Sigma_n|x^*x_n| < \infty$ for each $x^* \in X^*$; hence, if $\Sigma|x^*x_n| < \infty$ for each $x^* \in X^*$, then $\Sigma\|x_n\| < \infty$. But this is tantamount to the identity operator on X being absolutely 1-summing.

Exercises

1. *Hilbert-Schmidt operators and absolutely 2-summing operators.* Let E and F be Hilbert spaces with complete orthonormal systems $(e_i)_{i\in I}$ and $(f_j)_{j\in J}$, respectively. An operator $T: E \to F$ is called a *Hilbert-Schmidt operator* if

$$\sum_{I,J} |(Te_i, f_j)|^2 < \infty,$$

where $(\,,\,)$ will be used to denote the inner product.

(i) Show that $\Sigma_I\|Te_i\|^2 = \Sigma_{I,J}|(Te_i, f_j)|^2 = \Sigma_J\|T^*f_j\|^2$ and conclude that the quantity $\Sigma_{I,J}|(Te_i, f_j)|^2$ is independent of the complete orthonormal systems $(e_i)_{i\in I}, (f_j)_{j\in J}$. Naturally, we consider for a Hilbert-Schmidt operator T the functional $(\Sigma_{I,J}|(Te_i, f_j)|^2)^{1/2}$ and call this functional the *Hilbert-Schmidt norm* of T, denoted by $\sigma(T)$.

(ii) Every finite-rank bounded linear operator from E to F is a Hilbert-Schmidt operator, and every Hilbert-Schmidt operator is the limit in Hilbert-Schmidt norm of a sequence of finite-rank operators. Consequently, since $\|T\| \le \sigma(T)$, every Hilbert-Schmidt operator is compact. Notice that not every compact operator $S: E \to F$ is a Hilbert-Schmidt.

(iii) Every absolutely 2-summing operator $T: E \to F$ is a Hilbert-Schmidt operator with $\pi_2(T) \ge \sigma(T)$. [*Hint*: You might notice that as a consequence of (i), T is a Hilbert-Schmidt operator precisely when $\Sigma_I\|Te_i\|^2 < \infty$ for each complete orthonormal system $(e_i)_{i\in I}$ in E.]

(iv) If $T: E \to F$ is a Hilbert-Schmidt operator, then T can be realized in the form

$$Tx = \Sigma_n \lambda_n \langle x, e_n\rangle f_n,$$

where $(\lambda_n) \in l_2$, (e_n) is an orthonormal sequence in E and (f_n) is an orthonormal sequence in F, and $\|(\lambda_n)\|_2 = \sigma(T)$.

(v) Every Hilbert-Schmidt operator $T: E \to F$ is absolutely 2-summing with $\sigma(T) \ge \pi_2(T)$.

2. $\pi_p(X; Y) \subseteq \pi_q(X; Y)$, $1 \le p < q < \infty$. Show that if $1 \le p < q < \infty$ and T is absolutely p-summing, then T is absolutely q-summing with $\pi_q(T) \le \pi_p(T)$.

3. *Composition of absolutely summing operators.* Suppose $r, s > 1$ and $1/r + 1/s = 1$. If $R \in \Pi_r(X; Y)$ and $S \in \Pi_s(Y; Z)$, then $SR \in \Pi_1(X; Z)$, and
$$\underset{1}{\pi}(SR) \le \underset{r}{\pi}(R)\underset{s}{\pi}(S).$$

4. *The composition of absolutely 2-summing operators.* If $G: X \to Y$ and $A: Y \to Z$ are absolutely 2-summing, then $AG: X \to Z$ is nuclear (i.e., can be written in the form $AGx = \Sigma_n \lambda_n x_n^*(x) z_n$, where $(\lambda_n) \in l_1$, $(\|x_n^*\|) \in c_0$ and $(\|z_n\|) \in c_0$).

5. *Absolutely p-summing operators on* c_0. A bounded linear operator $T: E \to F$ is called *p-nuclear* ($p \ge 1$) whenever T can be written in the form $Tx = \Sigma_{n=1}^{\infty} x_n^*(x) y_n$, where $(x_n^*) \subseteq E^*$ and $(y_n) \subseteq F$ satisfy
$$\sum_n \|x_n^*\|^p < \infty \quad \text{and} \quad \sup_{\|y^*\| \le 1} \left(\sum_n |y^* y_n|^{p'} \right)^{1/p'} < \infty.$$
Here $1/p + 1/p' = 1$, and in case $p = 1$, the condition on the sequence (y_n) just requires that $\|y_n\| \to 0$.
Show that any absolutely p-summing operator $T: c_0 \to X$ is p-nuclear.

Notes and Remarks

In case $p = 1$ or 2, the absolutely p-summing operators were introduced and studied by A. Grothendieck (1956) in his infamous resumé. For general p, A. Pietsch (1967) is responsible for the initial study of the class of absolutely p-summing operators. It is to Pietsch that we owe the final form of the Grothendieck-Pietsch domination theorem, though Grothendieck's contribution in this regard is not to be slighted. Who is to be given the lion's share of credit is not at issue; rather, it is the result that counts, and the domination theorem is a basic one at that. Introducing measures where none were apparent is the theme of the theorem; the effects in Banach space theory (and abstract analysis in general) are only now beginning to be felt. We refer the reader to Pelczynski's (1976) lectures on applications of summing operators in the study of spaces of analytic functions or to J. Diestel's (1980) remarks regarding the absolutely 2-summing operators for a hint at the power provided by the machinery of the theory of absolutely p-summing operators.

Incidentally, our proof of the domination theorem is probably due to B. Maurey; we "discovered" it after several sessions of reading papers by him in various volumes of the Maurey-Schwartz seminar notes. It is practically the same as the proof found in Lindenstrauss-Tzafriri I.

As mentioned in the text, we have followed Grothendieck's approach to the Dvoretsky-Rogers theorem. Their original proof proceeded from the

Dvoretsky-Rogers lemma: *Let B be an n-dimensional normed linear space; then there exist points $x_1, \ldots, x_n$ of norm one in B such that for each $i \leq n$ and all real $t_1, \ldots, t_n$,*

$$\left\| \sum_{j=1}^{i} t_j x_j \right\| \leq \left(1 + \sqrt{\frac{i(i-1)}{n}}\right) \left(\sum_{j=1}^{i} t_j^2 \right)^{1/2}.$$

Their proof is particularly recommended to the geometrically minded students. From this lemma, Dvoretsky and Rogers were able to build, for any preassigned nonnegative sequence (t_n) in l_2, an unconditionally convergent series $\Sigma_n x_n$ in the infinite-dimensional Banach space for which $\|x_n\| = t_n$.

A decade after the Dvoretsky-Rogers lemma had been discovered, A. Dvoretsky returned to this topic and formulated his famous spherical sections theorem: *For each infinite-dimensional normed linear space F and each $n \geq 1$ and each $\varepsilon > 0$ there is a one-to-one linear mapping T of l_2^n into F such that $\|T\| \|T^{-1}\| < 1 + \varepsilon$.* This result has had a profound effect upon the directions taken by Banach space theory and, with developments related to the theory of absolutely p-summing operators, has played an important role in the disposition of numerous old problems in Banach space theory.

The Dvoretsky ε-spherical sections theorem was the object of an extensive study by T. Figiel, V. Milman, and J. Lindenstrauss (1977). By-products of their efforts include a new proof of the Dvoretsky-Rogers theorem and the easiest existing proof of the spherical sections theorem.

Exercise 1 is mentioned in passing by Grothendieck; a much finer thing can be said and will be said in the exercises following Chapter VII. Exercise 5 is due to C. Stegall and J. R. Retherford (1972); their paper is filled with important connections between operator theory and the classification of Banach spaces. Exercise 3 is a very special case of a result of A. Pietsch (1967), and Exercise 4 was known to A. Grothendieck (1956).

Related to issues raised in this chapter is the notion of an absolutely (p, q)-summing operator and particularly the work of B. Maurey and A. Pelczynski (1976), who give criteria for the composition of (p_i, q_i)-summing operators to be compact.

Bibliography

Diestel, J. 1980. Measure theory in action: absolutely 2-summing operators. Proceedings of the Conference on Measure Theory and its Applications, Northern Illinois University, DeKalb, Ill.

Dvoretsky, A. 1961. Some results on convex bodies and Banach spaces. Proc. Internat. Sympos. Linear Spaces, Jerusalem: Hebrew University, pp. 123–160.

Dvoretsky, A. and Rogers, C. A. 1950. Absolute and unconditional convergence in normed spaces. *Proc. Nat. Acad. Sci. USA*, **36**, 192–197.

Figiel, T., Lindenstrauss, J., and Milman, V. 1977. The dimension of almost spherical sections of convex sets. *Acta Math.*, **139**, 53–94.

Grothendieck, A. 1956. Resumé de la théorie métrique des produits tensoriels topologiques. *Bol. Soc. Mat. São Paolo*, **8**, 1–79.

Maurey, B. and Pelczynski, A. 1976. A criterion for compositions of (p, q)-absolutely summing operators to be compact. *Studia Math.*, **54**, 291–300.

Pelczynski, A. 1976. *Banach Spaces of Analytic Functions and Absolutely Summing Operators*. CBMS Regional Conference Series in Mathematics, volume 30.

Pietsch, A. 1967. Absolute p-summierende Abbildungen in normierten Räumen, *Studia Math.*, **28**, 333–353.

Stegall, C. and Retherford, J. R. 1972. Fully nuclear and completely nuclear operators with applications to $\mathscr{L}_1$- and $\mathscr{L}_\infty$-spaces. *Trans. Amer. Math. Soc.*, **163**, 457–492.

CHAPTER VII
The Classical Banach Spaces

To this juncture, we have dealt with general theorems concerning the nature of sequential convergence and convergence of series in Banach spaces. Many of the results treated thus far were first derived in special cases, then understood to hold more generally. Not too surprisingly, along the path to general results many important theorems, special in their domain of applicability, were encountered. In this chapter, we present more than a few such results.

There are three main objectives we hope to achieve in this chapter. First, we hope to reveal something of the character of Banach spaces that have likely already been encountered by the student and provide insight into just how the weak and norm topologies interact with familiar concepts in these more familiar acquaintances. Again, the classical Banach spaces play a central role in the development of general Banach space theory; coming to grips with their special properties is of paramount importance if one is to appreciate how and why this is so. Lastly, many of the more interesting phenomena to be discussed in these deliberations require some deeper understanding of the geometry of the classical spaces before these phenomena can be recognized as natural.

Weak and Pointwise Convergence of Sequences in $C(\Omega)$

The heart and soul of this section are each devoted to proving the following two theorems.

Theorem 1. *Let Ω be any compact Hausdorff space, and let (f_n) be a sequence of continuous scalar-valued functions defined on Ω.*

1. *In order that (f_n) be weakly convergent in $C(\Omega)$ to $f \in C(\Omega)$, it is necessary and sufficient that $\sup_n \|f_n\|_\infty < \infty$ and $f(\omega) = \lim_n f_n(\omega)$ for each $\omega \in \Omega$.*

2. *In order that* (f_n) *be a weak Cauchy sequence in* $C(\Omega)$, *it is necessary and sufficient that* $\sup_n \|f_n\|_\infty < \infty$ *and* $\lim_n f_n(\omega)$ *exist for each* $\omega \in \Omega$.

Theorem 2 (Baire's Classification Theorem).

1. *Let* Ω *be any topological space each closed subset of which is of the second category in itself. Then any bounded scalar-valued function on* Ω *which is the pointwise limit of a sequence of continuous scalar-valued functions on* Ω *has a point of continuity in each nonvoid closed subset of* Ω *(relative, of course, to the closed subset).*
2. *Let* Ω *be a separable metric space and* f *be a bounded scalar-valued function defined on* Ω. *If* f *has a point of continuity in each nonvoid closed subset of* Ω *(relative to the closed subset), then there exists a uniformly bounded sequence of continuous scalar-valued functions on* Ω *converging pointwise to* f.

The proof of part 1 of Theorem 1 is easy. We need to recall that the members of $C(\Omega)^*$ act on $C(\Omega)$ like integration via regular Borel measures on Ω. This in mind, suppose $f, f_n \in C(\Omega)$ $(n \geq 1)$ satisfy $f(\omega) = \lim_n f_n(\omega)$ for each $\omega \in \Omega$, where $\sup_n \|f_n\|_\infty < \infty$. Each regular Borel measure μ is a linear combination of (at most) four probability regular Borel measures (thanks to the Hahn-Jordan decomposition theorem). Therefore, to check that $f = \lim_n f_n$(weakly), it is enough to check that $\int f\, d\mu = \lim_n \int f_n\, d\mu$ holds for regular Borel probability measures μ, and this is clear from Lebesgue's bounded convergence theorem. On the converse side, we notice that weak convergence of a sequence (f_n) implies boundedness in any Banach space; so $f = \text{weak}\lim_n f_n$ in $C(\Omega)$ ensures $\sup_n \|f_n\|_\infty < \infty$. Further, for each $\omega \in \Omega$ the point charge (or point evaluation or point mass or Dirac δ-functional) δ_ω, whose value at $f \in C(\Omega)$ is

$$\delta_\omega(f) = f(\omega),$$

is clearly in $C(\Omega)^*$; that $\lim_n f_n(\omega) = f(\omega)$, for every $\omega \in \Omega$, clearly follows from this and with it part 1.

The proof of part 2 is similar to that of part 1. In fact, if $\lim_n f_n(\omega)$ exists for each $\omega \in \Omega$, where (f_n) is a uniformly bounded sequence of continuous functions defined on Ω, then

$$\lim_n \int_\Omega f_n\, d\mu = \int_\Omega \lim_n f_n\, d\mu$$

holds for each $\mu \in C(\Omega)^*$, by Lebesgue's bounded convergence theorem, and so (f_n) is weakly Cauchy in $C(\Omega)$. Conversely, weakly Cauchy sequences are always bounded in norm, and a careful test again with the δ_ω shows that weakly Cauchy things in $C(\Omega)$ are pointwise Cauchy (hence, convergent).

Theorem 1 has an easy proof and many applications. The proof of Theorem 2 lies deeper, and its applications are correspondingly more subtle.

Baire's category theorem is at the base of our considerations with the aim being the proof of the following result of which Theorem 2 (part 1) is an easy consequence.

Let Ω be a topological space, each closed subset of which is of the second category in itself, and (f_n) be a (uniformly bounded) sequence of continuous scalar-valued functions converging pointwise on Ω. Then the set of points $\omega \in \Omega$, where (f_n) is equicontinuous, is a dense $\mathscr{G}_\delta$-subset of Ω.

Let's see why this is so.

Take any $\varepsilon > 0$ and let $U(\varepsilon)$ be the (open) subset of Ω consisting of all those ω for which there is an open set $D(\omega)$ in Ω containing ω such that if $\omega', \omega'' \in D(\omega)$, then $|f_n(\omega') - f_n(\omega'')| < \varepsilon$ holds for all n. Plainly, as ε decreases, so too do the sets $U(\varepsilon)$. We claim that $U(\varepsilon)$ is dense in Ω for each ε. Of course, the points of $\cap_{m=1}^\infty U(1/m)$ are precisely the points of equicontinuity of the sequence (f_n); so once our claim has been established, we will be done with the present task.

Let O be any open set in Ω. Let

$$E_{n,m} = \left\{ \omega \in \bar{O} : |f_m(\omega) - f_n(\omega)| \leq \frac{\varepsilon}{6} \right\}$$

and let

$$F_p = \bigcap_{m,n \geq p} E_{n,m}.$$

Since each f_n is continuous, all the sets $E_{n,m}$ and F_p are closed subsets of Ω. Moreover, the assumption that $\lim_n f_n(\omega)$ exists for each $\omega \in \Omega$ (and hence for each $\omega \in \bar{O}$) certainly lets us conclude that

$$\bigcup_p F_p = \bar{O}.$$

Well! There must be a p so that F_p has nonempty interior (in $\bar{O}$). This (relative) interior necessarily intersects O in an open subset of Ω—call it V; we may choose V small enough that any point of V belongs to F_p, too. Let $\omega_0 \in F_p \cap V$. Of course, for $m, n \geq p$ we have $|f_m(\omega) - f_n(\omega)| \leq \varepsilon/6$ for all $\omega \in V$. Hence, for $n \geq p$ we have

$$|f_n(\omega_0) - f_n(\omega)| \leq |f_n(\omega_0) - f_p(\omega_0)| + |f_p(\omega_0) - f_p(\omega)| + |f_p(\omega) - f_n(\omega)|$$

$$\leq \frac{\varepsilon}{6} + \frac{\varepsilon}{6} + \frac{\varepsilon}{6} = \frac{\varepsilon}{2},$$

so long as $\omega \in V$. We can achieve strict inequality by shrinking V a bit; this shrinking can be done since f_p is continuous. Notice that there are only finitely many n smaller than p; so (after possibly p shrinkings) we can find an open subset V about ω_0 (contained in $F_p \cap O$) such that for any $n \geq 1$ and any $\omega \in V$

$$|f_n(\omega_0) - f_n(\omega)| < \frac{\varepsilon}{2}.$$

Of course, such things force V to be a part of $U(\varepsilon)$; $U(\varepsilon) \cap O$ is not empty, and this (because of O's arbitrariness) yields $U(\varepsilon)$'s density.

What of Theorem 2 (part 2)? Let f be a bounded real-valued function (we leave to the reader's imagination what variations in theme must be sought after in the complex case) defined on the *separable* metric space Ω having a point of continuity in each nonvoid closed subset (relative to that closed subset) of Ω. We will show that there is a sequence (f_n) of continuous real-valued functions defined on Ω for which $f(\omega) = \lim_n f_n(\omega)$ holds for each $\omega \in \Omega$. This we do in two steps: first, we show that for each real number y, the set $[f > y] = \{\omega \in \Omega : f(\omega) > y\}$ is an $\mathscr{F}_\sigma$-set in Ω; then (building on our first step and the faith inherent therein) we'll show that any such f must be the uniform limit of a sequence of functions each of the first Baire class (i.e., pointwise limit of a sequence of continuous functions).

Step 1. For each real number y, $[f > y]$ is an $\mathscr{F}_\sigma$-set in Ω.

Suppose that $z > y$. Take any $\omega \in \Omega$. Either $\omega \in [f > y]$ or $\omega \in [f < z]$. For any nonempty closed subset F of Ω, there is a point $\omega_F \in F$ at which $f|_F$ is continuous. If $\omega_F \in [f > y]$, then there is an open set $U(\omega_F)$ in Ω containing ω_F such that $F \cap U(\omega_F) \subseteq [f > y]$. Should ω_F find itself in $[f < z]$, then there would be an open set $U(\omega_F)$ in Ω containing ω_F such that $F \cap U(\omega_F) \subseteq [f < z]$. Whatever the situation may be, each nonempty closed subset F of Ω contains a proper closed subset $F_1(= F \setminus U(\omega_F))$ such that $F \setminus F_1$ is contained entirely in either $[f > y]$ or $[f < z]$.

Can F_1 be nonvoid? Well, yes! But if F_1 is nonempty, then there is a closed set F_2 properly contained in F_1 such that $F_1 \setminus F_2$ is contained in either $[f > y]$ or $[f < z]$.

Can F_2 be nonvoid? If so, there is a closed set F_3 properly contained in F_2 such that $F_2 \setminus F_3$ is contained in either $[f > y]$ or $[f < z]$.

Proceeding in this manner we generate a transfinite sequence ($F_\xi : \xi <$ the first uncountable ordinal) of closed subsets of Ω (with $F_0 = \Omega$) for which whenever F_ξ is nonempty, $F_{\xi+1}$ is a closed proper subset of F_ξ for which $F_\xi \setminus F_{\xi+1}$ is a subset of either $[f > y]$ or of $[f < z]$; in case η is a limit ordinal, we have $F_\eta = \cap_{\xi < \eta} F_\xi$.

Here is where we use our hypotheses. Ω is assumed to be a separable metric space. Therefore, there is a first F_η after which $F_\eta = F_{\eta+1} = \cdots$. By construction, F_η is empty. Therefore,

$$\Omega = \bigcup_{\xi < \eta} (F_\xi \setminus F_{\xi+1}),$$

where each $F_\xi \setminus F_{\xi+1}$ is contained in either $[f < y]$ or $[f > z]$. *Each of the sets $F_\xi \setminus F_{\xi+1}$ is an $\mathscr{F}_\sigma$!*

What have we done? We have represented Ω as the union of two $\mathscr{F}_\sigma$-subsets: one formed by taking the (countable) union of those $F_\xi \setminus F_{\xi+1}$ contained in $[f > y]$; the other formed by taking the (countable) union of those $F_\xi \setminus F_{\xi+1}$ contained in $[f < z]$.

Let $z_n \searrow y$. For each n, set $\Omega = A_n \cup B_n$, where A_n and B_n are $\mathscr{F}_\sigma$-subsets of Ω such that $A_n \subseteq [f > y]$ and $B_n \subseteq [f < z_n]$. $A = \cup_n A_n$, $B = \cap_n B_n$, and $C = \cap_n [f < z_n]$ are sets worth watching. First, $C = [f \leq y]$; so $C \cup [f > y] = \Omega$. Further, $A \cup B$ is a decomposition of Ω into disjoint sets satisfying $A = \cup_n A_n \subseteq [f > y]$ and $B = \cap_n B_n \subseteq \cap_n [f < z_n] = C$. It follows that $A = [f > y]$, and so $[f > y]$ is an $\mathscr{F}_\sigma$-set.

Step 1 has been taken.

The argument above can be modified to show that for each real y, $[f < y]$ is an $\mathscr{F}_\sigma$-subset of Ω, too.

Step 2. If f is a bounded real-valued function defined on metric space Ω for which $[f < y]$ and $[f > y]$ are $\mathscr{F}_\sigma$-sets regardless of the choice of real number y, then f is of the first Baire class. We'll sneak up on this one bit by bit.

To start, notice that the indicator function c_F of a closed subset F of a metric space Ω is of the first Baire class (think about it). Moreover, if S is an $\mathscr{F}_\sigma$ in the metric space Ω, then there is a (bounded) function g of the first Baire class such that $S = [g > 0]$; indeed, if $S = \cup_n F_n$ (F_n closed, $F_n \subseteq F_{n+1}$), then $g = \Sigma_n 2^{-n} c_{F_n}$ is the absolute sum of bounded functions of the first Baire class (and so of the same first Baire class) with $[g > 0] = S$.

To work! Take a bounded real-valued function f on Ω for which $[f > y]$ and $[f < z]$ are $\mathscr{F}_\sigma$-sets regardless of y, z. Suppose for this argument that $0 < f(\omega) < 1$ holds for any $\omega \in \Omega$. Take an $n \geq 1$. For $m = 0, 1, \ldots, n-1$, look at the $\mathscr{G}_\delta$-sets $[f \leq m/n]$, $[(m+1)/n \leq f]$; for each we can find bounded real-valued functions g'_m, g''_m of the first Baire class such that

$$\left[f \leq \frac{m}{n}\right] = [g'_m \leq 0] \quad \text{and} \quad \left[\frac{m+1}{n} \leq f\right] = [g''_m \leq 0].$$

The functions

$$g_m = \frac{\sup(g''_m, 0)}{\sup(g'_m, 0) + \sup(g''_m, 0)}$$

are also bounded, of the first Baire class, and satisfy

$$g_0(\omega) = g_1(\omega) = \cdots = g_{m-1}(\omega) = 0, \qquad g_{m+1}(\omega) = \cdots = g_{n-1}(\omega) = 1$$

whenever $m/n \leq f(\omega) \leq (m+1)/n$ with $g_m(\omega)$ somewhere between 0 and 1. Consequently,

$$g(\omega) \equiv n^{-1}(g_0(\omega) + \cdots + g_{n-1}(\omega))$$

is a bounded function of the first Baire class defined on Ω within $1/n$ of f throughout Ω. f is a uniform limit of such as g. f is itself of the first Baire class.

All's well that ends well.

The Classical Nonreflexive Sequence Spaces

Some Special Features of c_0, l_1, l_∞.

Presently we derive a few of the most basic structural properties of the nonreflexive sequence spaces c_0, l_1, and l_∞; we also discuss in some detail the dual of l_∞. Again, our main purpose is to gain insight into the very special nature of the spaces c_0, l_1, and l_∞. The properties on which we concentrate are categorical (i.e., homological) in nature and as such find frequent application in matters sequential.

Our first result says that l_∞ is "injective."

Theorem 3 (R. S. Phillips). *Let Y be a linear subspace of the Banach space X and suppose $T: Y \to l_\infty$ is a bounded linear operator. Then T may be extended to a bounded linear operator $S: X \to l_\infty$ having the same norm as T.*

PROOF. A bit of thought brings one to observe that the operator T must be of the form

$$Ty = (y_n^* y)$$

for some bounded sequence (y_n^*) in Y^*. If we let x_n^* be a Hahn-Banach extension of y_n^* to all of X, then the operator

$$Sx = (x_n^* x)$$

does the trick. □

Supposing l_∞ to be a closed linear subspace of a Banach space X, we can extend the identity operator $I: l_\infty \to l_\infty$ to an operator $S: X \to l_\infty$ with $\|S\| = 1$. The operator S is naturally a norm-one projection of X onto l_∞, thus providing us with an alternative description of Phillips's theorem: *l_∞ is complemented by a norm-one projection in any superspace.*

c_0 enjoys a similar property to that displayed by l_∞, at least among its separable superspaces.

Theorem 4 (A. Sobczyk). *Whenever c_0 is a closed linear subspace of a* **separable** *Banach space X, there is a bounded linear projection P from X onto c_0.*

PROOF (W. Veech). Let e_n^* denote the nth coordinate functional in $l_1 = c_0^*$; for each n, let x_n^* be a Hahn-Banach extension of e_n^* to all of X.

Look at $F = \{x^* \in B_{X^*}: x^* \text{ vanishes on } c_0\}$. Any weak* limit point of $\{x_n^*\}$ belongs to F; indeed, if x^* be such a limit point of $\{x_n^*\}$, then the value of x^* at any unit vector e_m must be arbitrarily closely approximated by infinitely many of the numbers $\{x_n^*(e_m): n \in N\}$—only one of which is not zero.

Define $d: X^* \times X^* \to [0, \infty)$ by

$$d(x^*, y^*) = \sum_n 2^{-n}|(x^* - y^*)(x_n)|,$$

where (x_n) is a sequence dense in S_X. Notice that d generates a topology on X^* that agrees on B_{X^*} with the weak* topology.

As noted above, any weak* limit point of the sequence (x_n^*) is in F. With the metrizability of F in our hands, we can restate this in the following (at first glance obscure) fashion: given any subsequence (y_n^*) of (x_n^*) there is a subsequence (z_n^*) of (y_n^*) which is weak* convergent to a point of F. Alternatively, the sequence (d_n) of real numbers given by $d_n = d$-distance of x^* to F has the property that each of its subsequences has a null subsequence. The result: $\lim_n d\text{-distance}(x_n^*, F) = 0$.

For each n pick a $z_n^* \in F$ close enough to x_n^* (in the d-metric) that $\lim_n d(z_n^*, x_n^*) = 0$. This just says that $0 = \text{weak}^* \lim_n (x_n^* - z_n^*)$. Now define $P: X \to c_0$ by $Px = (x_n^* x - z_n^* x)$; P is the sought-after projection. □

Similar to the case of l_∞ the "separable injectivity" of c_0 has another side to it: *if Y is a linear subspace of a* **separable** *Banach space X and $T: Y \to c_0$ is a bounded linear operator, then there is a bounded linear operator $S: X \to c_0$ extending T to all of X.* To see why this is so, we use the Phillips theorem to extend $T: Y \to l_\infty$ to a bounded linear operator $R: X \to l_\infty$. The separability of X implies that of the closed linear span Z of $RX \cup c_0$. But now Sobczyk's theorem ensures the existence of a bounded linear projection $P: Z \to c_0$ of Z onto c_0. Let $S = PR$.

We turn now to a brief look at l_1. It too possesses some striking mapping properties. In the case of l_1, the "projectivity" of l_1 comes about because of the strength of its norm. Face it: the norm of a vector $\Sigma_n t_n e_n$ in l_1 is as big as it can be $(\|\Sigma_n t_n e_n\| = \Sigma_n |t_n|)$ if respect for the triangle inequality and the "unit" vector is to be preserved. As a consequence of this, we note the following theorem.

Theorem 5. *If X is any Banach space and $T: X \to l_1$ is a bounded linear operator of X onto l_1 then X contains a complemented subspace that is isomorphic to l_1. Moreover, among the separable infinite-dimensional Banach spaces, the above assertion characterizes l_1 isomorphically.*

As is only fair, we start with the proof of the first assertion. Suppose $T: X \to l_1$ is as advertised. By the open mapping theorem there is a bounded sequence (x_n) in X such that $Tx_n = e_n$. Consider the bounded linear operator $S: l_1 \to X$ that takes e_n to x_n—the existence of a unique such S is obvious. Clearly, $TS: l_1 \to l_1$ is naught else but the identity on l_1 and "factors" through X. It follows that $ST: X \to X$ is a bounded linear operator whose square $[STST = S(TS)T = ST]$ is itself and whose range is isomorphic to the closed linear span of the x_n. But if (t_n) is any scalar sequence with only finitely many nonzero terms, then

$$\begin{aligned}\sum_n |t_n| &= \left\|\sum_n t_n e_n\right\|_1 \\ &= \left\|\sum_n t_n T x_n\right\| \\ &= \left\|T\sum_n t_n x_n\right\| \\ &\le \|T\|\left\|\sum_n t_n x_n\right\| \\ &= \|T\|\left\|\sum_n t_n S e_n\right\| \\ &= \|T\|\left\|S\sum_n t_n e_n\right\| \\ &\le \|T\|\,\|S\|\left\|\sum_n t_n e_n\right\| \\ &= \|T\|\,\|S\|\sum_n |t_n|.\end{aligned}$$

It follows that the closed linear span of the x_n is isomorphic to l_1, and the first assertion has been demonstrated.

To prove the second, we need a couple of facts about l_1 that are of interest in themselves. The first is a real classic, due to Banach and Mazur: *every separable Banach space X admits of a continuous linear operator $Q: l_1 \to X$ of l_1 onto X.* In fact, if we let (x_n) be a sequence in B_X that is dense in B_X, then we can define the operator $Q: l_1 \to X$ by $Qe_n = x_n$; it is again a consequence of the strength of l_1's norm that Q is a well-defined bounded linear operator. If $x \in B_X$ is given, then we can find an x_{n_1} so that

$$\|x - x_{n_1}\| \le \frac{1}{2\cdot 2},$$

and

$$\|2(x - x_{n_1})\| \le \tfrac{1}{2}.$$

Next pick $n_2 > n_1$ such that

$$\|2(x - x_{n_1}) - x_{n_2}\| \le \frac{1}{2\cdot 2^2},$$

so that

$$\|4(x - x_{n_1}) - 2x_{n_2}\| \le \frac{1}{2^2}.$$

Continue in this vein with the kth choice producing an $n_k > n_{k-1} > \cdots > n_1$ for which

$$\left\|2^{k-1}(x - x_{n_1}) - 2^{k-2}x_{n_2} - \cdots - 2x_{n_{k-1}} - x_{n_k}\right\| \leq \frac{1}{2^k}.$$

The result is that

$$\left\|x - 2^{1-k}x_{n_k} - 2^{2-k}x_{n_{k-1}} - \cdots - \tfrac{1}{2}x_{n_2} - x_{n_1}\right\| \leq \frac{1}{2^k}.$$

It follows from this that the vector

$$e = e_{n_1} + \tfrac{1}{2}e_{n_2} + \tfrac{1}{4}e_{n_3} + \cdots + \frac{1}{2^k}e_{n_{k+1}} + \cdots \in l_1$$

is carried by Q right onto x.

Returning to our second assertion, we see now that if X is a separable infinite-dimensional Banach space with the property that X is isomorphic to a complemented subspace of any separable space of which X is a quotient, then X is isomorphic to a complemented subspace of l_1. What are the complemented infinite-dimensional subspaces of l_1? Well, all of them are isomorphic to l_1. Of course, this takes proof, and we set forth to prove this now. We follow the direction of Pelczynski in this matter.

The first thing to show is the following.

Theorem 6 (Pelczynski). *Every infinite-dimensional closed linear subspace of l_1 contains a complemented subspace of l_1 that is isomorphic to l_1.*

Proof. Let Z be an infinite-dimensional closed linear subspace of l_1.

Choose any z_1 in Z having norm one. Let k_1 be chosen so that the contribution of the coordinates of z_1 past k_1 to the norm of z_1 amounts to no more than $\frac{1}{4}$.

Since Z is infinite dimensional, there is a z_2 in Z of norm one the first k_1 coordinates of which are zero. Let k_2 be chosen so that the contribution of the coordinates of z_2 beyond k_2 to the norm of z_2 amounts to no more than $\frac{1}{8}$.

Again, since Z is infinite dimensional, there is a z_3 in Z of norm one the first k_2 of whose coordinates are zero. Let k_3 be chosen so that the contribution of the coordinates of z_3 past k_3 to the norm of z_3 amount to no more than $\frac{1}{16}$.

The inductive step is clear.

Agree that $k_0 = 0$. Let

$$b_n = \sum_{j=k_{n-1}+1}^{k_n} z_{n,j}e_j,$$

where $z_{n,j}$ denotes the jth coordinate of z_n and e_j denotes the jth unit vector. Notice that the closed linear span $[b_n]$ of $\{b_n\}$ is isometric to l_1 and is the range of a norm-one projection P. Moreover we have $\|z_n - b_n\|$ not exceeding 2^{-n-1} for any n. Let (b_n^*) be the sequence in $[b_n]^*$ biorthogonal to (b_n); we have

$$\|b_n^*\| = \frac{1}{\|b_n\|} \le \frac{1}{\|z_n\| - \|z_n - b_n\|} = \frac{1}{1-2^{-n-1}}.$$

Consider the operator $T: l_1 \to l_1$ defined by

$$Tx = x - Px + \sum_n b_n^*(Px) z_n.$$

Since $Px \in [b_n]$ and (b_n^*) is biorthogonal to (b_n) with $[b_n]$ isometric to l_1, we see that $(b_n^* Px) \in l_1$; it follows that T is well-defined, bounded, and linear. Moreover, if $x \in l_1$ and $\|x\|_1 \le 1$, then

$$\|x - Tx\| = \left\| Px - \sum_n b_n^*(Px) z_n \right\| \le \|P\| \sum_n \|b_n^*\| \|b_n - z_n\|$$

$$\le \sum_n \frac{2^{-n-1}}{1-2^{-n-1}} = \sum_n (2^{n+1}-1)^{-1} < 1.$$

Therefore, $\|I - T\| < 1$. It follows easily from this that T^{-1} exists as a bounded linear operator on l_1; i.e., T is an isomorphism of l_1 onto itself. To see what T^{-1} looks like, just consider the equation

$$[I-(I-T)] \sum_n (I-T)^n = I,$$

and you can see that $T^{-1} = \Sigma_n (I-T)^n$. Clearly, T takes $[b_n]$ onto $[z_n]$; so $[z_n]$ is isomorphic to l_1. Finally, $Q = TPT^{-1}$ is a bounded linear projection of l_1 onto $[z_n] \subseteq Z$. □

With Theorem 6 in hand we are ready to finish the proof of the second assertion of Theorem 5. Before proceeding with this task, we establish some notational conventions. Suppose (X_n) is a sequence of Banach spaces. Then $(\Sigma_n X_n)_1$ denotes the Banach space of all sequences (x_n), where $x_n \in X_n$, for each n, $\|(x_n)\| = \Sigma_n \|x_n\| < \infty$. It is plain that if each X_n is isomorphic to l_1 with a common bound for the norms of the isomorphisms, then $(\Sigma_n X_n)_1$ is isomorphic to l_1. Sometimes $(\Sigma_n X_n)_1$ is denoted by $(X_1 \oplus X_2 \oplus \cdots)_1$. Also, if X and Y are Banach spaces, then $X \times Y$ is isomorphic to $(X \oplus Y)_1$. Now we finish off Theorem 5.

Let X be an infinite-dimensional complemented subspace of l_1 (recall this is what we have been able to conclude about any Banach space X with the property that it is complemented in any space of which it is a quotient). We will assume that the symbol "$\sim$" will signal the existence of an isomorphism between the left- and right-hand extremities. If Y is a complement of X, then $l_1 \sim (X \oplus Y)_1$. By Theorem 6, there are closed linear subspaces Z_1 and Z of X that are complemented in l_1 such that $X \sim (Z_1 \oplus Z)_1$ and $Z_1 \sim l_1$. The punch line comes from the "Pelczynski decomposition method"; all of the following are easy to see:

$$\begin{aligned}
l_1 \sim (X \oplus Y)_1 &\sim (Z_1 \oplus Z \oplus Y)_1\\
&\sim ((l_1 \oplus Z)_1 \oplus Y)_1\\
&\sim (l_1 \oplus Z \oplus Y)_1\\
&\sim (l_1 \oplus l_1 \oplus \cdots \oplus Z \oplus Y)_1\\
&\sim ((X \oplus Y)_1 \oplus (X \oplus Y)_1 \oplus \cdots \oplus Z \oplus Y)_1\\
&\sim ((X \oplus X \oplus \cdots)_1 \oplus (Y \oplus Y \oplus \cdots)_1 \oplus Z \oplus Y)_1\\
&\sim ((X \oplus X \oplus \cdots)_1 \oplus (Y \oplus Y \oplus \cdots)_1 \oplus Z)_1\\
&\sim ((X \oplus Y)_1 \oplus (X \oplus Y)_1 \oplus \cdots \oplus Z)_1\\
&\sim (l_1 \oplus l_1 \oplus \cdots \oplus Z)_{l_1}\\
&\sim (l_1 \oplus Z)_1\\
&\sim (Z_1 \oplus Z)_1\\
&\sim X.
\end{aligned}$$

This completes the proof of Theorem 5.

The fanciest of our footwork is done. We have seen that both c_0 and l_∞ share injective-type properties while l_1's strength of norm ensures that every separable Banach space occurs as a quotient of l_1 (with Theorem 5 telling us that l_1 is the smallest such space in some sense). We will in the next few sections follow up on more sequentially oriented properties of these spaces, but it seems that this is as likely a place as any to discuss one more space that naturally enters the study of the spaces c_0, l_1, and l_∞: ba, the dual of l_∞. Curiously it will be through the study of ba that two of the most striking sequential properties of these spaces will be unearthed.

Take an $x^* \in l_\infty^*$. Then for each $\Delta \subseteq N$, the characteristic function of Δ, c_Δ belongs to l_∞, and so we can evaluate $x^*(c_\Delta)$. It is easy to see that $x^*(c_\Delta)$ is an additive function of Δ; furthermore, given any pairwise disjoint subsets $\Delta_1, \Delta_2, \ldots, \Delta_n$ of N we have

$$\begin{aligned}
\sum_{i=1}^{n} |x^* c_{\Delta_i}| &= \sum_{i=1}^{n} x^* c_{\Delta_i} \operatorname{sgn} x^* c_{\Delta_i}\\
&= x^*\left(\sum_{i=1}^{n} \operatorname{sgn} x^* c_{\Delta_i} \cdot c_{\Delta_i} \right)\\
&\leq \|x^*\|,
\end{aligned}$$

because $\|\sum_{i=1}^{n} \operatorname{sgn} x^* c_{\Delta_i} \cdot c_{\Delta_i}\|_\infty \leq 1$. So members of l_∞^* lead us naturally to finitely additive measures whose total variation is bounded. This natural intrusion of finitely additive measures into the study of l_∞ (through duality) is worth spending some time on; it is even worth exploring in some generality.

Suppose Ω is a set and Σ is a σ-field of subsets of Ω. Denote by $B(\Sigma)$ the Banach space of bounded, Σ-measurable scalar-valued functions defined on Ω with the supremum norm $\|\cdot\|_\infty$ and denote by $\text{ba}(\Sigma)$ the Banach space of bounded additive scalar-valued measures defined on Σ with variational norm $\|\cdot\|_1$. We plan to show that $B(\Sigma)^* = \text{ba}(\Sigma)$ with the action of a $\mu \in \text{ba}(\Sigma)$ given by means of integration. It is plain that a computation virtually identical to that of the previous paragraph gives a member of $\text{ba}(\Sigma)$ for each member of $B(\Sigma)^*$, namely, $\mu(E) = x^*(c_E)$. Moreover, $\|\mu\|_1 \le \|x^*\|$. Next, if $\mu \in \text{ba}(\Sigma)$, we can define an integral $\int d\mu$ in such a way that every $f \in B(\Sigma)$ can be integrated. How? Start with a simple function $f = \sum_{i=1}^n a_i c_{A_i}$, where $a_1, a_2, \dots, a_n$ are scalars and $A_1, A_2, \dots, A_n$ are disjoint members of Σ; then $\int f\,d\mu$ is defined in the only sensible way:

$$\int f\,d\mu = \int \sum_{i=1}^n a_i c_{A_i}\,d\mu = \sum_{i=1}^n a_i \mu(A_i).$$

Observe that if $\|f\|_\infty \le 1$, then

$$\begin{aligned}\left|\int f\,d\mu\right| &= \left|\sum_{i=1}^n a_i \mu(A_i)\right| \\ &\le \sum_{i=1}^n |a_i||\mu(A_i)| \\ &\le \sup_{1\le i\le n} |a_i| \sum_{i=1}^n |\mu(A_i)| \\ &\le \|\mu\|_1.\end{aligned}$$

It is now clear that $\int d\mu$ acts in a linear continuous fashion on the simple functions modeled on Σ endowed with the supremum norm. As such it can be uniquely extended to the uniform closure of this class in a norm-preserving fashion; the uniform closure of these simple functions is just $B(\Sigma)$. Whatever the extension is, we call its value at an $f \in B(\Sigma)$ "$\int f\,d\mu$."

To summarize, start with an $x^* \in B(\Sigma)^*$, define $\mu \in \text{ba}(\Sigma)$ by $\mu(A) = x^*(c_A)$, and note that $\|\mu\|_1 \le \|x^*\|$. Observe that μ generates $\int d\mu$, which precisely reproduces the values of x^*. Moreover, $\int d\mu$ as a functional has functional norm no more than $\|\mu\|_1$. We have proved the following theorem.

Theorem 7. *The dual of $B(\Sigma)$ is identifiable with the space* $\text{ba}(\Sigma)$ *under the correspondence*

$$x^* \in B(\Sigma)^* \leftrightarrow \mu \in \text{ba}(\Sigma)$$

given by

$$x^* f = \int f\,d\mu.$$

Furthermore, $\|x^*\| = \|\mu\|_1$.

The alert reader will notice that the above representation theorem is really quite formal and cannot be expected to produce much of value unless we go quite a bit deeper into the study of finitely additive measures. This calls for a few words about bounded additive measures; in other words, we digress for a bit. We hope to make one overriding point: a scalar-valued measure being bounded and additive is very like a countably additive measure and is not (at least for the purposes we have in mind) at all pathological.

For instance, suppose $\mu \in \mathrm{ba}(\Sigma)$ and let (A_n) be a sequence of disjoint members of Σ. For each n we have

$$\sum_{k=1}^{n} |\mu(A_k)| \leq \|\mu\|_1$$

so that

$$\sum_n |\mu(A_n)| \leq \|\mu\|_1$$

and $\Sigma_n \mu(A_n)$ is an absolutely convergent series. The point is that μ adds up disjoint sets—even countably many of them—it just may not be judicious enough to add up to the most pleasing sum.

In reality the fact that $\Sigma_n \mu(A_n)$ and $\mu(\cup_n A_n)$ might disagree is not μ's "lack of judgment" but a failure on the part of the underlying σ-field Σ. Suppose, for the sake of this discussion, that μ has only nonnegative values. Then for any sequence (A_n) of pairwise disjoint members of Σ we have

$$\sum_n \mu(A_n) \leq \mu\left(\bigcup_n A_n\right).$$

That strict inequality above might occur is due to the "featherbedding" nature of unions in Σ. If we look at the proper model for the algebra Σ, then on that model μ is countably additive. This statement bears scrutiny.

Recall the Stone representation theorem. It says that *for any Boolean algebra $\mathscr{A}$ there is a totally disconnected compact Hausdorff space $\Omega_{\mathscr{A}}$ for which the Boolean algebra $\mathscr{S}(\mathscr{A})$ of simultaneously closed and open subsets of $\Omega_{\mathscr{A}}$ is isomorphic* (*as a Boolean algebra*) *to $\mathscr{A}$*.

Start with Σ, pass to Ω_Σ, then to $\mathscr{S}(\Sigma)$. μ has an identical twin $\hat{\mu}$ working on $\mathscr{S}(\Sigma)$, but $\hat{\mu}$ has better working conditions than μ. In fact, if (K_n) is a sequence of disjoint members of $\mathscr{S}(\Sigma)$ whose union K belongs to $\mathscr{S}(\Sigma)$, then (since K is compact and each K_n is open) only a finite number of the K_n are nonvoid! $\hat{\mu}$ is countably additive on $\mathscr{S}(\Sigma)$ and so has a unique (regular) countably additive extension to the σ-field of subsets of Ω_Σ generated by $\mathscr{S}(\Sigma)$.

What happened? To begin with, if we have a sequence (A_n) of disjoint members of Σ and we look at $\cup_n A_n = A$, each A_i and A correspond to a K_i and K in $\mathscr{S}(\Sigma)$. The isomorphism between Σ and $\mathscr{S}(\Sigma)$ tells us that K is the supremum in the algebra $\mathscr{S}(\Sigma)$ of the K_n. *However*, K is *not* (necessarily) the union of the K_n. No; in fact, Stone showed that whenever you take a family $\{K_\alpha\}_{\alpha \in A}$ of "clopen" sets in the Boolean algebra $\mathscr{S}(\Sigma)$, then the supremum of the K_α—should such exist in $\mathscr{S}(\Sigma)$—must be the *closure* of the union. It follows that $K = \overline{\cup_n K_n}$ over in Ω_Σ. Returning to our A_n, we see that μ's value at the union of the A_n included not only $\Sigma_n \mu(A_n)$ but in a phantom fashion $\hat{\mu}(K \setminus \cup_n K_n)$. This justifies the claim of featherbedding on part of union in Σ.

What about integrals with respect to members of ba(Σ)? They respect uniform convergence and even some types of pointwise convergence. Of course, one cannot expect them to be like Lebesgue integrals without countable additivity. On the other hand, one is only after integrating members of $B(\Sigma)$ wherein uniform convergence is the natural mode of convergence; so this is not too great a price to pay.

Finally, it ought to be pointed out that members of ba(Σ) are like countably additive measures: if $\mu \in$ ba(Σ) and (A_n) is a sequence of disjoint members of Σ, then there is a subsequence (B_n) of (A_n) such that μ is countably additive on the σ-field $\mathscr{B}$ generated by the B_n. Why is this? Suppose μ has all its values between 0 and 1. Let K and N be infinite disjoint subsets of the set $\mathbf{N}$ of natural numbers. Then either $\Sigma_{n \in K}\mu(A_n)$ or $\Sigma_{n \in N}\mu(A_n)$ is less than or equal to $\frac{1}{2}$. Whichever the case, call the infinite subset N_1 and let B_1 be A_{n_1} where n_1 is the first member of N_1. Now break $N_1 \setminus \{n_1\}$ into two disjoint infinite subsets K and N; either $\Sigma_{n \in K}\mu(A_n) \le \frac{1}{4}$ or $\Sigma_{n \in N}\mu(A_n) \le \frac{1}{4}$. Whichever the case, call the indexing set N_2 and let B_2 be A_{n_2}, where n_2 is the first index occurring in N_2. Repeat this procedure, and a bit of thought will show that the resulting sequence (B_n) satisfies our claim for it.

l_∞^*, Schur's Theorem about l_1, and the Orlicz-Pettis Theorem (Again).

We saw in the preceding section that l_∞^* is not quite so unwieldy as might be guessed. In this section a few of the truly basic limiting theorems regarding l_∞^* are derived. They include the Nikodym-Grothendieck boundedness theorem, Rosenthal's lemma and Phillips's lemma. From this list we show that in l_1 the weak and the norm convergences of sequences coincide—an old fact discovered by Schur in 1910. Then we derive the Orlicz-Pettis theorem much as Orlicz and Pettis did in the 1930s.

Throughout this discussion Ω is a set, Σ is a σ-field of subsets of Ω, and ba(Σ) $= B(\Sigma)^*$ is the space of bounded, finitely additive scalar-valued measures defined on Σ. For $\mu \in$ ba(Σ) the variation $|\mu|$ is the member of ba(Σ) whose value at a member E of Σ is given by

$$|\mu|(E) = \sup\left\{ \Sigma |\mu(E_i)| \right\},$$

where the supremum is taken over all finite collections $\{E_1, \ldots, E_n\}$ of pairwise disjoint members of Σ contained in E. Of course, $|\mu|(\Omega)$ is just the variational norm $\|\mu\|_1$ of μ. It is noteworthy that for any $E \in \Sigma$ we have

$$\sup_{\substack{F\in\Sigma\\ F\subseteq E}} |\mu(F)| \le |\mu|(E) \le 4 \sup_{\substack{F\in\Sigma\\ F\subseteq E}} |\mu(F)|.$$

The left side holds trivially, whereas the right follows from considering for a fixed $E \in \Sigma$ and a given partition π of E into a finite number of disjoint members of Σ the real and imaginary parts of each value of μ on members of π and checking the positive and negative possibilities of each.

Let us start our more serious discussion with a fundamental bounding principle.

Nikodym-Grothendieck Boundedness Theorem. *Suppose $\mathscr{F} = \{\mu_t : t \in T\}$ is a family of members of* ba(Σ) *satisfying*

$$\sup_t |\mu_t(E)| < \infty$$

for each $E \in \Sigma$. Then

$$\sup_{t,\, E\in\Sigma} |\mu_t(E)| < \infty.$$

PROOF. Should the conclusion fail, there would be a sequence (μ_n) of members of $\mathscr{F}$ for which

$$\sup_{n,\, E\in\Sigma} |\mu_n(E)| = \infty.$$

Suppose such is the case.

Observe: *If $\rho > 0$, then there is an n and a partition $\{E, F\}$ of Ω into disjoint members of Σ such that both $|\mu_n(E)|, |\mu_n(F)| > \rho$.* In fact, choose n and E so that $E \in \Sigma$ and

$$|\mu_n(E)| > \sup_k |\mu_k(\Omega)| + \rho.$$

Then

$$\begin{aligned}|\mu_n(\Omega \setminus E)| &= |\mu_n(E) - \mu_n(\Omega)| \\ &\ge |\mu_n(E)| - |\mu_n(\Omega)| > \rho.\end{aligned}$$

Now let n_1 be the first positive integer for which there is a partition $\{E, F\}$ of Ω into disjoint members of Σ for which

$$|\mu_{n_1}(E)|, |\mu_{n_1}(F)| > 2.$$

One of the quantities

$$\sup_{n,\, B\in\Sigma} |\mu_n(E \cap B)|, \qquad \sup_{n,\, B\in\Sigma} |\mu_n(F \cap B)|$$

is infinite. If the first, set $B_1 = E$ and $G_1 = F$; otherwise, set $B_1 = F$ and $G_1 = E$.

Let $n_2 > n_1$ be the first such positive integer for which there is a partition $\{E, F\}$ of B_1 into disjoint members of Σ such that

$$|\mu_{n_2}(E)|, |\mu_{n_2}(F)| > |\mu_{n_2}(G_1)| + 3.$$

One of the quantities

$$\sup_{n, B \in \Sigma} |\mu_n(E \cap B)|, \qquad \sup_{n, B \in \Sigma} |\mu_n(F \cap B)|$$

is infinite. Should it be the first of these, set $B_2 = E$ and $G_2 = F$; otherwise, set $B_2 = F$ and $G_2 = E$.

Continue.

We obtain a sequence (G_k) of pairwise disjoint members of Σ and a strictly increasing sequence (n_k) of positive integers such that for each $k > 1$

$$|\mu_{n_k}(G_k)| > \sum_{j=1}^{k-1} |\mu_{n_k}(G_j)| + k + 1.$$

Relabel (μ_{n_k}) by (μ_k).

Partition the set $\mathbb{N}$ of natural numbers into infinitely many disjoint infinite subsets $N_1, N_2, \ldots$. The additivity of $|\mu_1|$ gives

$$\sum_{k=1}^{\infty} |\mu_1|\left(\bigcup_{n \in N_k} G_n\right) \le |\mu_1|\left(\bigcup_n G_n\right)$$
$$\le |\mu_1|(\Omega).$$

It follows that there is a subsequence (G_{k_i}) of $(G_k)_{k \ge 2}$ such that

$$|\mu_1|\left(\bigcup_{i=1}^{\infty} G_{k_i}\right) < 1.$$

Repeat the above argument; this time work with $|\mu_{k_1}|$ instead of $|\mu_1|$ and $(G_{k_i})_{i \ge 2}$ instead of $(G_k)_{k \ge 2}$. You'll find a subsequence $(G_{k_{i_j}})$ of $(G_{k_i})_{i \ge 2}$ such that

$$|\mu_{k_1}|\left(\bigcup_{j=1}^{\infty} G_{k_{i_j}}\right) < 1.$$

Repeat with $|\mu_{k_{i_1}}|$ replacing $|\mu_{k_1}|$ and $(G_{k_{i_j}})_{j \ge 2}$ in lieu of $(G_{k_i})_{i \ge 2}$.

Let G_{m_i} denote the first member of the ith subsequence so generated $(m_1 = 1, m_2 = k_1, m_3 = k_{i_1}, \ldots)$. Then for each j,

$$|\mu_{m_j}|\left(\bigcup_{i=j+1}^{\infty} G_{m_i}\right) < 1.$$

If we let

$$D = \bigcup_{j=1}^{\infty} G_{m_j},$$

then

$$
\begin{aligned}
|\mu_{m_j}(D)| &= \left|\mu_{m_j}\left(\bigcup_{i=1}^{j-1} G_{m_i} \cup G_{m_j} \cup \bigcup_{i=j+1}^{\infty} G_{m_i}\right)\right| \\
&\geq |\mu_{m_j}(G_{m_j})| - \left|\mu_{m_j}\left(\bigcup_{i=1}^{j-1} G_{m_i}\right)\right| - \left|\mu_{m_j}\left(\bigcup_{i=j+1}^{\infty} G_{m_i}\right)\right| \\
&\geq |\mu_{m_j}(G_{m_j})| - \sum_{i=1}^{j-1} |\mu_{m_j}(G_{m_i})| - |\mu_{m_j}|\left(\bigcup_{i=j+1}^{\infty} G_{m_i}\right) \\
&\geq m_j \uparrow \infty,
\end{aligned}
$$

a contradiction. □

Rosenthal's lemma is our next stop. It provides the sharpest general disjointification principle there is.

Rosenthal's Lemma. *Let $(\mu_n) \subseteq \text{ba}(\Sigma)$ be uniformly bounded. Then given $\varepsilon > 0$ and a sequence (E_n) of disjoint members of Σ there is an increasing sequence (k_n) of positive integers for which*

$$|\mu_{k_n}|\left(\bigcup_{j \neq n} E_{k_j}\right) < \varepsilon$$

for all n.

Proof. We may assume that $\sup_m |\mu_m|(\cup_n E_n) \leq 1$.

Partition $\mathbf{N}$ into an infinite number of infinite (disjoint) subsets (N_k). If for some p there is no $k \in N_p$ with

$$|\mu_k|\left(\bigcup_{\substack{j \neq k \\ j \in N_p}} E_j\right) \geq \varepsilon,$$

then for each $k \in N_p$ we have

$$|\mu_k|\left(\bigcup_{\substack{j \neq k \\ j \in N_p}} E_j\right) < \varepsilon.$$

Enumerating N_p will produce the sought-after subsequence. What if no such p arises? Well, then it must be that for each p there's a $k_p \in N_p$ for which

$$|\mu_{k_p}|\left(\bigcup_{\substack{j \neq k_p \\ j \in N_p}} E_j\right) \geq \varepsilon.$$

Notice that

$$|\mu_{k_p}|\left(\bigcup_n E_{k_n}\right) + |\mu_{k_p}|\left(\bigcup_n E_n \setminus \bigcup_n E_{k_n}\right) = |\mu_{k_p}|\left(\bigcup_n E_n\right) \leq 1,$$

which, since

$$\bigcup_{\substack{j \neq k_p \\ j \in N_p}} E_j \subseteq \bigcup_n E_n \setminus \bigcup_n E_{k_n},$$

gives us

$$|\mu_{k_p}|\left(\bigcup_n E_{k_n}\right) \leq 1 - \varepsilon$$

for all p.

Repeat the above argument starting this time with the sequences $\mu_n' = \mu_{k_n}$ and $E_n' = E_{k_n}$; our starting point now will be the inequality

$$|\mu_n'|\left(\bigcup_n E_n'\right) \leq 1 - \varepsilon.$$

Proceeding as above, either we arrive immediately at a suitable subsequence or extract a subsequence (j_{k_n}) of (k_n) for which another ε can be shaved off the right side of the above inequality making

$$|\mu_{j_{k_p}}|\left(\bigcup_n E_{j_{k_n}}\right) \leq 1 - 2\varepsilon$$

hold for all p. □

Whatever the first n is that makes $1 - n\varepsilon < 0$, the above procedure must end satisfactorily by n steps or face the possibility that $0 \leq 1 - n\varepsilon < 0$.

From Rosenthal's lemma and the Nikodym-Grothendieck boundedness theorem we derive another classic convergence theorem pertaining to l_∞^*.

Phillips's Lemma. *Let* $\mu_n \in \mathrm{ba}(2^{\mathbf{N}})$ *satisfy* $\lim_n \mu_n(\Delta) = 0$ *for each* $\Delta \subseteq \mathbf{N}$. *Then*

$$\lim_n \sum_j |\mu_n(\{j\})| = 0.$$

PROOF. The Nikodym-Grothendieck theorem tells us that $\sup_n \|\mu_n\| < \infty$, and so the possibility of applying Rosenthal's lemma arises.

Were the conclusion of Phillips's lemma not to hold, it would be because for some $\delta > 0$ and some subsequence [which we will still refer to as (μ_n)] of (μ_n) we have

$$\sum_j |\mu_n(\{j\})| \geq 6\delta$$

for all n.

Let F_1 be a finite subset of $\mathbb{N}$ for which

$$|\mu_1(F_1)| > \delta.$$

Using the fact that $(\mu_n(\Delta))$ is null for each Δ, choose $n_2 > n_1 = 1$ so that

$$\sum_{j \in F_1} |\mu_{n_2}(\{j\})| < \delta.$$

Next, let F_2 be a finite subset of $\mathbb{N} \setminus F_1$ for which

$$\begin{aligned} |\mu_{n_2}(F_2)| &\geq \tfrac{1}{5} \sum_{j \notin F_1} |\mu_{n_2}(\{j\})| \\ &\geq \tfrac{1}{5}\Big(\sum_j |\mu_{n_2}(\{j\})| - \sum_{j \in F_1} |\mu_{n_2}(\{j\})| \Big) \\ &\geq \tfrac{1}{5}(6\delta - \delta) = \delta. \end{aligned}$$

Using the fact that $(\mu_n(\Delta))$ is null for each Δ, choose $n_3 > n_2$ so that

$$\sum_{j \in F_1 \cup F_2} |\mu_{n_3}(\{j\})| < \delta.$$

Let F_3 be a finite subset of $\mathbb{N} \setminus (F_1 \cup F_2)$ for which

$$\begin{aligned} |\mu_{n_3}(F_3)| &\geq \tfrac{1}{5} \sum_{j \notin F_1 \cup F_2} |\mu_{n_3}(\{j\})| \\ &\geq \tfrac{1}{5}\Big(\sum_j |\mu_{n_3}(\{j\})| - \sum_{j \in F_1 \cup F_2} |\mu_{n_3}(\{j\})| \Big) \\ &> \tfrac{1}{5}(6\delta - \delta) = \delta. \end{aligned}$$

Our procedure should now be clear. We extract a subsequence (ν_n) of (μ_n) and a sequence (F_n) of pairwise disjoint finite subsets of $\mathbb{N}$ for which given n

$$\sum_n |\nu_n(\{j\})| \geq 6\delta,$$

$$\sum_{j \in F_1 \cup \cdots \cup F_{n-1}} |\nu_n(\{j\})| < \delta,$$

and

$$|\nu_n(F_n)| \geq \delta.$$

Rosenthal's lemma allows us to further prune (ν_n) and (F_n) so as to attain

$$|\nu_n|\Big(\bigcup_{k \neq n} F_k \Big) < \frac{\delta}{2}.$$

On so refining, we see that

$$\begin{aligned}\left|\nu_n\left(\bigcup_m F_m\right)\right| &= \left|\nu_n(F_n)+\nu_n\left(\bigcup_{m\neq n} F_m\right)\right|\\ &\geq |\nu_n(F_n)| - |\nu_n|\left(\bigcup_{m\neq n} F_m\right)\\ &> \delta - \frac{\delta}{2} = \frac{\delta}{2}.\end{aligned}$$

□

Schur's Theorem. *In l_1, weak and norm convergences of sequences coincide.*

PROOF. Each $x \in l_1$ defines a $\mu_x \in \mathrm{ba}(2^N) = l_\infty^* = l_1^{**}$ by looking at x's image μ_x under the natural imbedding of l_1 into l_1^{**}, where for any Δ, $\mu_x(\Delta)$ is given by

$$\mu_x(\Delta) = \sum_{n\in\Delta} x(n), \qquad x = (x(n)) \in l_1.$$

Should (x_k) be a weakly null sequence in l_1, then the corresponding sequence $(\mu_k = \mu_{x_k})$ in $\mathrm{ba}(2^N)$ satisfies

$$\begin{aligned}\lim_n \mu_n(\Delta) &= \lim_n \sum_{m\in\Delta} x_n(m)\\ &= \lim_n \chi_\Delta(x_n) = 0.\end{aligned}$$

Phillips's lemma now tells us that

$$0 = \lim_n \sum_j |\mu_n(\{j\})| = \lim_n \sum_j |x_n(j)| = \lim_n \|x_n\|_1.$$

□

Okay, it is time for the Orlicz-Pettis theorem again—only this time we prove it in much the same way Orlicz and Pettis did it in the first place using Schur's theorem except that we use Phillips's lemma.

PROOF OF THE ORLICZ-PETTIS THEOREM. As usual, there is some initial footwork making clear that if anything could go wrong with the Orlicz-Pettis theorem, it would happen where a weakly subseries convergent series $\Sigma_n x_n$ could be found for which $\|x_n\| \geq \varepsilon > 0$ holds for all n. This proof shows that whenever $\Sigma_n x_n$ is weakly subseries convergent, there is a subsequence (x_{n_k}) of (x_n) that is norm null.

Whatever goes on with the series $\Sigma_n x_n$, all the action happens in the closed linear span $[x_n]$ of the x_n; so we may as well assume that X is separable. For each n choose an $x_n^* \in B_{X^*}$ such that $x_n^* x_n = \|x_n\|$. Since X is separable, B_{X^*} is weak* compact and metrizable (the proof of this will be given later in detail; however, a look at step 1 of the proof of the Eberlein-Smulian theorem ought to be convincing of this fact). Therefore, there is a subsequence (y_n^*) of (x_n^*) which is weak* convergent, say, to y_0^*;

let (y_n) be the corresponding subsequence of (x_n). $\Sigma_n y_n$ is weakly subseries convergent. Therefore, (y_n) is weakly null. It follows that for very large n, $(y_n^* - y_0^*)(y_n)$ is very close to $\|y_n\|$. Since the series Σy_n is weakly subseries convergent, for any $\Delta \subseteq N$ the series $\Sigma_{n\in\Delta} y_n$ converges weakly to some $\sigma_\Delta \in X$. Define $\mu_n \in l_\infty^*$ at $\Delta \subseteq N$ by

$$\mu_n(\Delta) = \left(y_n^* - y_0^*\right)(\sigma_\Delta).$$

Because $y_0^* = \text{weak}^* \lim y_n^*$, $\lim_n \mu_n(\Delta) = 0$ for each $\Delta \subseteq N$. Phillips's lemma concludes that $\lim_n \Sigma_k |\mu_n(\{k\})| = 0$. But $(y_n^* - y_0^*)(y_n) = \mu_n(\{n\})$ with the left side being a good approximation of $\|y_n\|$ for n big and the right side being a good approximation of 0 for n big. Enough said. □

Weak Compactness in $\text{ca}(\Sigma)$ and $L_1(\mu)$

Let Ω be a set and Σ be a σ-field of subsets of Ω. Denote by $\text{ca}(\Sigma)$ the linear subspace of $\text{ba}(\Sigma)$ consisting of the countably additive measures on Σ. It is clearly the case that $\text{ca}(\Sigma)$ is a closed linear subspace of $\text{ba}(\Sigma)$ if the latter is normed by $\|\mu\|_\infty = \sup\{|\mu(E)|: E \in \Sigma\}$; from this and the inequality $\|\mu\|_\infty \leq \|\mu\|_1 = \text{variation of } \mu = |\mu|(\Omega) \leq 4\|\mu\|_\infty$, we see that $(\text{ca}(\Sigma), \|\cdot\|_1)$ is a Banach space. Further, it is a standard exercise that $|\mu| \in \text{ca}(\Sigma)$ whenever $\mu \in \text{ca}(\Sigma)$.

It is our purpose in this section to derive criteria for weak compactness in $\text{ca}(\Sigma)$. On doing so, we will derive the classical conditions for a subset of $L_1(\mu)$ to be weakly compact and recognize both $\text{ca}(\Sigma)$ and $L_1(\mu)$ as Banach spaces in which weakly Cauchy sequences are weakly convergent. The Kadec-Pelczynski theorem, recognizing the role of l_1's unit vector basis in nonweakly convergent sequences in $L_1(\mu)$, will be given its due attention, and the Dieudonné-Grothendieck criterion for weak compactness in $\text{rca}(\Sigma)$ will be established. Here $\text{rca}(\Sigma)$ denotes the space of *regular* members of $\text{ca}(\Sigma)$, where Σ is the Borel σ-field of subsets of a compact Hausdorff space Ω. A well-known consequence of this and Phillips's lemma, i.e., weak* convergent sequences in l_∞^* are weakly convergent, will finish this section.

We begin our discussion with an idea of Saks. Take a nonnegative $\lambda \in \text{ca}(\Sigma)$. For $A, B \in \Sigma$ define the pseudo λ-distance between A and B by

$$d_\lambda(A, B) = \lambda(A \Delta B),$$

where $A\Delta B = (A\setminus B)\cup(B\setminus A)$ is the symmetric difference of A and B. The seed of Saks's idea is in the following easily proved result.

Theorem 8. *(Σ, d_λ) is a* complete *pseudometric space on which the operations $(A, B) \to A \cup B$, $(A, B) \to (A \cap B)$, and $A \to A^c$ are all continuous (the first two as functions of two variables).*

Saks's program is to study convergence of sequences of countably additive measures on Σ by means of viewing the measures as continuous functions on pseudometric spaces of the (Σ, d_λ) ilk; particularly useful in this connection is the completeness of (Σ, d_λ) since it brings to mind the Baire category technique, a technique mastered by none more thoroughly than Saks.

Completeness being so crucial to the implementation of Saks's scheme, we would be remiss if we didn't say at least a few words toward the proof of the completeness aspect of Theorem 8 (other assertions can be safely left to the enjoyment of the careful reader).

To see that (Σ, d_λ) is complete, notice that for $A, B \in \Sigma$,

$$d_\lambda(A, B) = \|c_A - c_B\|_{L_1(\lambda)}.$$

Therefore, if (A_n) is a d_λ-Cauchy sequence in Σ, (c_A) is norm Cauchy in $L_1(\lambda)$, hence convergent in $L_1(\lambda)$-mean to some $f \in L_1(\lambda)$. Passing to an appropriate subsequence will convince you that f is itself of the form c_A for some $A \in \Sigma$. Of course, A is the d_λ-limit of (A_n).

Naturally, if $\mu \in \text{ba}(\Sigma)$ is continuous on (Σ, d_λ), we say μ is λ continuous. Notice that λ-continuity of μ automatically implies μ is itself in ca(Σ). In this connection it is noteworthy that the λ continuity of $\mu \in \text{ba}(\Sigma)$ is just saying that μ satisfies the condition: for each $\varepsilon > 0$ there is a $\delta > 0$ such that $|\mu(E) - \mu(F)| \leq \varepsilon$ whenever $|\lambda(E) - \lambda(F)| = |\lambda(E \Delta F)| \leq \delta$; in particular, whenever $\lambda(E) \leq \delta$, then $|\mu(E)| \leq \varepsilon$, and so μ is absolutely continuous with respect to λ. The converse is also true; i.e., if μ is absolutely continuous with respect to λ, then μ is a continuous function on (Σ, d_λ).

Suppose $\mathscr{K}$ is a family of finitely additive scalar-valued measures defined on Σ. We say (for the moment) that $\mathscr{K}$ is *equi-λ-continuous at* $E \in \Sigma$ if for each $\varepsilon > 0$ there is a $\delta > 0$ such that if $F \in \Sigma$ and $d_\lambda(E, F) \leq \delta$, then $|\mu(E) - \mu(F)| \leq \varepsilon$ for all $\mu \in \mathscr{K}$; uniformly *equi-λ-continuous on* Σ if for each $\varepsilon > 0$ there is a $\delta > 0$ such that given $E, F \in \Sigma$ with $d_\lambda(E, F) \leq \delta$, then $|\mu(E) - \mu(F)| \leq \varepsilon$ for all $\mu \in \mathscr{K}$; *uniformly countably additive* provided for each decreasing sequence (E_n) of members of Σ with $\cap_n E_n = \varnothing$ and each $\varepsilon > 0$ there is an N_ε such that $|\mu(E_n)| \leq \varepsilon$ for n beyond N_ε and all $\mu \in \mathscr{K}$.

The momentary excess of verbiage is eliminated by the next theorem.

Theorem 9. *Let $\mathscr{K}$ be a family of finitely additive scalar-valued measures defined on Σ. Then the following are equivalent (TFAE):*

1. *$\mathscr{K}$ is equi-λ-continuous at some $E \in \Sigma$.*
2. *$\mathscr{K}$ is equi-λ-continuous at $\varnothing$.*
3. *$\mathscr{K}$ is uniformly equi-λ-continuous on Σ.*

Moreover 1 to 3 imply that $\mathscr{K}$ is uniformly countably additive.

PROOF. Suppose 1 holds. Let $\varepsilon > 0$ be given and choose $\delta > 0$ so that should $B \in \Sigma$ be within δ of E, then $|\mu(B) - \mu(E)| \leq \varepsilon$ for all $\mu \in \mathscr{K}$.

Notice that if $A \in \Sigma$ and $\lambda(A) \le \delta$, then $\lambda((E \cup A)\Delta E) \le \lambda(A) \le \delta$ and $\lambda((E \setminus A)\Delta E) \le \lambda(A) \le \delta$. It follows that if $A \in \Sigma$ and $\lambda(A) \le \delta$, then

$$\begin{aligned} |\mu(A)| &= |\mu(A \cup E) - \mu(E \setminus A)| \\ &\le |\mu(A \cup E) - \mu(E)| + |\mu(E) - \mu(E \setminus A)| \\ &\le \varepsilon + \varepsilon = 2\varepsilon \end{aligned}$$

for all $\mu \in \mathscr{K}$. This is 2.

Next, if $C, D \in \Sigma$ and

$$\lambda(C \setminus D) + \lambda(D \setminus C) = \lambda(C \Delta D) \le \delta,$$

then for all $\mu \in \mathscr{K}$ we have

$$\begin{aligned} |\mu(C) - \mu(D)| &= |\mu(C \setminus D) - \mu(D \setminus C)| \\ &\le |\mu(C \setminus D)| + |\mu(D \setminus C)| \\ &\le 2\varepsilon + 2\varepsilon = 4\varepsilon, \end{aligned}$$

and now 3 is in hand.

The last assertion follows from 3 and λ's countable additivity.

From now on a $\mathscr{K}$ satisfying 1 to 3 will be called *uniformly* λ*-continuous*; sometimes this is denoted by $\mathscr{K} \underset{\text{unif}}{\ll} \lambda$ and sometimes by

$$\lim_{\lambda(E) \to 0} \mu(E) = 0 \quad \text{uniformly for } \mu \in \mathscr{K}. \qquad \square$$

A bit more about uniformly countably additive families is in order.

Theorem 10. *Let $\mathscr{K} \subseteq \text{ca}(\Sigma)$. Then TFAE:*

1. *If (E_n) is a sequence of disjoint members of Σ, then for each $\varepsilon > 0$ there is an n_ε such that for $m \ge n \ge n_\varepsilon$,*

$$\left| \sum_{i=n}^{m} \mu(E_i) \right| \le \varepsilon$$

   for all $\mu \in \mathscr{K}$.

2. *If (E_n) is a sequence of disjoint members of Σ, then for each $\varepsilon > 0$ there is an n_ε such that for $n \ge n_\varepsilon$,*

$$\left| \sum_{i=n}^{\infty} \mu(E_i) \right| \le \varepsilon$$

   for all $\mu \in \mathscr{K}$.

3. *If (E_n) is a sequence of disjoint members of Σ, then for each $\varepsilon > 0$ there is an n_ε such that for $n \ge n_\varepsilon$,*

$$|\mu(E_n)| \le \varepsilon$$

   for all $\mu \in \mathscr{K}$.

4. *If (E_n) is a monotone increasing sequence in Σ, then for each $\varepsilon > 0$ there is an n_ε such that if $m, n \geq n_\varepsilon$, then*
$$|\mu(E_m) - \mu(E_n)| \leq \varepsilon$$
for all $\mu \in \mathscr{K}$.
5. *If (E_n) is a monotone decreasing sequence in Σ, then for each $\varepsilon > 0$ there is an n_ε such that if $m, n \geq n_\varepsilon$, then*
$$|\mu(E_m) - \mu(E_n)| \leq \varepsilon$$
for all $\mu \in \mathscr{K}$.
6. *$\mathscr{K}$ is uniformly countably additive on Σ.*

Proof. The proof is purely formal and proceeds as with one measure at a time with the phrase "for all $\mu \in \mathscr{K}$" carefully tacked on; for this reason we go through the proof that 3 implies 1, leaving the details of the other parts of proof to the imagination of the reader.

Suppose (E_n) is a sequence of disjoint members of Σ for which 1 failed; then there would be an $\varepsilon > 0$ such that for any N there would be $m_N \geq n_N \geq N$ with an accompanying $\mu_N \in \mathscr{K}$ for which
$$\left|\mu_N\left(\bigcup_{i=n_N}^{m_N} E_i\right)\right| = \left|\sum_{i=n_N}^{m_N} \mu_N(E_i)\right| \geq \varepsilon.$$
Take $N = 1$ and choose $m_1 \geq n_1 \geq 1$ in accordance with the above quagmire. Let $F_1 = \cup_{i=n_1}^{m_1} E_i$. Let $\nu_1 = \mu_1$.

Next, take $N = m_1 + 1$ and choose $m_2 \geq n_2 \geq N$, again according to the dictates above. Let F_2 be $\cup_{i=n_2}^{m_2} E_i$, and let $\nu_2 = \mu_N$.

Our procedure is clear; we generate a sequence (F_k) of pairwise disjoint members of Σ along with a corresponding sequence (ν_k) in $\mathscr{K}$ for which
$$|\nu_k(F_k)| \geq \varepsilon,$$
thereby denying 3. □

Formalities out of the way, we recall from the first section that we proved the following: *let (X, d) be a complete (pseudo) metric space, and let (f_n) be a sequence of continuous scalar-valued functions defined on X. Suppose that for each $x \in X$, $\lim_n f_n(x)$ exists. Then $\{x \in X : (f_n)$ is equicontinuous at $x\}$ is a set of the second category in X.*

An almost immediate consequence of this is the following classical result.

Vitali-Hahn-Saks Theorem. *Let (μ_n) be a sequence in* ca(Σ) *each term of which is λ-continuous, where λ is a nonnegative member of* ca(Σ). *Assume that $\lim_n \mu_n(E) = \mu(E)$ exists for each $E \in \Sigma$. Then $\{\mu_n\}$ is uniformly λ-continuous and μ is both λ-continuous and countably additive.*

Proof. Viewing the μ_n as functions on the complete pseudometric space (Σ, d_λ), we can apply the cited result from the first section. The equicontinuity of the family $\{\mu_n\}$ on a set of second category implies its equicontinuity at some $E \in \Sigma$ and brings Theorem 9 into play. □

Another "oldie-but-goodie":

Nikodym's Convergence Theorem. *Suppose* (μ_n) *is a sequence from* $\text{ca}(\Sigma)$ *for which*

$$\lim_n \mu_n(E) = \mu(E)$$

exists for each $E \in \Sigma$. *Then* $\{\mu_n\}$ *is uniformly countably additive and* $\mu \in \text{ca}(\Sigma)$.

Proof. Consider the absolutely convergent series

$$\sum_n \frac{|\mu_n|(\cdot)}{(1 + \|\mu_n\|_1)2^n}$$

in $\text{ca}(\Sigma)$; its sum $\lambda \in \text{ca}(\Sigma)$ is nonnegative, and together $\{\mu_n\}$ and λ fit perfectly in the hypotheses of the Vitali-Hahn-Saks theorem. Its conclusion suits $\{\mu_n\}$ and μ well. □

Weak convergence in $\text{ca}(\Sigma)$? No, we haven't forgotten!

Theorem 11. *A sequence* (μ_n) *in* $\text{ca}(\Sigma)$ *converges weakly to* $\mu \in \text{ca}(\Sigma)$ *if and only if for each* $E \in \Sigma$, $\mu(E) = \lim_n \mu_n(E)$.

Proof. Since the functional $\nu \to \nu(E)$ belongs to $\text{ca}(\Sigma)^*$ for each $E \in \Sigma$, the necessity of $\mu(E) = \lim_n \mu_n(E)$ for each $E \in \Sigma$ is clear.

Suppose for the sake of argument that $\mu(E) = \lim_n \mu_n(E)$ holds for each $E \in \Sigma$. Now Nikodym's boundedness theorem allows us to conclude that the μ_n are uniformly bounded, and so $\sup_n |\mu_n|(\Omega) < \infty$. It follows that the series

$$\sum_n \frac{|\mu_n|(\cdot)}{2^n}$$

is absolutely convergent in the Banach space $\text{ca}(\Sigma)$; let λ be its sum. For each n there is an $f_n \in L_1(\lambda)$ such that

$$\mu_n(E) = \int_E f_n \, d\lambda,$$

this thanks to the Radon-Nikodym theorem. Similarly, there is an $f \in L_1(\lambda)$ such that

$$\mu(E) = \int_E f \, d\lambda$$

for each $E \in \Sigma$. Since $\sup_n |\mu_n|(\Omega) < \infty$ and $\|f_n\|_1 = |\mu_n|(\Omega)$, (f_n) is an L_1-bounded sequence. Since $\mu(E) = \lim_n \mu_n(E)$ holds for each $E \in \Sigma$,

$$\int fg\, d\lambda = \lim_n \int f_n g\, d\lambda$$

holds for each simple function g. But the collection of all simple g is dense in $L_\infty(\lambda)$ so that an easy $\varepsilon/2 + \varepsilon/2 = \varepsilon$ argument shows

$$\int fh\, d\lambda = \lim_n \int f_n h\, d\lambda$$

holding for all $h \in L_\infty(\lambda) = L_1(\lambda)^*$. It follows that (f_n) converges weakly to f in $L_1(\lambda)$, which in turn implies that (μ_n) converges weakly to μ in ca(Σ). □

Immediate from the above is the following corollary.

Corollary. *A sequence (f_n) in $L_1(\lambda)$ converges weakly to f in $L_1(\lambda)$ if and only if $\int_E f\, d\lambda = \lim_n \int_E f_n\, d\lambda$ for each $E \in \Sigma$.*

In tandem with the Vitali-Hahn-Saks-Nikodym convergence principles the above proofs suggest the following important theorem.

Theorem 12. *Weakly Cauchy sequences in* ca(Σ) *are weakly convergent. Consequently, for any $\lambda \in \text{ca}^+(\Sigma)$, weakly Cauchy sequences in $L_1(\lambda)$ are weakly convergent.*

PROOF. Let (μ_n) be a weakly Cauchy sequence in ca(Σ). Since each $E \in \Sigma$ determines the member $\nu \to \nu(E)$ of ca$(\Sigma)^*$, $\lim_n \mu_n(E) = \mu(E)$ exists for each $E \in \Sigma$. The Vitali-Hahn-Saks-Nikodym clique force μ to be a member of ca(Σ). The just-established criteria for weak convergence in ca(Σ) make μ the weak limit of (μ_n).

The second assertion follows from the first on observing again that for $\lambda \in \text{ca}^+(\Sigma)$, $L_1(\lambda)$ is a closed subspace of ca(Σ). □

We are closing in on weak compactness criteria for both ca(Σ) and $L_1(\lambda)$. The next lemma will bring these criteria well within our grasp.

Lemma. *Let $\mathscr{A}$ be an algebra of sets generating Σ and suppose $\{\mu_n\}$ is a uniformly countably additive family for which $\lim_n \mu_n(E)$ exists for each $E \in \mathscr{A}$. Then $\lim_n \mu_n(E)$ exists for each $E \in \Sigma$.*

PROOF. Look at $\Lambda = \{E \in \Sigma : \lim_n \mu_n(E) \text{ exists}\}$. By hypothesis, $\mathscr{A} \subseteq \Lambda$. We claim that Λ is a monotone class; from this it follows that $\Lambda = \Sigma$, proving the lemma.

Let (E_m) be a monotone sequence of members of Λ with $E_m \to E$. By the uniform countable additivity of the μ_n,

$$\mu_n(E) = \lim_m \mu_n(E_m)$$

uniformly in n; one need only glance at and believe in Theorem 10 to see this. So given $\varepsilon > 0$ there is an m such that

$$|\mu_n(E_m) - \mu_n(E)| \leq \varepsilon$$

for all n. But $(\mu_n(E_m))_n$ converges; so there is an N_ε such that if $p, q \geq N_\varepsilon$ then

$$|\mu_p(E_m) - \mu_q(E_m)| \leq \varepsilon.$$

Plainly

$$|\mu_p(E) - \mu_q(E)| \leq 3\varepsilon$$

should p, q exceed N_ε. It follows that $(\mu_n(E))$ is a convergent sequence. □

Theorem 13. *Let $\mathscr{K}$ be a subset of* ca(Σ). *Then TFAE:*

1. *$\mathscr{K}$ is relatively weakly compact.*
2. *$\mathscr{K}$ is bounded and uniformly countably additive.*
3. *$\mathscr{K}$ is bounded and there is a $\lambda \in$ ca$^+(\Sigma)$ such that $\mathscr{K}$ is uniformly λ-continuous.*

PROOF. Suppose $\mathscr{K}$ is relatively weakly compact. Then there is an $M > 0$ such that $\|\mu\|_1 \leq M$ for all $\mu \in \mathscr{K}$. We claim that *given $\varepsilon > 0$ there is a finite set $\{\mu_1, \ldots, \mu_n\} \subseteq \mathscr{K}$ and a $\delta > 0$ such that $|\mu_1|(E), |\mu_2|(E), \ldots, |\mu_n|(E) \leq \delta$ implies $|\mu(E)| \leq \varepsilon$ for all $\mu \in \mathscr{K}$*. Indeed if this were not the case, then there would be a bad $\varepsilon > 0$ for which no such finite set or $\delta > 0$ exists. Take any $\mu_1 \in \mathscr{K}$. There must be $E_1 \in \Sigma$ and $\mu_2 \in \mathscr{K}$ for which

$$|\mu_1|(E_1) \leq \tfrac{1}{2}, \qquad |\mu_2(E_1)| > \varepsilon.$$

Further there must be $E_2 \in \Sigma$ and $\mu_3 \in \mathscr{K}$ such that

$$|\mu_1|(E_2), |\mu_2|(E_2) \leq \tfrac{1}{4} \quad \text{and} \quad |\mu_3(E_2)| > \varepsilon.$$

Continuing in this fashion, we get a sequence (μ_n) in $\mathscr{K}$ and a sequence (E_n) in Σ such that

$$|\mu_1|(E_n), \ldots, |\mu_n|(E_n) \leq 2^{-n} \quad \text{and} \quad |\mu_{n+1}(E_n)| > \varepsilon. \tag{1}$$

Passing to a subsequence, we can arrange that (μ_n) converges weakly to some $\mu \in$ ca(Σ); if (n_k) denotes the indices of this extracted subsequence and we replace E_m by $E_{n_{m+1}-1}$, then for the weakly convergent sequence we can assume (1) as well. Let $\lambda = \Sigma_n 2^{-n}|\mu_n|$; by the Vitali-Hahn-Saks theorem, $\{\mu_n\}$ is uniformly λ-continuous. But $\lambda(E_m)$ tends to 0. Therefore, $\lim_m \mu_n(E_m) = 0$ uniformly in n, a hard thing to do in light of $|\mu_{n+1}(E_n)| > \varepsilon$ for all n; that is, we reach a contradiction.

Our claim is established; we now use the claim to show how 1 implies 3. From the claim we see that there is a sequence (ν_n) in $\mathscr{K}$ such that if $|\nu_n|(E) = 0$ for all n, then $|\mu(E)| = 0$ for all $\mu \in \mathscr{K}$. If we look at $\lambda =$

$\Sigma_n 2^{-n}|\nu_n| \in \text{ca}^+(\Sigma)$, then it is plain that each μ in $\mathscr{K}$ is λ-continuous. Were $\mathscr{K}$ *not* uniformly λ-continuous, there would exist an $\varepsilon > 0$, a sequence (E_n) of members of Σ, and a sequence (μ_n) from $\mathscr{K}$ such that even though $0 = \lim_n \lambda(E_n)$, $|\mu_n(E_n)| \geq \varepsilon$ for all n. Passing to a subsequence, we could as well assume the sequence (μ_n) is weakly convergent to a $\mu \in \text{ca}(\Sigma)$. But this would say something which, in view of Theorem 11 and the Vitali-Hahn-Saks theorem, is not possible.

It follows that 1 implies 3.

Since Theorem 9 tells us that 3 implies 2, we aim for 1 with 2 in hand. Suppose $\mathscr{K}$ is bounded and uniformly additive. Take a sequence (μ_n) from $\mathscr{K}$, and let $\lambda = \Sigma_n 2^{-n}|\mu_n| \in \text{ca}^+(\Sigma)$. For each n, let f_n be the Radon-Nikodym derivative of μ_n with respect to λ. Since each f_n is the pointwise limit of a sequence of simple functions, there is a countable collection $\Gamma_n \subseteq \Sigma$ such that f_n is measurable with respect to the σ-field Σ_n generated by Γ_n. Look at $\cup_n \Gamma_n = \Gamma$, and let $\mathscr{A}$ be the algebra generated by Γ. Both Γ and $\mathscr{A}$ are countable. An easy diagonal argument produces a subsequence (μ'_n) of (μ_n) that converges on each member of $\mathscr{A}$. It follows from our lemma that (μ'_n) converges on each member of the σ-field $\sigma(\mathscr{A})$ generated by $\mathscr{A}$. Therefore, (f'_n) converges weakly in $L_1(\lambda, \sigma(\mathscr{A}))$, a subspace of $L_1(\lambda)$. Hence, (f'_n) converges weakly in $L_1(\lambda)$ and so (μ'_n) converges weakly in ca(Σ). The Eberlein-Šmulian theorem comes to our rescue to conclude that $\mathscr{K}$ must be relatively weakly compact. □

An immediate consequence of the above corollary and the Radon-Nikodym theorem is the following.

Theorem *(Dunford-Pettis). Let $\lambda \in \text{ca}^+(\Sigma)$ and $\mathscr{K}$ be a subset of $L_1(\lambda)$. Then TFAE*:

1. *$\mathscr{K}$ is relatively weakly compact.*
2. *$\mathscr{K}$ is bounded and the indefinite integrals of members of $\mathscr{K}$ are uniformly countably additive.*
3. $\sup_{f \in \mathscr{K}} \|f\|_1 < \infty$, *and given $\varepsilon > 0$ there is a $\delta > 0$ such that if $\lambda(A) \leq \delta$, then $\int_A |f|\, d\lambda \leq \varepsilon$ for all $f \in \mathscr{K}$.*

Corollary (Kadec-Pelczynski). *Suppose $\mathscr{K}$ is a nonweakly compact bounded subset of $L_1(\lambda)$, where λ is a nonnegative member of* ca(Σ). *Then $\mathscr{K}$ contains a sequence (f_n) which is equivalent to the unit vector basis of l_1.*

Proof. By the Dunford-Pettis theorem, we know that the measures $\{\int_{(\cdot)} f\, d\lambda : f \in \mathscr{K}\}$ are not uniformly countably additive on Σ. Therefore, there is a sequence (f_n) in $\mathscr{K}$, a disjoint sequence (E_n) in Σ, and a $\delta > 0$ such that for all n,

$$\int_{E_n} |f_n|\, d\lambda > \delta.$$

By Rosenthal's lemma we can (pass to an appropriate subsequence so as to) assume that

$$\int_{E_n} |f_n| \, d\lambda > \delta \quad \text{and} \quad \int_{\bigcup_{j \neq n} E_j} |f_n| \, d\lambda < \frac{\delta}{2}.$$

If $(\gamma_n) \in l_1$, then

$$\begin{aligned}\left\| \sum_{n=1}^{\infty} \gamma_n f_n \right\|_1 &\geq \left\| \sum_{n=1}^{\infty} \gamma_n f_n \chi_{\cup_m E_m} \right\|_1 \\ &\geq \sum_{n=1}^{\infty} \int_{E_n} |\gamma_n f_n| \, d\lambda - \left\| \sum_{n=1}^{\infty} \gamma_n f_n \chi_{\cup_{m \neq n} E_m} \right\|_1 \\ &\geq \delta \sum_n |\gamma_n| - \frac{\delta}{2} \sum_n |\gamma_n| \\ &= \frac{\delta}{2} \sum_n |\gamma_n|.\end{aligned}$$

□

As one can quickly gather from the above corollary, the Dunford-Pettis criterion is a powerful tool in the study of L_1 and its subspaces; when combined with some ideas from basis theory, this power is displayed in a stunning dichotomy for subspaces of $L_1[0,1]$, also discovered by Kadec and Pelczynski. An exposition of this dichotomy, following closely along the original path cleared by its discoverers, is our next task.

Theorem (Kadec-Pelczynski). *Let X be a nonreflexive subspace of $L_1[0,1]$. Then X contains a subspace complemented in L_1 and isomorphic to l_1.*

To help us get started, we first provide a way of producing complemented copies of l_1 inside $L_1[0,1]$.

Lemma. *Let (f_n) be a sequence from $L_1[0,1]$, and suppose that for each $\varepsilon > 0$ there is an n_ε such that the set $\{t : |f_{n_\varepsilon}(t)| \geq \varepsilon \|f_{n_\varepsilon}\|_1\}$ has measure $< \varepsilon$. Then (f_n) has a subsequence (g_n) such that $(g_n / \|g_n\|)$ is a basic sequence equivalent to l_1's unit vector basis and for which the closed linear span $[g_n]$ of the g_n is complemented in $L_1[0,1]$.*

Proof of lemma. We first take care to see just what the set $\{t : |f(t)| \geq \varepsilon \|f\|_1\}$ having measure $< \varepsilon$ entails. Call this set E. Then

$$\begin{aligned}\int_E \frac{|f(t)|}{\|f\|} \, dt &= \int_0^1 \frac{|f(t)|}{\|f\|} \, dt - \int_{E^c} \frac{|f(t)|}{\|f\|} \, dt \\ &= 1 - \int_{[|f(t)| < \varepsilon \|f\|]} \frac{|f(t)|}{\|f\|} \, dt > 1 - \varepsilon.\end{aligned}$$

Therefore, under the hypotheses of the lemma, we can find E_1 and n_1 so that

$$\lambda(E_1) < \frac{1}{4^2}$$

and

$$\int_{E_1} \frac{|f_{n_1}(t)|}{\|f_{n_1}\|} dt > 1 - \frac{1}{4^2}.$$

Next, applying the hypotheses again and keeping the absolute continuity of integrals in mind, we can find E_2 and $n_2 > n_1$ so that

$$\lambda(E_2) < \frac{1}{4^3},$$

$$\int_{E_2} \frac{|f_{n_2}(t)|}{\|f_{n_2}\|} dt > 1 - \frac{1}{4^3},$$

and

$$\int_{E_2} \frac{|f_{n_1}(t)|}{\|f_{n_1}\|} dt < \frac{1}{4^3}.$$

Continually applying such tactics, we generate a subsequence (g_n) of (f_n) and sets E_n so that

$$\int_{E_n} \frac{|g_n(t)|}{\|g_n\|} dt > 1 - \frac{1}{4^{n+1}}$$

and

$$\int_{E_n} \sum_{k=1}^{n-1} \frac{|g_k(t)|}{\|g_k\|} dt < \frac{1}{4^{n+1}}.$$

Now we disjointify: let

$$A_n = E_n \setminus \bigcup_{k=n+1}^{\infty} E_k$$

and set

$$h_n(t) = \frac{g_n(t)}{\|g_n\|} \chi_{A_n}.$$

Some computations:

$$\left\| \frac{g_n}{\|g_n\|} - h_n \right\| \le \int_{A_n^c} \frac{|g_n(t)|}{\|g_n\|}\,dt$$

$$\le \int_{E_n^c} \frac{|g_n(t)|}{\|g_n\|}\,dt + \int_{E_n \setminus A_n} \frac{|g_n(t)|}{\|g_n\|}\,dt$$

$$\le \frac{1}{4^{n+1}} + \sum_{k=n+1}^{\infty} \int_{E_k} \frac{|g_n(t)|}{\|g_n\|}\,dt$$

$$< \frac{1}{4^{n+1}} + \sum_{k=n+1}^{\infty} \frac{1}{4^{k+1}} < \frac{1}{4^n}.$$

Therefore,

$$1 \ge \|h_n\|_1 = \int_{A_n} \frac{|g_n(t)|}{\|g_n\|}\,dt$$

$$\ge \int_{E_n} \frac{|g_n(t)|}{\|g_n\|}\,dt - \sum_{k=n+1}^{\infty} \int_{E_k} \frac{|g_n(t)|}{\|g_n\|}\,dt$$

$$\ge 1 - \frac{1}{4^{n+1}} - \sum_{k=n+1}^{\infty} \frac{1}{4^{k+1}}$$

$$> 1 - \frac{1}{4^n}.$$

So,

$$\left\| \frac{g_n}{\|g_n\|} - \frac{h_n}{\|h_n\|} \right\| \le \left\| \frac{g_n}{\|g_n\|} - h_n \right\| + \left\| h_n - \frac{h_n}{\|h_n\|} \right\|$$

$$\le \frac{1}{4^n} + (1 - \|h_n\|) \le \frac{2}{4^n}.$$

Some reflections:

The h_n are disjointly supported nonzero members of $L_1[0,1]$; therefore, $(h_n/\|h_n\|)$ is a basic sequence in $L_1[0,1]$ equivalent to the unit vector basis of l_1, $[h_n]$ is complemented in $L_1[0,1]$ by means of a norm-one projection P, and the coefficient functionals φ_n^* of (h_n) extend to members of $L_1[0,1]^*$ having norm one. All this was noticed in our earlier work.

Our computations alert us to the proximity of the g_n, on normalization, to the h_n, on normalization. In particular,

$$\sum_n \|P\|\,\|\varphi_n^*\| \left\| \frac{g_n}{\|g_n\|} - \frac{h_n}{\|h_n\|} \right\| \le \sum_n \frac{2}{4^n} < 1.$$

An appeal to Theorem 12 of Chapter V concludes the proof of the lemma. □

Now for the proof of the main theorem we start with the nonweakly compact closed unit ball B_X of X. Let $0 < \mu \leq 1$ and set, for any $f \in L_1[0,1]$,

$$\alpha(f,\mu) = \sup\left\{\int_E |f(t)|\,dt : \lambda(E) = \mu\right\}.$$

If $\alpha_X(\mu) = \sup_{f \in B_X} \alpha(f,\mu)$, then the nonreflexivity of X is reflected by the conclusion that

$$\alpha^* = \lim_{\mu \to 0} \alpha_X(\mu) > 0.$$

Therefore, we can choose $f_n \in B_X$, measurable sets $E_n \subseteq [0,1]$, and $\mu_n > 0$ such that

$$\lim_n \mu_n = 0,$$

$$\int_{E_n} |f_n(t)|\,dt = \mu_n,$$

and

$$\lim_n \alpha(f_n, \mu_n) = \alpha^*.$$

Consider the functions f_n' given by

$$f_n'(t) = f_n(t)\chi_{E_n}(t).$$

Notice that given $\varepsilon > 0$ there is an n_ε so that the set $\{t : |f_{n_\varepsilon}'(t)| \geq \varepsilon \|f_{n_\varepsilon}'\|_1\} < \varepsilon$; in other words, we have established the hypotheses of our lemma. Rewarded with the conclusions of that lemma, we can find an increasing sequence (k_n) of positive integers such that the sequence (f_{k_n}') satisfies the following: first, $(f_{k_n}'/\|f_{k_n}'\|)$ is a basic sequence equivalent to the unit vector basis of l_1, and second, the closed linear span $[f_{k_n}']$ of the f_{k_n}' is complemented in $L_1[0,1]$ (and, of course, isomorphic to l_1).

Let $g_n = f_{k_n}$, $g_n' = f_{k_n}'$, and $g_n'' = g_n - g_n'$.

Of course, $\{g_n'' : n \geq 1\}$ is relatively weakly compact in $L_1[0,1]$; so with perhaps another turn at extracting subsequences, we may assume (g_n'') is weakly convergent. Now notice that we've located a sequence g_n in B_X that can be expressed in the form

$$g_n = g_n' + g_n'',$$

where (g_n') spans a complemented l_1 in $L_1[0,1]$ and (g_n'') is weakly convergent. It is important to keep in mind that neither the g_n' nor the g_n'' need find themselves in X. Regardless, we show that some suitable modification of the g_n', when normalized, are close enough to X to ensure the applicability of Theorem 12 of Chapter V, thereby establishing the existence in X of an isomorphic copy of l_1 that is complemented in $L_1[0,1]$.

Since (g_n'') is weakly convergent, $(g_{2n}'' - g_{2n+1}'')$ is weakly null; thanks to Mazur's theorem, there is a sequence (h_n'') of convex combinations of

$(g''_{2n} - g''_{2n+1})$ tending to zero in norm. We may assume h''_n to be of the following form: for some $k_1 < k_2 < \cdots < k_n < \cdots$,

$$h''_m = \sum_{i=k_m}^{k_{m+1}-1} a_i^{(m)}(g''_{2i} - g''_{2i+1}),$$

where, naturally, $a_{k_m}^{(m)} + \cdots + a_{k_{m+1}}^{(m)} = 1$ and all the a are ≥ 0. It is important to keep tabs on the vectors

$$h'_m = \sum_{i=k_m}^{k_{m+1}-1} a_i^{(m)}(g'_{2i} - g'_{2i+1})$$

and

$$h_m = \sum_{i=k_m}^{k_{m+1}-1} a_i^{(m)}(g_{2i} - g_{2i+1}).$$

In particular, we should notice that

$$h_m = h'_m + h''_m,$$

that due to the nature of the sums involved in the definition of h'_m, the closed linear span of the h'_m is a complemented copy of l_1 found inside the closed linear span of the g'_n, itself a complemented subspace of $L_1[0,1]$, and that each h_m belongs to X. What is important here is the fact that

$$0 = \lim_n \|h''_m\| = \lim_m \|h_m - h'_m\|.$$

It follows that passing to a subsequence of the h_m, we can force $(\|h_m - h'_m\|)$ to tend to zero as quickly as we need to. How quickly ought we shoot for? Well, quickly enough to apply Theorem 12 of Chapter V. A word of warning in this connection. The h'_m span a complemented copy of l_1 in $L_1[0,1]$, but only on normalization do we get the vector basis of this copy; not to worry, since $\|h_m\|$ is close to $\|h'_m\|$ for m large enough. This remark in hand, Theorem 12 of Chapter V ought to be applied easily.

If Ω is a compact Hausdorff space, then $C(\Omega)^*$ can be identified with the space $\text{rca}(\Sigma)$ of regular Borel measures defined on Ω. Recall that a measure $\mu \in \text{rca}(\Sigma)$ precisely when for any Borel set B in Ω, $|\mu|(B) = \sup\{|\mu|(K): K$ is a compact subset of $B\}$.

Theorem 14 (Dieudonné-Grothendieck). *Let Ω be a compact Hausdorff space and Σ be the σ-field of Borel subsets of Ω. Suppose $\mathscr{K}$ is a bounded subset of* $\text{rca}(\Sigma)$.

In order for $\mathscr{K}$ to be relatively weakly compact, it is both necessary and sufficient that given a sequence (O_n) of disjoint open subsets of Ω, then

$$\lim_n \mu(O_n) = 0$$

uniformly for $\mu \in \mathscr{K}$.

PROOF. Necessity is clear from the uniform countable additivity of relatively weakly compact subsets of ca(Σ).

To establish sufficiency we will mount a two pronged attack by proving that should the bounded set $\mathscr{K}$ satisfy $\lim_n \mu(O_n) = 0$ uniformly for any sequence (O_n) of disjoint open sets in Ω, then

1. *Given $\varepsilon > 0$ and a compact set $K \subseteq \Omega$ there is an open set U of Ω containing K for which*
$$|\mu|(U \setminus K) \leq \varepsilon$$
for all $\mu \in \mathscr{K}$.
2. *Given $\varepsilon > 0$ and an open set $U \subseteq \Omega$ there exists a compact set $K \subseteq U$ for which*
$$|\mu|(U \setminus K) \leq \varepsilon$$
for all $\mu \in \mathscr{K}$.

In tandem 1 and 2 will then be used to derive the relative weak compactness of $\mathscr{K}$. Before proceeding, it is worthwhile to make a couple of points: first, conditions 1 and 2 obviously say that $\mathscr{K}$ is "uniformly regular" with 1 expressing uniform outer regularity and 2 expressing uniform inner regularity; second, although we do not pursue this here, each of 1 and 2 *by itself* is equivalent to the relative weak compactness of $\mathscr{K}$, and so their appearance in the present proof ought not to be viewed as at all accidental. On with the proof.

Suppose 1 fails. Then there is a compact set K_0 and an $\varepsilon_0 > 0$ such that for any open set V containing K_0 we can find a $\mu_V \in \mathscr{K}$ for which
$$|\mu_V|(V \setminus K_0) > \varepsilon_0.$$
Starting with Ω we know that there is a $\mu_1 \in \mathscr{K}$ such that
$$|\mu_1|(\Omega \setminus K_0) > \varepsilon_0.$$
Since μ_1 is regular there is a compact set $K_1 \subseteq \Omega \setminus K_0$ such that
$$|\mu_1(K_1)| > \frac{\varepsilon_0}{2}.$$
Notice that K_0 and K_1 are disjoint compact sets so there are disjoint open sets Y and Z that contain K_0 and K_1, respectively. By regularity, Z can be chosen to satisfy
$$|\mu_1|(Z \setminus K_1) \leq \frac{\varepsilon_0}{4}.$$
Let $U_1 = Z$ and $V_1 = Y$. Then
$$U_1 \subseteq \Omega \setminus V_1, \qquad K_0 \subseteq V_1 \subseteq V_0 = \Omega, \qquad K_1 \subseteq U_1 = Z.$$
It follows that
$$|\mu_1|(U_1 \setminus K_1) \leq \frac{\varepsilon_0}{4}$$

so that

$$|\mu_1(U_1)| \geq |\mu_1(K_1)| - |\mu_1|(U_1 \setminus K_1)$$
$$\geq \frac{\varepsilon_0}{2} - \frac{\varepsilon_0}{4} = \frac{\varepsilon_0}{4}.$$

Back to the well once again. V_1 is an open set containing K_0. Hence there is a $\mu_2 \in \mathscr{K}$ such that

$$|\mu_2|(V_1 \setminus K_0) > \varepsilon_0.$$

Regularity of μ_2 gives a compact set $K_2 \subseteq V_1 \setminus K_0$ such that

$$|\mu_2(K_2)| > \frac{\varepsilon_0}{2}.$$

K_0 and K_2 are disjoint compact sets, and so they can be enveloped in disjoint open sets Y and Z; again the regularity of μ_2 allows us also to assume that

$$|\mu_2|(Z \setminus K_2) \leq \frac{\varepsilon_0}{4}.$$

Let $U_2 = Z \cap V_1$ and $V_2 = Y \cap V_1$. Then

$$U_2 \subseteq V_1 \setminus V_2, \qquad K_0 \subseteq V_2 \subseteq V_1, \qquad K_2 \subseteq U_2 \subseteq Z.$$

So

$$|\mu_2|(U_2 \setminus K_2) \leq |\mu_2|(Z \setminus K_2) \leq \frac{\varepsilon_0}{4}$$

and

$$|\mu_2(U_2)| \geq |\mu_2(K_2)| - |\mu_2|(U_2 \setminus K_2)$$
$$\geq \frac{\varepsilon_0}{2} - \frac{\varepsilon_0}{4} = \frac{\varepsilon_0}{4}.$$

Our procedure is clearly producing a sequence (U_n) of disjoint open sets and a corresponding sequence (μ_n) of members of $\mathscr{K}$ for which $|\mu_n(U_n)| \geq \varepsilon_0/4$, a contradiction.

Before establishing 2, we make a fuel stop:

2′. *Given $\varepsilon > 0$ there exists a compact set $K \subseteq \Omega$ such that*

$$|\mu|(\Omega \setminus K) \leq \varepsilon$$

for all $\mu \in \mathscr{K}$.

If not, then there exists $\varepsilon_0 > 0$ such that for any compact set $K \subseteq \Omega$ there is a $\mu_K \in \mathscr{K}$ with

$$|\mu_K|(\Omega \setminus K) > \varepsilon_0.$$

Starting with the compact set $\varnothing$, there is a $\mu_1 \in \mathscr{K}$ for which

$$|\mu_1|(\Omega) > \varepsilon_0.$$

Since μ_1 is regular, there is a compact set $K_1 \subseteq \Omega$ for which

$$|\mu_1(K_1)| > \frac{\varepsilon_0}{2}.$$

By 1 there is an open set V_1 containing K_1 for which

$$|\mu|(V_1 \setminus K_1) \le \frac{\varepsilon_0}{4}\frac{1}{2}$$

for all $\mu \in \mathscr{K}$. Let U_1 be an open set satisfying

$$K_1 \subseteq U_1 \subseteq \overline{U_1} \subseteq V_1.$$

Then

$$|\mu|(\overline{U_1} \setminus K_1) \le \frac{\varepsilon_0}{4}\frac{1}{2}$$

for all $\mu \in \mathscr{K}$ and

$$\begin{aligned} |\mu_1(U_1)| &\ge |\mu_1(K_1)| - |\mu_1|(\overline{U_1} \setminus K_1) \\ &> \frac{\varepsilon_0}{2} - \frac{\varepsilon_0}{8} > \frac{\varepsilon_0}{4}. \end{aligned}$$

Next, there is a $\mu_2 \in \mathscr{K}$ such that

$$|\mu_2|(\Omega \setminus K_1) > \varepsilon_0.$$

It follows that

$$\begin{aligned} |\mu_2|(\Omega \setminus \overline{U_1}) &\ge |\mu_2|(\Omega \setminus K_1) - |\mu_2|(\overline{U_1} \setminus K_1) \\ &> \varepsilon_0 - \frac{\varepsilon_0}{8}. \end{aligned}$$

Since μ_2 is regular, there is a compact set $K_2 \subseteq \Omega \setminus \overline{U_1}$ such that

$$|\mu_2(K_2)| \ge \frac{1}{2}\left(\varepsilon_0 - \frac{\varepsilon_0}{8}\right).$$

Of course, K_2 and $\overline{U_1}$ are disjoint; so there is an open set V_2 for which

$$K_2 \subseteq V_2 \subseteq \overline{V_2} \subseteq \Omega \setminus \overline{U_1}$$

and for which (using 1)

$$|\mu|(V_2 \setminus K_2) < \frac{\varepsilon_0}{4}\frac{1}{2^2}$$

for all $\mu \in \mathscr{K}$. Now pick an open set U_2 for which

$$K_2 \subseteq U_2 \subseteq \overline{U_2} \subseteq V_2.$$

Notice that $\overline{U_1}$ and $\overline{U_2}$ are disjoint,

$$|\mu|(\overline{U_2} \setminus K_2) < \frac{\varepsilon_0}{4}\frac{1}{2^2}$$

for all $\mu \in \mathscr{K}$, and

$$\begin{aligned} |\mu_2(U_2)| &\ge |\mu_2(K_2)| - |\mu_2|(\overline{U_2} \setminus K_2) \\ &> \frac{1}{2}\left(\varepsilon_0 - \frac{\varepsilon_0}{8}\right) - \frac{\varepsilon_0}{4}\frac{1}{2^2} \ge \frac{\varepsilon_0}{4}. \end{aligned}$$

The inductive procedure should be clear. Again, we produce a sequence (U_n) of disjoint open sets and a corresponding sequence (μ_n) of members of $\mathscr{K}$ for which $|\mu_n(U_n)| \geq \varepsilon_0/4$, again, a contradiction.

Now to establish 2, we know by 2′ that given $\varepsilon > 0$ there is a compact set $F \subseteq \Omega$ such that

$$|\mu|(\Omega \setminus F) \leq \varepsilon$$

for all $\mu \in \mathscr{K}$. Given an open set U, $(\Omega \setminus U) \cap F$ is compact, and so by 1 there's an open set V containing $(\Omega \setminus U) \cap F$ for which

$$|\mu|(V \setminus [(\Omega \setminus U) \cap F]) \leq \varepsilon$$

holds for all $\mu \in \mathscr{K}$. Let $K = F \setminus V$. Then $K \subseteq U \cap F$ and

$$|\mu|(U \setminus K) \leq |\mu|(\Omega \setminus F) + |\mu|(V \setminus [(\Omega \setminus U) \cap F]) \leq 2\varepsilon.$$

1 and 2 have been established. Now we show that $\mathscr{K}$ is relatively weakly compact. By the Eberlein-Smulian theorem we can restrict our attention to the case where $\mathscr{K}$ can be listed in a single sequence (μ_n). each term of which can be assumed to be λ-continuous with respect to a fixed $\lambda \in \mathrm{rca}^+(\Sigma)$. Let f_n denote the Radon-Nikodym derivative $d\mu_n/d\lambda$ of μ_n with respect to λ; (f_n) is a bounded sequence in $L_1(\lambda)$.

Were $\mathscr{K}$ not relatively weakly compact, then we could find an $\varepsilon > 0$, a subsequence (g_n) of (f_n), and a sequence (B_n) of Borel sets in Ω for which

$$\lambda(B_n) \leq \frac{1}{2^{n+1}}$$

yet

$$\int_{B_n} |g_n(\omega)| \, d\lambda(\omega) \geq \varepsilon$$

for all n. By regularity, we can enlarge the B_n slightly to open U_n and obtain

$$\lambda(U_n) \leq \frac{1}{2^n}$$

and

$$\int_{U_n} |g_n| \, d\lambda \geq \varepsilon$$

for all n. Looking at $V_n = \cup_{m=n}^{\infty} U_m$, we get a decreasing sequence (V_n) of open sets with

$$\lim_n \lambda(V_n) = 0$$

and

$$\int_{V_n} |g_n| \, d\lambda \geq \varepsilon$$

for each n. By 2 we have that for each n there is a compact set $K_n \subseteq V_n$ such that

$$\int_{V_n \setminus K_n} |g_k|\, d\lambda \leq \frac{\varepsilon}{2^{n+1}}$$

for all k. Looking at $F_n = K_1 \cap \cdots \cap K_n \subseteq K_n \subseteq V_n$, we see that

$$\int_{F_n} |g_n|\, d\lambda = \int_{V_n} |g_n|\, d\lambda - \int_{V_n \setminus F_n} |g_n|\, d\lambda,$$

which, since $V_n \setminus F_n \subseteq (V_1 \setminus K_1) \cup \cdots \cup (V_n \setminus K_n)$, is less than or equal to

$$\int_{V_n} |g_n|\, d\lambda - \sum_{k=1}^{n} \int_{V_k \setminus K_k} |g_n|\, d\lambda \geq \varepsilon - \sum_{k=1}^{n} \frac{\varepsilon}{2^{k+1}} > \frac{\varepsilon}{2}.$$

The sequence (F_n) is a decreasing sequence of compact sets whose λ measures tend to 0; consequently, $\cap_n F_n$ is a compact set of λ-measure 0. By 1 there is an open set W containing $\cap_n F_n$ such that for all k,

$$\int_{W \setminus \cap_n F_n} |g_k|\, d\lambda \leq \frac{\varepsilon}{4}.$$

Since $\lambda(\cap_n F_n) = 0$,

$$\int_W |g_k|\, d\lambda \leq \frac{\varepsilon}{4}$$

for all k. But W is an open set containing $\cap_n F_n$; so there is an m such that $F_m \subseteq W$ from which we have

$$\frac{\varepsilon}{2} \leq \int_{F_m} |g_m|\, d\lambda \leq \int_W |g_m|\, d\lambda \leq \frac{\varepsilon}{4},$$

another contradiction. At long last we're home free! □

One striking application of the Dieudonné-Grothendieck criterion in tandem with the Phillips lemma is to the study of l_∞. To describe this next result of Grothendieck, we need to notice the following about l_∞: *l_∞ is isometrically isomorphic to $C(K)$, where K is the Stone space of the Boolean algebra $2^{\mathbf{N}}$ of all subsets of the natural numbers.* This is an easy consequence of the Stone representation theorem and the Stone-Weierstrass theorem. After all, in the notation of the second section, l_∞ is just $B(2^{\mathbf{N}})$. Now the map that takes a simple function $\Sigma_{i=1}^n a_i c_{A_i}$ in $B(2^{\mathbf{N}})$ to the function $\Sigma_{i=1}^n a_i c_{\hat{A}_i}$ (where $A \to \hat{A}$ is the Stone representation of $2^{\mathbf{N}}$ as the algebra of clopen subsets of K) in $C(K)$ is a well-defined linear isometry on the simple functions. The domain of the map is dense, and its range is also (thanks to K's total disconnectedness and the Stone-Weierstrass theorem). Therefore, the isometry extends to a linear isometry of $B(2^{\mathbf{N}}) = l_\infty$ *onto* $C(K)$.

Theorem 15 (Grothendieck). *In l_∞^*, weak* convergent sequences are weakly convergent.*

Proof. Let K be as described in the paragraph preceding the statement of the theorem and suppose that (μ_n) is weak* null in $C(K)^*$. We show that $\{\mu_n\}$ is relatively weakly compact. From this it follows that (μ_n) is weakly null.

Suppose such is not the case. Then the Dieudonné-Grothendieck criterion provides us with an $\varepsilon > 0$ and a sequence (O_n) of open disjoint subsets of K and a subsequence (ν_n) of (μ_n) for which we always have

$$|\nu_n(O_n)| \geq \varepsilon.$$

Using the regularity of the ν_n and K total disconnectedness, we can (and do) assume that the O_n are clopen as well. Now we take note and make use of the fact that the Boolean algebra $2^{\mathbb{N}}$ is σ-complete in the sense that every countable collection of elements in $2^{\mathbb{N}}$ has a least upper bound therein; this σ-completeness is of course shared by the Stone algebra of clopen subsets of K and allows us to unravel the procedure described in the second section, The Classical Nonreflexive Sequence Spaces. More precisely we can define $\check{\nu}_n \in \mathrm{ba}(2^{\mathbb{N}})$ by

$$\check{\nu}_n(\Delta) = \nu_n\Big(\sup_{k \in \Delta} O_k\Big)$$

for any $\Delta \subseteq \mathbb{N}$. Since (ν_n) is weak* null and $\sup_{k \in \Delta} O_k$ is a clopen set in K for any $\Delta \subseteq \mathbb{N}$,

$$0 = \lim_n \nu_n\big(C_{\sup_{k \in \Delta} O_k}\big) = \lim_n \check{\nu}_n(\Delta)$$

for any Δ. It follows from the Phillips lemma that

$$\begin{aligned} 0 &= \lim_n \sum_k |\check{\nu}_n(\{k\})| \\ &= \lim_n \sum_k |\nu_n(O_k)|, \end{aligned}$$

a contradiction. □

Weakly Convergent Sequences and Unconditionally Convergent Series in $L_p[0,1]$ $(1 \leq p < \infty)$

In this section we present a couple of the finer aspects of "Sequences and series in Banach spaces" in case the terms live in $L_p[0,1]$ for $1 \leq p < \infty$. We give complete proofs of the pertinent facts only in case $1 \leq p \leq 2$; what happens (and why) in case $p > 2$ is outlined in the exercises. To be frank, this latter case causes only minor difficulties once the case $1 \leq p \leq 2$ is understood. In addition, the situation in which unconditionally convergent series in $L_p[0,1]$, for $1 \leq p \leq 2$, find themselves is one of the central themes of present-day "Sequences and series in Banach spaces"; so special attention to this case seems appropriate.

We start this section with the beautiful Khinchine inequalities, proceed directly to Orlicz's theorem about the square summability of unconditionally convergent series in $L_p[0,1]$ for $1 \le p \le 2$, pass to a proof of Banach and Saks that weakly convergent sequences in $L_p[0,1]$ for $1 < p \le 2$ admit subsequences whose arithmetic means are norm convergent and close with Szlenk's complementary result to the same effect as that of Banach and Saks in case $p = 1$.

Recall the definition of the Rademacher functions; each acts on $[0,1]$ and has values in $[-1,1]$. The first r_1 is just 1 everywhere. The second, r_2, is 1 on $[0,\frac{1}{2})$ and -1 on $[\frac{1}{2},1]$; r_3 is 1 on $[0,\frac{1}{4})$ and $[\frac{1}{2},\frac{3}{4})$ but -1 on $[\frac{1}{4},\frac{1}{2})$ and $[\frac{3}{4},1]$. Get the picture? Okay.

Theorem (Khinchine's Inequalities). *Let $(r_n)_{n\ge 1}$ denote the sequence of Rademacher functions. Then for each $1 \le p < \infty$ there is a constant $k_p > 0$ such that*

$$k_p^{-1}\Big(\sum_i a_i^2\Big)^{1/2} \le \Big\|\sum_i a_i r_i\Big\|_p \le k_p\Big(\sum_i a_i^2\Big)^{1/2}$$

holds, for any finite sequence (a_i) of reals.

PROOF. The Rademacher functions are orthonormal over $[0,1]$ and belong to $L_\infty[0,1]$ with sup norm 1. Consequently, we need only show the existence of constants in the following situations:

(i) If $2 \le p$, then we need to show there is a $K > 0$ such that $\|\Sigma_i a_i r_i\|_p \le K(\Sigma_i a_i^2)^{1/2}$.

(ii) If $1 \le p < 2$, then we need to show there is a $k > 0$ such that $k(\Sigma_i a_i^2)^{1/2} \le \|\Sigma_i a_i r_i\|_p$.

Let's establish Khinchine's inequalities for $p \ge 2$ then. Again the monotonicity of the L_p-norms lets us concentrate on p an even integer, say $p = 2l$, where $l \ge 1$ is a whole number. Look at

$$S_n = \sum_{i=1}^n a_i r_i$$

and take the integral over $[0,1]$ of its pth power. We write down what results; the reader is advised to reflect on what's written down in light of the binomial (and multinomial) formula!

$$\begin{aligned}\int_0^1 S_n^p(t)\,dt &= \int_0^1 S_n^{2l}(t)\,dt \\ &= \int\Big(\sum_{i=1}^n a_i r_i\Big)^{2l} \\ &= \sum A_{\alpha_1,\ldots,\alpha_j} a_{i_1}^{\alpha_1}\cdots a_{i_j}^{\alpha_j}\int r_{i_1}^{\alpha_1}\cdots r_{i_j}^{\alpha_j},\end{aligned}$$

$\alpha_1, \ldots, \alpha_j$ are positive integers,

$$\sum \alpha_j = 2l,$$

$$A_{\alpha_1,\ldots,\alpha_j} = \frac{(\alpha_1 + \cdots + \alpha_j)!}{(\alpha_1)! \cdots (\alpha_j)!},$$

and $i_1, \ldots, i_j$ are different integers between 1 and n. Thinking about the fact that the r_i (under the last integral sign above) are the Rademacher functions, we see that the form of $\int S_n^p$ is really considerably simpler than at first guessed, namely,

$$\int S_n^p = \int S_n^{2l} = \sum A_{2\beta_1,\ldots,2\beta_j} a_{i_1}^{2\beta_1} \cdots a_{i_j}^{2\beta_j},$$

since $\int r_{i_1}^{\alpha_1} \cdots r_{i_j}^{\alpha_j}$ is 0 or 1 depending on the existence of odd powers $\alpha_1, \ldots, \alpha_j$ or nonexistence thereof. Of course, in this form of $\int S_n^p$, we know that $\beta_1, \ldots, \beta_j$ are positive integers and $\Sigma \beta_j = l$. Writing $\int S_n^p$ again, we have

$$\int S_n^p = \sum \frac{A_{2\beta_1,\ldots,2\beta_j}}{A_{\beta_1,\ldots,\beta_j}} A_{\beta_1,\ldots,\beta_j} a_{i_1}^{2\beta_1} \cdots a_{i_j}^{2\beta_j}.$$

We wish to apply Hölder's inequality; so we estimate the ratios $A_{2\beta_1,\ldots,2\beta_j}/A_{\beta_1,\ldots,\beta_j}$:

$$\begin{aligned}
\frac{A_{2\beta_1,\ldots,2\beta_j}}{A_{\beta_1,\ldots,\beta_j}} &= \frac{(2l)!}{(2\beta_1)! \cdots (2\beta_j)!} \frac{(\beta_1)! \cdots (\beta_j)!}{(l)!} \\
&= \frac{(2l)(2l-1)\cdots(l+1)(l)!(\beta_1)!\cdots(\beta_j)!}{(2\beta_1)\cdots(\beta_1+1)(\beta_1)!\cdots(2\beta_j)\cdots(\beta_j+1)(\beta_j)!(l)!} \\
&= \frac{(2l)(2l-1)\cdots(l+1)}{(2\beta_1)\cdots(\beta_1+1)\cdots(2\beta_j)\cdots(\beta_j+1)} \\
&\le \frac{(2l)(2l-1)\cdots(l+1)}{2^{\beta_1}\cdots 2^{\beta_j}} = \frac{(2l)(2l-1)\cdots(l+1)}{2^{\beta_1+\cdots+\beta_j}} \\
&\le \frac{l^l}{2^l}.
\end{aligned}$$

This gives

$$\int S_n^{2l} \le \left(\frac{l}{2}\right)^l \left(\sum_{i=1}^n a_i^2\right)^l.$$

From this we see that

$$\left(\int S_n^p\right)^{1/p} = \left(\int S_n^{2l}\right)^{1/2l}$$
$$\le \sqrt{\frac{l}{2}}\left(\sum_{i=1}^{n} a_i^2\right)^{1/2}$$

and (i) has been established.

To establish (ii), we have need of an old friend: Liapounov's inequality. Recall what this says: *If $f \ge 0$ belongs to all the $L_p[0,1]$ for $p > 0$, then* $\log \int_0^1 f^p$ *is a* convex *function of $p > 0$.*

Liapounov's inequality is an interesting consequence of Hölder's inequality that ought to be worked out by the reader. Let's get on with (ii). We are only concerned with $1 \le p < 2$. Pick λ_1 and λ_2 so that $\lambda_1, \lambda_2 > 0$, $\lambda_1 + \lambda_2 = 1$ and $p\lambda_1 + 4\lambda_2 = 2$. Then

$$\left(\sum_{i=1}^{n} a_i^2\right) = \|S_n\|_2^2$$
$$\le \|S_n\|_p^{p\lambda_1}\|S_n\|_4^{4\lambda_2} \qquad \text{(by Liapounov's inequality)}$$
$$\le \|S_n\|_p^{p\lambda_1}\left[\sqrt{2}\left(\sum_{i=1}^{n} a_i^2\right)^{1/2}\right]^{4\lambda_2},$$

by (i). On dividing both sides by the appropriate quantity we get (with careful use of $p\lambda_1 + 4\lambda_2 = 2$)

$$4^{-\lambda_2/p\lambda_1}\left(\sum a_i^2\right)^{1/2} \le \|S_n\|_p$$

and with it (ii). □

Let $1 \le p < \infty$ and suppose that $\sum_k f_k$ is unconditionally convergent in $L_p(0,1)$. Let r_n denote the nth Rademacher function. By the bounded multiplier test if (α_n) is a sequence of numbers with $|\alpha_n| \le 1$ for all n, then there is a $K > 0$ such that

$$\left\|\sum_{k=1}^{n} \alpha_k f_k\right\|_p \le K$$

for all n. It follows that if $0 \le t \le 1$, then

$$\int_0^1 \left|\sum_{k=1}^{n} r_k(t) f_k(s)\right|^p ds = \left\|\sum_{k=1}^{n} r_k(t) f_k\right\|_p^p \le K^p$$

holds for all n. Khinchine's inequality alerts us to the fact that there is an

$A > 0$ such that for any scalar sequence (β_n) we have

$$\left(\sum_{k=1}^{n} \beta_k^2\right)^{1/2} \le A \left\| \sum_{k=1}^{n} \beta_k r_k \right\|_p .$$

Combining these observations, we conclude that

$$\begin{aligned}\int_0^1 \left[\sum_{k=1}^{n} f_k^2(s)\right]^{p/2} ds &\le \int_0^1 A^p \left\| \sum_{k=1}^{n} f_k(s) r_k \right\|_p^p ds \\ &= A^p \int_0^1 \int_0^1 \left| \sum_{k=1}^{n} f_k(s) r_k(t) \right|^p dt\, ds \\ &= A^p \int_0^1 \int_0^1 \left| \sum_{k=1}^{n} f_k(s) r_k(t) \right|^p ds\, dt \\ &\le A^p \int_0^1 K^p\, dt = A^p K^p .\end{aligned}$$

Summarizing we get the following fact.

General fact. If $\Sigma_k f_k$ is unconditionally convergent in $L_p(0,1)$, then there is a $C > 0$ such that

$$\int_0^1 \left(\sum_k f_k^2(t)\right)^{p/2} dt \le C.$$

Our way is paved to prove the following.

Theorem *(Orlicz). Let $1 \le p \le 2$ and suppose that $\Sigma_n f_n$ is unconditionally convergent in $L_p(0,1)$. Then*

$$\sum_k \|f_k\|_p^2 < \infty .$$

PROOF. Let n be any positive integer. Then if $\frac{1}{2} + 1/q = 1/p$, we have

$$\begin{aligned}\left(\sum_{k=1}^{n} |f_k(t)|^p d_k\right)^{1/p} &= \left\|\left(f_k(t) d_k^{1/p}\right)\right\|_{l_p} \\ &\le \left\|\left(f_k(t)\right)\right\|_{l_2} \left\|\left(d_1^{1/p}, \ldots, d_n^{1/p}, 0, 0, \ldots\right)\right\|_{l_q}\end{aligned}$$

for any $d_1, \ldots, d_n \ge 0$. A bit of computation shows that $q = 2p/(2-p)$. On taking pth powers, we have for the same d that

$$\sum_{k=1}^{n} |f_k(t)|^p d_k \le \left(\sum_{k=1}^{n} |f_k(t)|^2\right)^{p/2} \left(\sum_{k=1}^{n} d_k^{2/(2-p)}\right)^{(2-p)/2}$$

Term-by-term integration gives

$$\int_0^1 \sum_{k=1}^n |f_k(t)|^p d_k \, dt \le \left(\sum_{k=1}^n d_k^{2/(2-p)} \right)^{(2-p)/2} \int_0^1 \left(\sum_{k=1}^n |f_k(t)|^2 \right)^{p/2} dt$$
$$\le \left(\sum_{k=1}^n d_k^{2/(2-p)} \right)^{(2-p)/2} C,$$

where C was awarded us as an upper bound for the integral involved through the graces of our general fact. It now follows that if r is conjugate to $2/(2-p)$, then

$$\sum_{k=1}^n \left(\int_0^1 |f_k(t)|^p \, dt \right)^r \le C^r.$$

Computing what exactly it means for r to be such as it is, we see that $r = 2/p$ and so

$$\sum_{k=1}^n \|f_k\|_p^2 = \left(\sum_{k=1}^n \int_0^1 |f_k(t)|^p \, dt \right)^{2/p} \le C^r < \infty.$$

The arbitrary nature of n and the fixed nature of C force

$$\sum_n \|f_n\|_p^2 < \infty.$$

□

Theorem (Banach-Saks). *Suppose $1 < p \le 2$. If (f_n) is a weakly null sequence in $L_p[0,1]$, then (f_n) admits a subsequence (f_{k_n}) for which $\|\sum_{i=1}^n f_{k_i}\| = O(n^{1/p})$.*

PROOF. Since (f_n) is weakly null, we may as well assume that each f_n has norm ≤ 1. Because $1 < p \le 2$, it is easy to convince yourself that there is a constant $A > 0$ for which

$$|a+b|^p \le |a|^p + p|a|^{p-1} b \operatorname{sign}(a) + A|b|^p \tag{2}$$

holds regardless of the real numbers a, b considered. Now, let $S_1 = f_{k_1} = f_1$. Choose $k_2 > k_1$ so that

$$\left| \int_0^1 |f_{k_1}(t)|^{p-1} \operatorname{sign}(f_{k_1}(t)) f_{k_2}(t) \, dt \right| \le \frac{1}{p}.$$

Let $S_2 = f_{k_1} + f_{k_2}$. Choose $k_3 > k_2$ so that

$$\left| \int_0^1 |S_2(t)|^{p-1} \operatorname{sign}(S_2(t)) f_{k_3}(t) \, dt \right| \le \frac{1}{p}.$$

The path to the choice of (k_n) ought to be clear. Notice that

$$\|S_n\|_p^p \le \|S_{n-1}\|_p^p + p \int_0^1 |S_{n-1}(t)|^{p-1} \operatorname{sign}(S_{n-1}(t)) f_{k_n}(t) \, dt$$
$$+ A \|f_{k_n}\|_p^p,$$

as follows from inequality (2) in the obvious fashion,

$$\leq \|S_{n-1}\|_p^p + 1 + A.$$

Running through this last inequality a few times gives

$$\left\|\sum_{i=1}^{n} f_{k_i}\right\|_p^p = \|S_n\|_p^p \leq \|S_1\|_p^p + C(n-1) \leq Cn$$

for some $C > 1$. This, though, is what we wanted, since it tells us that

$$\left\|\sum_{i=1}^{n} f_{k_i}\right\|_p \leq C^{1/p} n^{1/p}. \qquad \square$$

Corollary. *If $1 < p \leq 2$, then any bounded sequence in $L_p[0,1]$ admits a subsequence whose arithmetic means converge in norm.*

Proof. If (g_n) is a bounded sequence in $L_p[0,1]$, then (g_n) admits a subsequence (g_{m_n}) weakly convergent to some $g_0 \in L_p[0,1]$—this thanks to the reflexivity of the L_p in question. On replacing (g_n) by $(g_{m_n} - g_0)$, we find ourselves with a sequence $(f_n) = (g_{m_n} - g_0)$ that satisfies the hypotheses of the Banach-Saks theorem. The conclusion of that theorem speaks of a subsequence (f_{k_n}) of (f_n) for which

$$\left\|\sum_{l=1}^{n} f_{k_l}\right\| \leq Cn^{1/p}$$

for all n and some $C > 0$ independent of n. What does this mean for $(g_{m_{k_n}} - g_0)$? Well,

$$\left\|\left(\frac{1}{n}\sum_{l=1}^{n} g_{m_{k_l}} - g_0\right)\right\|_p = \left\|\frac{1}{n}\sum_{l=1}^{n}(g_{m_{k_l}} - g_0)\right\|_p$$

$$= \frac{1}{n}\left\|\sum_{l=1}^{n} f_{k_l}\right\|_p \leq Cn^{(1-p)/p}.$$

What about $p = 1$? Again, weakly null sequences admit of subsequences whose arithmetic means converge in norm to zero. Here, however, we have the opportunity to use the weak-compactness criteria developed in the third section, Weak compactness in ca(Σ) and $L_1(\mu)$—an opportunity not to be denied.

First, we provide a small improvement on the argument given in the Banach-Saks theorem in case of $L_2[0,1]$ to prove that *if (f_n) is a weakly null sequence in $L_2[0,1]$, then there is a subsequence (f_{k_n}) for which*

$$\lim_n \sup_{j_1 < \cdots < j_n} \left\|\frac{1}{n}\sum_{i=1}^{n} f_{k_{j_i}}\right\|_2 = 0.$$

As before, we assume $\|f_n\|_2 \leq 1$ for all n and that (f_n) is weakly null. Let

$k_1 = 1$ and choose $k_2 > k_1$ so that

$$\int_0^1 f_{k_1}(t) f_{k_2}(t)\, dt < \tfrac{1}{2}.$$

Next, let $k_3 > k_2$ with

$$\int_0^1 f_{k_1}(t) f_{k_3}(t)\, dt, \int_0^1 f_{k_2}(t) f_{k_3}(t)\, dt < \tfrac{1}{3}.$$

Again, let $k_4 > k_3$ be chosen so that

$$\int_0^1 f_{k_1}(t) f_{k_4}(t)\, dt, \int_0^1 f_{k_2}(t) f_{k_4}(t)\, dt, \int_0^1 f_{k_3}(t) f_{k_4}(t)\, dt < \tfrac{1}{4}.$$

The extraction procedure should be clear. Let $j_1 < \cdots < j_n$. Then

$$\begin{aligned}\left\| \sum_{i=1}^n f_{k_{j_i}} \right\|_2^2 &= \sum_{i=1}^n \|f_{k_{j_i}}\|_2^2 + 2 \sum_{1 \le l < i \le n} \int_0^1 f_{n_{l_i}} f_{n_{i_j}} \\ &\le n + 2 \sum_{i=2}^n \sum_{l=1}^{i-1} (j_i)^{-1} = n + 2 \sum_{i=2}^n \frac{i-1}{j_i} \\ &\le n + 2n = 3n.\end{aligned}$$

The assertion we are after follows quickly from this. □

Diagonal Lemma. *Let (f_n) be a weakly null sequence in $L_1[0,1]$. Then for each $\varepsilon > 0$ there is a subsequence (g_n) of (f_n) such that*

$$\overline{\lim_k} \sup_{n_1 < \cdots < n_k} \left\| \frac{1}{k} \sum_{i=1}^k g_{n_i} \right\|_1 \le \varepsilon.$$

PROOF. We may assume that $\|f_n\|_1 \le 1$ for all n. Let m, n be positive integers and set

$$E_{m,n} = \{ t : |f_n(t)| \ge m \};$$

on so doing, notice that (if λ denotes Lebesgue measure)

$$|f_n|_1 = \int_0^1 |f_n(t)| \ge \int_{E_{m,n}} |f_n(t)| \ge m \lambda(E_{m,n}),$$

or

$$\lambda(E_{m,n}) \le \frac{1}{m}.$$

The set $\{ f_n : n \ge 1 \}$ is relatively weakly compact in $L_1[0,1]$; so the Dunford-Pettis criterion supplies us with an $\eta > 0$ for which $\int_E |f_n(t)|\, dt \le \varepsilon/3$ whenever $\lambda(E) \le \eta$. In tandem with the simple calculation made above, we find that there is an m_0 so that

$$\int_{E_{m_0,n}} |f_n(t)| dt \le \frac{\varepsilon}{3}$$

for all n. Define $\bar{f}_n$ by

$$\bar{f}_n(t) = \begin{cases} f_n(t) & \text{if } t \in E_{m_0, n}, \\ 0 & \text{otherwise.} \end{cases}$$

It is plain that $f_n(t) - \bar{f}_n(t)$ is just $f_n C_{[|f_n(t)| \le m_0]}$, and so $f_n - \bar{f}_n \in m_0 B_{L_2[0,1]}$. Thanks to the weak compactness of $m_0 B_{L_2[0,1]}$, we can find an increasing sequence (n_k) of positive integers and an $h \in m_0 B_{L_2[0,1]}$ such that

$$h = \text{weak} \lim_{k \to \infty} f_{n_k} - \bar{f}_{n_k};$$

our earlier remarks let us assume even more, namely,

$$\lim_k \sup_{i_1 < \cdots < i_k} \left\| \frac{1}{k} \sum_{j=1}^k \left(f_{n_{i_j}} - \bar{f}_{n_{i_j}}\right) - h \right\|_2 = 0.$$

Since $(f_{n_k} - \bar{f}_{n_k})$ converges weakly to h in $L_2[0,1]$, the same holds true in $L_1[0,1]$; but (f_{n_k}) is weakly null in $L_1[0,1]$; so $-h = \text{weak}\lim_k \bar{f}_{n_k}$. It now follows from the fact that

$$\|\bar{f}_{n_k}\|_1 = \int_{E_{m_0, n_k}} |f_{n_k}(t)|\,dt \le \varepsilon/3 \text{ for all } k, \text{ that } \|h\|_1 \le \varepsilon/3.$$

Putting all the parts together, we have that if k is big enough, then

$$\sup_{i_1 < \cdots < i_k} \left\| \frac{1}{k} \sum_{j=1}^k \left(f_{n_{i_j}} - \bar{f}_{n_{i_j}}\right) - h \right\|_1 \le \sup_{i_1 < \cdots < i_k} \left\| \frac{1}{k} \sum_{j=1}^k \left(f_{n_{i_j}} - \bar{f}_{n_{i_j}}\right) - h \right\|_2 \le \frac{\varepsilon}{3},$$

which tells us that for the same large k,

$$\begin{aligned} &\sup_{i_1 < \cdots < i_k} \left\| \frac{1}{k} \sum_{j=1}^k f_{n_{i_j}} \right\|_1 \\ &\quad \le \sup_{i_1 < \cdots < i_k} \left(\left\| \frac{1}{k} \sum_{j=1}^k \left(f_{n_{i_j}} - \bar{f}_{n_{i_j}}\right) - h \right\|_1 + \frac{1}{p} \sum_{j=1}^k \|f_{n_{i_j}}\|_1 + \|h\|_1 \right) \\ &\quad \le \varepsilon. \end{aligned}$$

$(g_k = f_{n_k})$ is our subsequence. □

Theorem (Szlenk). *In $L_1[0,1]$, every weakly convergent sequence has a subsequence whose arithmetic means are norm convergent.*

Proof. We suppose that (f_n) is a weakly null sequence in $B_{L_1[0,1]}$. If the many virtues claimed of the diagonal lemma are to be believed, then we can find a sequence $(n_k(l))_{l=1}^{\infty}$ of strictly increasing sequences of positive integers, each subsequent sequence a subsequence of its predecessor such that for any l we have

$$\overline{\lim_k} \sup_{i_1 < \cdots < i_k} \left\| \frac{1}{k} \sum_{j=1}^k f_{n_{i_j}} \right\|_1 \le \frac{1}{l}.$$

Of course, the subsequence we are looking for is precisely the sequence (g_n) whose mth item is $f_{n_m}(m)$. Let's check it out: If $k > l$, then

$$\left\| \frac{1}{k} \sum_{j=1}^{k} g_j \right\|_1 \le \left\| \frac{1}{k} \sum_{j=1}^{l} g_j \right\|_1 + \frac{k-l}{l} \left\| \frac{1}{k-l} \sum_{j=l+1}^{k} g_j \right\|_1$$

$$\le \left\| \frac{1}{k} \sum_{j=1}^{l} g_j \right\| + \left\| \frac{1}{k-l} \sum_{j=1}^{k-l} f_{n_{l+j}}(l+j) \right\|_1 .$$

Now

$$\overline{\lim_{k \to \infty}} \left\| \frac{1}{k} \sum_{j=1}^{k} g_j \right\|_1 \le \overline{\lim_{k \to \infty}} \left\| \frac{1}{k} \sum_{j=1}^{l} g_j \right\| + \overline{\lim_{k \to \infty}} \left\| \frac{1}{k-l} \sum_{j=1}^{k-l} f_{n_{l+j}}(l+j) \right\|_1 .$$

The first of these dominating terms ($\overline{\lim}_{k \to \infty} \|(1/k)\Sigma_{j=1}^{l} g_j\|$) is 0; after all, $k > l$ fixes l but lets k go wild! The second dominating term is no more than $1/l$ because of the diagonal lemma. But this gives us

$$\overline{\lim_{k \to \infty}} \left\| \frac{1}{k} \sum_{j=1}^{k} g_j \right\|_1 \le \frac{1}{l}$$

regardless of what the l is. It must be that

$$\lim_{k \to \infty} \left\| \frac{1}{k} \sum_{j=1}^{k} g_j \right\|_1 = 0. \qquad \square$$

Exercises

1. *The Dunford-Pettis property.* A Banach space X enjoys the *Dunford-Pettis property* if given weakly null sequences (x_n) and (x_n^*) in X and X^*, respectively, then $\lim_n x_n^*(x_n) = 0$.

   (i) A Banach space X has the Dunford-Pettis property if and only if for any Banach space Y, each weakly compact linear operator $T: X \to Y$ is completely continuous, i.e., takes weakly convergent sequences in X to norm convergent sequences in Y.

   (ii) For any compact Hausdorff space Ω, $C(\Omega)$ has the Dunford-Pettis property. (*Hint*: Think of Egorov's theorem.)

   (iii) If X^* has the Dunford-Pettis property, then so does X.

2. *Operators on c_0.* The bounded linear operators from c_0 to a Banach space X correspond precisely to the weakly unconditionally Cauchy series in X.

   (i) A bounded linear operator $T: c_0 \to X$ is weakly compact if and only if the series $\Sigma_n Te_n$ is weakly subseries convergent.

   (ii) A bounded linear operator $T: c_0 \to X$ is compact if and only if the series $\Sigma_n Te_n$ is norm subseries convergent.

3. *Operators into l_1.* The bounded linear operators from a Banach space X into l_1 correspond precisely to the sequences (x_n^*) in X^* for which $\Sigma_n |x_n^* x| < \infty$, for each $x \in X$, i.e., the weakly unconditionally Cauchy series in X^*.

   (i) An operator $T: X \to l_1$ is weakly compact if and only if the series $\Sigma_n x_n^*$ is weakly subseries convergent in X^*.

   (ii) An operator $T: X \to l_1$ is compact if and only if the series $\Sigma_n x_n^*$ is norm subseries convergent in X^*.

4. *Operators into c_0.* The bounded linear operators from a Banach space X into c_0 correspond precisely to the weak* null sequences in X^*.

   (i) A bounded linear operator $T: X \to c_0$ is a weakly compact operator if and only if the sequence $(T^* e_n^*)$ is weakly null in X^*.

   (ii) A bounded linear operator $T: X \to c_0$ is a compact operator if and only if the sequence $(T^* e_n^*)$ is norm null in X^*.

   (iii) For X any of the spaces c_0, l_∞, l_p $(1 \le p < \infty)$, $L_p[0,1]$ $(1 \le p \le \infty)$, or $\mathrm{ba}(\Sigma)$ there exists a noncompact linear operator from X into c_0.

   (iv) Every bounded linear operator from l_∞ to c_0 is weakly compact; therefore, c_0 is not isomorphic to a complemented subspace of l_∞.

5. *Operators on l_1.* The bounded linear operators from l_1 to a Banach space X correspond precisely to the bounded sequences in X.

   (i) A bounded linear operator $T: l_1 \to X$ is weakly compact if and only if the set $\{Te_n : n \ge 1\}$ is relatively weakly compact.

   (ii) A bounded linear operator $T: l_1 \to X$ is compact if and only if the set $\{Te_n : n \ge 1\}$ is relatively norm compact.

6. *The sum operator: a universal nonweakly compact operator.*

   (i) The operator $\sigma: l_1 \to l_\infty$ defined by

$$\sigma((t_n)) = \left(\sum_{i=1}^{n} t_i \right)$$

   is a nonweakly compact bounded linear operator.

   (ii) A bounded linear operator $T: X \to Y$ is not weakly compact if and only if there exist bounded linear operators $S: l_1 \to X$ and $U: Y \to l_\infty$ such that $UTS = \sigma$. [*Hint*: Pelczynski's proof of the Eberlein-Smulian theorem gives an inkling of how to find a bounded sequence (x_n) whose image is basic and satisfies $y^* T x_n \ge \delta$ for some $y^* \in Y^*$. The operator S is induced by (x_n) and leaves little choice as to how U is to be defined.]

7. *A universal noncompact operator.*

   (i) The formal identity operator $i: l_1 \to l_\infty$ is a noncompact bounded linear operator that has the sum operator σ as a factor.

   (ii) A bounded linear operator $B: X \to Y$ is not compact if and only if there exist bounded linear operators $J: l_1 \to X$ and $W: Y \to l_\infty$ such that $WBJ = i$.

[*Hint*: In case B is weakly compact but not compact, aim the Bessaga-Pelczynski selection principle toward inducing a weakly compact operator from l_1 into X.]

8. *Edgar's ordering of Banach spaces, I.* Following G. A. Edgar, a partial ordering of Banach spaces can be defined: Given Banach spaces X and Y, we say that $X < Y$ if $X = \bigcap_{T \in \mathscr{L}(X;Y)} T^{**-1}(Y)$.

(i) Banach space X satisfies Mazur's condition (weak* sequentially continuous functionals on X^{**} are in X) if and only if $X < c_0$.

(ii) $c_0 < X$ if and only if c_0 is isomorphic to a subspace of X.

9. *Edgar's ordering of Banach spaces, II.*

(i) $X < l_\infty$ if and only if any $x^{**} \in X^{**}$ which is weak* continuous on bounded weak* separable subsets of X^* is in X.

(ii) $l_\infty < X$ if and only if l_∞ is isomorphic to a subspace of X.

10. *The Yosida-Hewitt decomposition theorem.* Let Ω be a set and Σ be a σ-field of subsets of Ω. Suppose $\mu, \nu \in \mathrm{ba}(\Sigma)$. Define

$$(\mu \vee \nu)(E) = \sup_{F \subset E,\, F \in \Sigma} \{\mu(F) + \nu(E \setminus F)\}.$$

(i) $\mu \vee \nu \in \mathrm{ba}(\Sigma)$. Further, if $\eta \in \mathrm{ba}(\Sigma)$ satisfies $\eta(E) \ge \mu(E)$ and $\eta(E) \ge \nu(E)$ for all $E \in \Sigma$, then $\eta(E) \ge (\mu \vee \nu)(E)$ for all $E \in \Sigma$.

(ii) If $\mu, \nu \in \mathrm{ca}(\Sigma)$, then $\mu \vee \nu \in \mathrm{ca}(\Sigma)$.

We say $\eta \in \mathrm{ba}(\Sigma)$ is *purely finitely additive* if given a countably additive μ on Σ for which $|\mu|(E) \le |\eta|(E)$ holds for each $E \in \Sigma$, then $\mu = 0$.

(iii) Let $\mu \in \mathrm{ba}(\Sigma)$. Then μ can be written in the form $\mu = \mu_c + \mu_{\mathrm{pfa}}$ where $\mu_c \in \mathrm{ca}(\Sigma)$ and μ_{pfa} is a purely finitely additive member of $\mathrm{ba}(\Sigma)$. [*Hint*: By considering μ,

$$\mu = \left(\frac{|\mu| + \mu}{2}\right) - \left(\frac{|\mu| - \mu}{2}\right),$$

as the difference of two members of $\mathrm{ba}^+(\Sigma)$, one need only consider nonnegative μ. Now let $\Gamma \subseteq \mathrm{ca}^+(\Sigma)$ be the set $\{\gamma \in \mathrm{ca}^+(\Sigma) : \gamma(E) \le \mu(E)$ for all $E \in \Sigma\}$. Choosing $\gamma_n \in \Gamma$ so that $\gamma_n(\Omega) \nearrow \sup\{\gamma(\Omega) : \gamma \in \Gamma\}$ and letting $\mu_n = \gamma_1 \vee \gamma_2 \vee \cdots \vee \gamma_n$, notice that $\lim_n \mu_n(E)$ exists for all $E \in \Sigma$.]

(iv) If Ω is the set $\mathbb{N}$ of natural numbers and Σ is the σ-field of all subsets of Ω, then any purely finitely additive measure on Σ vanishes on finite sets.

11. *Edgar's ordering of Banach spaces, III.*

(i) $X < l_1$ if and only if any $x^{**} \in X^{**}$ such that $x^{**}(\text{weak}^* \sum_n x_n^*) = \sum_n x^{**} x_n^*$, for each weakly unconditionally Cauchy series $\sum_n x_n^*$ in X^*, belongs to X.

(ii) $l_1 < X$ if and only if X is nonreflexive.

12. *Pelczynski's property V.* A Banach space X has *property V* whenever given any Banach space Y, every unconditionally converging operator $T: X \to Y$ is weakly compact.

(i) For any compact Hausdorff space Ω, $C(\Omega)$ has property V. (*Hint*: Use the Dieudonné-Grothendieck criterion and Gantmacher's theorem.)

(ii) If X has property V, then a subset K of X^* is relatively weakly compact whenever

$$\lim_n x^*x_n = 0 \quad \text{uniformly for } x^* \in K$$

for any weakly unconditionally Cauchy series $\Sigma_n x_n$. [*Hint*: The condition cited implies not only the boundedness of the linear operator $T: X \to l_\infty(K)$ defined by $(Tx)(x^*) = x^*x$ but also the fact that T is unconditionally converging.]

(iii) If X^* has property V, then weak* null sequences in X^{**} are weakly null. (*Hint*: Phillips's lemma is worth keeping in mind.)

(iv) The converse of (ii) also holds.

(v) If X has property V, the weakly Cauchy sequences in X^* are weakly convergent. (*Hint*: Schur's lemma is worth keeping in mind.)

13. *Relatively disjoint families of measures.* Let Ω be a set, Σ be a σ-field of subsets of Ω and $0 < \varepsilon < \delta$. A sequence (μ_n) in ca(Σ) is called (δ, ε)*-relatively disjoint* if $\sup_n |\mu_n|(\Omega) < \infty$ and there exists a sequence (E_n) of pairwise disjoint members of Σ such that for each n

$$|\mu_n|(E_n) > \delta \quad \text{and} \quad \sum_{m \neq n} |\mu_n|(E_m) < \varepsilon.$$

The sequence (μ_n) is called *relatively disjoint* if it is (δ, ε)-relatively disjoint for some choice of ε and δ.

Relatively disjoint sequences in ca(Σ) are basic sequences equivalent to the unit vector basis of l_1 with a closed linear span that is complemented in ca(Σ).

14. *Phillips's lemma and limited sets.* A subset B of a Banach space X is *limited* if $\lim_n x_n^* x = 0$ uniformly for $x \in B$ whenever (x_n^*) is a weak* null sequence in X^*.

(i) Limited sets are bounded.

(ii) Relatively compact sets are limited.

(iii) In separable Banach spaces limited sets are relatively compact.

(iv) The set $\{e_n : n \geq 1\}$ of unit coordinate vectors is limited in l_∞, but not in c_0.

15. *A theorem of Buck.* In a finite-dimensional Banach space, a bounded sequence each subsequence of which has norm convergent arithmetic means is itself convergent.

16. *Weakly null sequences in c_0.* If (x_n) is a weakly null sequence in c_0, then (x_n) has a subsequence each subsequence of which has norm-null arithmetic means.

17. $C[0,1]$ *fails the weak Banach-Saks property.*

(i) For fixed positive integer k, show that there exists a nonnegative sequence $(g_n^k)_{n\ge 1}$ in $B_{C[0,1]}$ such that
(a) $g_n^k(t) = 0$ if $t \notin ((k-1)/k, k/(k+1))$.
(b) $\lim_n g_n^k(t) = 0$, for all $t \in [0,1]$.
(c) If $n_1 < n_2 < \cdots < n_k$, then there is a $t_0 \in [0,1]$ such that $g_{n_1}^k(t_0) = g_{n_2}^k(t_0) = \cdots = g_{n_k}^k(t_0) = 1$.

(ii) Following (i), prove that the sequence (f_n) defined by

$$f_n = g_n^1 + g_n^2 + \cdots + g_n^n$$

is a weakly null sequence in $C[0,1]$ for which given $n_1 < n_2 < \cdots < n_m < n_{m+1} < \cdots < n_{2m}$ we have $(f_{n_1} + \cdots + f_{n_{2m}})(t) \ge \frac{1}{2}$ for all $t \in [0,1]$.

18. *Orlicz's theorem in L_p, $p > 2$.* Let $p > 2$.

(i) There is $M > 0$ so that for any real numbers a, b

$$|a|^p + pb|a|^{p-1}\operatorname{sgn} a + M|b|^p \le |a+b|^p.$$

(ii) There is $M > 0$ so that for any f, g in $L_p[0,1]$

$$\int_0^1 |f(t)|^p\,dt + M\int_0^1 |g(t)|^p\,dt \le \int_0^1 |f(t) + \sigma g(t)|^p\,dt$$

holds for some sign $\sigma = \pm 1$.

(iii) From (ii) derive Orlicz's theorem for $p > 2$, i.e., if $\Sigma_n f_n$ is an unconditionally convergent series in $L_p[0,1]$, then $\Sigma_n \|f_n\|_p^p < \infty$.

19. *The Banach-Saks theorem for L_p, $p > 2$.* Let $p > 2$ and denote by $[p]$ the greatest positive integer $\le p$.

(i) There are $A, B > 0$ such that for any real numbers a, b

$$|a+b|^p \le |a|^p + p|a|^{p-1} b \operatorname{sgn} a + A|b|^p + B\sum_{j=2}^{[p]} |a|^{p-j}|b|^j.$$

(ii) If (f_n) is a weakly null sequence in $L_p[0,1]$, then (f_n) has a subsequence (g_n) such that

$$\begin{aligned}\int_0^1 \left|\sum_{i=1}^n g_i(t)\right|^p dt &\le \int_0^1 \left|\sum_{i=1}^{n-1} g_i(t)\right|^p dt \\ &\quad + p\int_0^1 \left|\sum_{i=1}^{n-1} g_i(t)\right|^{p-1} \operatorname{sgn}\left(\sum_{i=1}^{n-1} g_i(t)\right) g_n(t)\,dt \\ &\quad + A\int_0^1 |g_n(t)|^p\,dt + B\sum_{j=2}^{[p]}\int_0^1 \left|\sum_{i=1}^{n-1} g_i(t)\right|^{p-j} |g_n(t)|^j\,dt.\end{aligned}$$

(iii) The subsequence (g_n) extracted in (ii) satisfies an estimate $\|\Sigma_{i=1}^n g_i\|_p \le M\sqrt{n}$ for some $M > 0$.

20. *Absolutely p-summing operators in Hilbert space.*

(i) Using Khinchine's inequalities, prove that the natural inclusion map $i: l_1 \to l_2$ is absolutely 1-summing.

(ii) Let H, K be Hilbert spaces. Recall that a bounded linear operator $T: H \to K$ is a Hilbert-Schmidt operator when T admits a representation in the form $Tx = \Sigma_n \lambda_n \langle x, h_n \rangle k_n$, where $(\lambda_n) \in l_2$, (h_n) is an orthonormal sequence in H and (k_n) is an orthonormal sequence in K. Show that every Hilbert-Schmidt operator has the natural inclusion map $i: l_1 \to l_2$ as a factor.

Consequently, every Hilbert-Schmidt operator is absolutely 1-summing and the absolutely p-summing operators from H to K coincide with the Hilbert-Schmidt class for every $1 \leq p \leq 2$.

21. *Weakly compact sets in $L_\infty[0,1]$ and the Dunford-Pettis property for $L_1[0,1]$.*

(i) Weakly compact sets in $L_\infty[0,1]$ are norm separable.

(ii) If $\mathscr{K}$ is a weakly compact subset of $L_\infty[0,1]$, then for each $\varepsilon > 0$ there is a measurable set $B \subseteq [0,1]$ whose complement has measure less than ε such that $\{fc_B : f \in \mathscr{K}\}$ is relatively norm compact in $L_\infty[0,1]$.

(iii) $L_1[0,1]$ has the Dunford-Pettis property.

22. *Cotype 2.* A Banach space E has *cotype 2* if there exists a $c(E) > 0$ such that given $x_1, x_2, \ldots, x_n \in E$; then

$$\left(\sum_{i=1}^{n} \|x_i\|^2\right)^{1/2} \leq c(E)\left(\int_0^1 \left\|\sum_{i=1}^{n} r_i(t) x_i\right\|^2 dt\right)^{1/2}$$

where (r_n) is the sequence of Rademacher functions.

(i) If E has cotype 2, then $\pi_p(X; E) = \pi_2(X; E)$ for any $2 \leq p < \infty$ and any Banach space X.

(ii) Hilbert spaces have cotype 2.

(iii) Let H, K be Hilbert spaces and $1 \leq p < \infty$. Then $\pi_p(H; K)$ coincides with the class of Hilbert-Schmidt operators from H to K.

(iv) If $1 \leq p \leq 2$, then $L_p(0,1)$ has cotype 2.

Notes and Remarks

Banach proved Theorem 1 in case Ω is a compact metric space; however, his proof carries over to the general case. Once the dual of $C(\Omega)$ is known as a space of measures, the weak convergence or weak Cauchyness of a sequence is easily recognized. Banach was in position to recognize this (at least in case

Ω is a compact metric space) being close to Saks and recognizing the relevance of integration theory.

Theorem 2 is due to R. Baire. It provides an elegant internal characterization of functions of the first Baire class. For an enjoyable read we recommend Hausdorff's discussion of Baire's classification scheme of the bounded Borel functions.

Theorem 3 is due to R. S. Phillips (1940) and can be found in his contribution "On linear transformations," a paper filled with still delicious tidbits. The injectivity of l_∞ is shared by other Banach spaces including $l_\infty(\Gamma)$- and $L_\infty(\mu)$-spaces. The complete characterization of spaces complemented by a norm-one projection in any superspace was obtained by L. Nachbin (1950), D. Goodner (1950), and J. L. Kelley (1952) in the real case and M. Hasumi (1958) in the complex case. Their result is *a Banach space X is complemented by a norm-one projection in any super space if and only if there exists an extremally disconnected compact Hausdorff space Ω such that X is isometrically isomorphic to $C(\Omega)$*. If you relax the demand that the projection be of norm one, then you are face to face with a long-standing open problem in Banach space theory: *Which Banach spaces are complemented in any superspace?*

Theorem 4 is a marvelous discovery of A. Sobczyk (1941). Naturally, Sobczyk's proof differs from the proof of Veech presented in the text. Another proof, due to A. Pełczynski and found in his "Projections in certain Banach spaces" (1960), is strongly recommended, too.

The statement of Theorem 4 actually characterizes c_0 among the separable Banach spaces; i.e., any infinite-dimensional separable Banach space complemented in any separable super space is isomorphic to c_0. That this is so is an admirable achievement of modern Banach space theory with the deciding blow being struck by M. Zippin (1977).

The fact that every separable Banach space is a quotient of l_1 is, as we've already noted in the text, due to S. Banach and S. Mazur (1933). The corollary fact that l_1 is the unique "projective" object among the separable infinite-dimensional Banach spaces seems to be due to J. Lindenstrauss. G. Köthe has extended the result to nonseparable spaces.

The description of $B(\Sigma)^*$ is due to T. H. Hildebrandt (1934) and, independently, G. Fichtenholtz and L. V. Kantorovich (1934). The paper of K. Yosida and E. Hewitt (1952) is *must* reading in coming to understand the exact nature of individual members of ba(Σ); Exercise 10 is due to Yosida and Hewitt.

Each of the results of the section on l_∞^*, Schur's theorem about l_1, and the Orlicz-Pettis theorem is a "name" theorem; each has earned its place as such. The Nikodym boundedness theorem *in* ca(Σ) was already referred to in Dunford and Schwartz as a "striking improvement of the principle of uniform boundedness" in that space. Grothendieck's generalization spent some years in surprising anonymity, although it appeared in his widely ignored Saõ Paulo lecture notes from the mid fifties.

Rosenthal's lemma was instrumental in H. P. Rosenthal's (1970) study of operators on $C(\Omega)$ spaces, where Ω is an extremally disconnected compact Hausdorff space. Using variations on a common theme of disjointification, Rosenthal showed that nonweakly compact operators on such $C(\Omega)$ fix a copy of l_∞ and that a dual containing a copy of $c_0(\Gamma)$ also contains a copy of $l_\infty(\Gamma)$. Exercise 13 is to be found in this study. We follow Kupka's approach in the text but recommend the reader treat himself to a reading of Drewnowski's generalization (1975) of the Rosenthal lemma.

Our presentation of many of the results of this chapter was inspired by an unpublished manuscript of J. Jerry Uhl, Jr., accompanied by many enjoyable conversations with that individual regarding this material. This is especially true of Phillips's lemma and Schur's theorem, two grand old interchange-of-limits jewels. Incidentally, the original objective of Phillips's lemma was part (d) of Exercise 14.

Everything that appears in the third section, Weak Compactness in ca(Σ) and $L_1(\mu)$, save the results of M. I. Kadec and A. Pelczynski (1962) is at least stated in Dunford and Schwartz, and to a greater or lesser extent we have followed the spirit of their presentation. It was R. E. Huff who pointed out the proof of Theorem 9 and its natural similarity to many of the "continuity at a point implies global continuity" style results that occur in topological algebra.

M. Fréchet introduced the metric d_λ and O. Nikodym took over the study of (Σ, d_λ). The upshot of Nikodym's efforts is the fundamental Nikodym convergence theorem.

G. Vitali (1907) showed that if (f_n) is a sequence of Lebesgue-integrable functions on $[0,1]$ which converge almost everywhere to f, then $\int_0^1 f(s)\,ds$ and $\lim_n \int_0^1 f_n(s)\,ds$ exist and are equal if and only if the indefinite integrals of the f_n are uniformly absolutely continuous with respect to Lebesgue measure. H. Hahn proved that if (f_n) is a sequence of Lebesgue-integrable functions on $[0,1]$ and if $\lim_n \int_E f_n(s)\,ds$ exists for every measurable set E, then the indefinite integrals are uniformly absolutely continuous and converge to a set function continuous with respect to Lebesgue measure. These set the stage for the Vitali-Hahn-Saks theorem proved in the generality set forth herein by S. Saks, by much the same method as employed here.

The weak sequential completeness of ca(Σ) and $L_1(\lambda)$ is an easy consequence of the Vitali-Hahn-Saks and Nikodym convergence theorems.

Theorem 13 is due in the main to V. M. Dubrovskii (1947); we follow Dunford and Schwartz in principle for our presentation. Naturally the Dunford-Pettis theorem can be found in their memoir "Linear operations on summable functions" (1940).

The paper of M. I. Kadec and A. Pelczynski (1962) analyzes the structure of subspaces of $L_p[0,1]$ for $p \geq 2$ in addition to containing the gems treated in the text. Among the noteworthy results contained in Kadec-Pelczynski is their proof that *Hilbertian subspaces of* $L_p[0,1]$ *are complemented whenever*

$p \geq 2$, and their discovery that *regardless of $p > 1$, if X is a complemented infinite-dimensional subspace of $L_p[0,1]$, then either X is isomorphic to l_2 or X contains a complemented subspace isomorphic to l_p.*

The Kadec-Pelczynski alternative for subspaces of $L_1[0,1]$ was substantially improved by H. P. Rosenthal (1970). In his quest to know all there is to know about subspaces of $L_1[0,1]$, Rosenthal discovered the following.

Theorem (Rosenthal). *Let X be a closed linear subspace of $L_1[0,1]$. X is reflexive if and only if X does not contain l_1^n uniformly; in which case, X is isomorphic to a subspace of $L_1[0,1]$ for some $1 < p \leq 2$.*

The proof of this theorem depends on some diabolically clever change-of-density arguments that evolve from the Grothendieck-Pietsch domination scheme. It was an analysis of Rosenthal's argument that, in part, put B. Maurey and G. Pisier on the right path toward their "Great Theorem."

The Dieudonné-Grothendieck theorem was proved in a special case by J. Dieudonné and given general treatment by A. Grothendieck in his *Canadian Journal of Mathematics* memoir (1953). It was there that the Dunford-Pettis property was first isolated and the results of Exercise 1 derived. Theorem 15 is also found in this basic contribution; spaces X with the property that weak* null sequences in X^* are weakly null are often called *Grothendieck spaces*.

Khinchine's inequalities are an old and venerable contribution due to A. Khinchine. It is only recently that S. Szarek and U. Haagerup found the best constants in these inequalities.

Our presentation of Orlicz's theorem follows W. Orlicz's original proof (1930); Exercise 18 indicates the modification necessary in case $p > 2$. Actually with a bit of tender love and care Orlicz's proof can be made to prove the following: *Suppose $\Sigma_n f_n$ is a series in $L_p[0,1]$ for which $\Sigma_n \varepsilon_n f_n$ converges for almost all sequence (ε_n) of signs in $\{\pm 1\}^N$. Then $\Sigma_n \|f_n\|_p^{\text{cotype } L_p[0,1]} < \infty$. Here cotype $L_p[0,1] = 2$ if $1 \leq p \leq 2$, whereas cotype $L_p[0,1] = p$ if $p > 2$.* In light of our first proof of the Orlicz-Pettis theorem, it seems fitting that this sharpening of Orlicz's theorems apparently involves some apparent relationship to the behavior of sums of independent random variables having values in a Banach space.

The application of Khinchine's inequalities to p-summing operators was first broached by A. Pelczynski (1967) and A. Pietsch (1967).

The Banach-Saks phenomenon in $L_p[0,1]$ for $1 \leq p < \infty$ has a curious tale accompanying it. In their original note Banach and Saks (1930) make special mention of the failure of the phenomenon in $L_1[0,1]$; indeed, they claim to produce a weakly null sequence in $L_1[0,1]$ without any subsequences having norm convergent arithmetic means. Of course, W. Szlenk's proof (1965) bares the Banach-Saks slip.

We cannot leave our discussion of the Banach-Saks-Szlenk theorem without recalling the now celebrated discovery of J. Komlós (1967): *Given a bounded sequence* (f_n) *in* $L_1[0,1]$ *there exists an* $f \in L_1[0,1]$ *and a subsequence* (g_n) *of* (f_n) *such that each subsequence* (h_n) *of* (g_n) *satisfies*

$$f = \lim_n \frac{1}{n} \sum_{k=1}^{n} h_k \quad \textit{almost everywhere}.$$

That $C[0,1]$ fails the so-called weak Banach-Saks property was first shown by J. Schreier; we take our proof (Exercise 17) from J. Bourgain's (1979) penetrating study of operators on $C(\Omega)$ that fix copies of $C(\alpha)$ for various ordinals α.

The uncovering of the sum operator as a universally nonweakly compact operator was the work of J. Lindenstrauss and A. Pelczynski (1968) while W. B. Johnson (1971) used this to show the universality of the formal identity $i: l_1 \to l_\infty$ as a noncompact operator.

Bibliography

Banach, S. and Mazur, S. 1933. Zur Theorie der linearen Dimension, *Studia Math.*, **4**, 100–112.

Banach, S. and Saks, S. 1930. Sur la convergence forte dans les champs L^p, *Studia Math.*, **2**, 51–57.

Bourgain, J. 1979. The Szlenk index and operators on $C(K)$-spaces, *Bull. Soc. Math. Belg.* **31**, 87–117.

Drewnowski, L. 1975. Un théoreme sur les opérateurs de $l_\infty(\Gamma)$, *C. R. Acad. Sc. Paris*, **281**, 967–969.

Dubrovskii, V. M. 1947. On some properties of completely additive set functions and their application to a generalization of a theorem of Lebesgue, *Mat. Sb.*, **20**, 317–329.

Dubrovskii, V. M. 1947. On the basis of a family of completely additive functions of sets and on the properties of uniform additivity and equi-continuity, *Dokl. Akad. Nauk SSSR*, **58**, 737–740.

Dubrovskii, V. M. 1948. On properties of absolute continuity and equi-continuity," *Dokl. Akad. Nauk SSSR*, **63**, 483–486.

Dunford, N. and Pettis, B. J. 1940. Linear operations on summable functions, *Trans. Amer. Math. Soc.*, **47**, 323–392.

Edgar, G. A. 1983. An ordering for Banach spaces. *Pac. J. Math.*, **108**, 83–98.

Fichtenholz, G. and Kantorovich, L. V. 1934. Sur les opérationes linéaires dans l'espace des fonctions bornées, *Studia Math.*, **5**, 69–98.

Goodner, D. B. 1950. Projections in normed linear spaces, *Trans. Amer. Math. Soc.*, **69**, 89–108.

Grothendieck, A. 1953. Sur les applications linéaires faiblement compactes d'espaces du type $C(K)$, *Canadian J. Math.*, **5**, 129–173.

Grothendieck, A. 1955. *Espaces Vectoriels Topologiques*. Soc. de Mat. de São Paulo.

Hahn, H. 1922. Über Folgen linearer Operationen, *Monatsh. Math. u. Phys.*, **32**, 3–88.

Hasumi, M. 1958. The extension property of complex Banach spaces, *Tôhoku Math. J.*, **10**, 135–142.

Hausdorff, F. 1957. *Set Theory*. Chelsea Publishing Co., New York.

Hildebrandt, T. H. 1934. On bounded functional operations. *Trans. Amer. Math. Soc.*, **36**, 868–875.

Johnson, W. B. 1971. A universal non-compact operator. *Colloq. Math.*, **23**, 267–268.
Kadec, M. I. and Pelczynski, A. 1962. Bases, lacunary sequences and complemented subspaces in L_p. *Studia Math.*, **21**, 161–176.
Kelley, J. L. 1952. Banach spaces with the extension property. *Trans. Amer. Math. Soc.*, **72**, 323–326.
Komlós, J. 1967. A generalization of a problem of Steinhaus. *Acta Math. Acad. Sci. Hung.*, **18**, 217–229.
Kupka, J. 1974. A short proof and generalization of a measure theoretic disjointization lemma. *Proc. Amer. Math. Soc.*, **45**, 70–72.
Lindenstrauss, J. and Pelczynski, A. 1968. Absolutely summing operators in $\mathscr{L}_p$ spaces and their applications. *Studia Math.*, **29**, 275–326.
Nachbin, L. 1950. A theorem of the Hahn-Banach type for linear transformations. *Trans. Amer. Math. Soc.*, **68**, 28–46.
Nikodym, O. M. 1933. Sur les familles bornées de fonctions parfaitement additives d'ensemble abstrait. *Monatsh. Math. u. Phys.*, **40**, 418–426.
Nikodym, O. M. 1933. Sur les suites convergentes de fonctions parfaitement additives d'ensemble abstrait. *Monatch. Math. u. Phys.*, **40**, 427–432.
Orlicz, W. 1930. Über unbedingte Konvergenz in Funktionräumen. *Studia Math.*, **1**, 83–85.
Pelczynski, A. 1960. Projections in certain Banach spaces. *Studia Math.*, **19**, 209–228.
Pelczynski, A. 1967. A characterization of Hilbert-Schmidt operators. *Studia Math.*, **28**, 355–360.
Phillips, R. S. 1940. On linear transformations. *Trans. Amer. Math. Soc.*, **48**, 516–541.
Pietsch, A. 1967. Absolut p-summierende Abbildungen in normierten Räumen. *Studia Math*, **28**, 333–353.
Rosenthal, H. P. 1970. On relatively disjoint families of measures, with some applications to Banach space theory. *Studia Math.*, **37**, 13–36.
Schreier, J. 1930. Ein Gegenbeispiel zur Theorie der schwachen Konvergenz. *Studia Math.*, **2**, 58–62.
Sobczyk, A. 1941. Projection of the space m on its subspace c_0. *Bull. Amer. Math. Soc.*, **47**, 938–947.
Szlenk, W. 1965. Sur les suites faiblements convergentes dans l'espace L. *Studia Math.*, **25**, 337–341.
Veech, W. A. 1971. Short proof of Sobczyk's theorem. *Proc. Amer. Math. Soc.*, **28**, 627–628.
Vitali, G. 1907. Sull'integrazione per serie. *Rend. Circ. Mat. Palermo*, **23**, 137–155.
Yosida, K. and Hewitt, E. 1952. Finitely additive measures. *Trans. Amer. Math. Soc.*, **72**, 46–66.
Zippin, M. 1977. The separable extension problem. *Israel J. Math.*, **26**, 372–387.

CHAPTER VIII

Weak Convergence and Unconditionally Convergent Series in Uniformly Convex Spaces

In this chapter, we prove three results too stunning not to be in the spotlight. These results are typical of the most attractive aspects of the theory of Banach spaces in that they are proved under easily stated, commonly understood hypotheses, are readily appreciated by Banach spacers and non-Banach spacers alike, and have proofs that bare their geometric souls.

The fundamentally geometric concept underlying each of the results is that of uniform convexity. Recall that a Banach space X is *uniformly convex* if given $\varepsilon > 0$ there is a $\delta > 0$ such that whenever $x, y \in S_X$ and $\|x - y\| = \varepsilon$, then $\|(x + y)/2\| \leq 1 - \delta$. An illustration should enlighten the reader as to the origin of the name.

Since the notion of uniform convexity involves keeping (uniform) control of *convex* combinations of points on the sphere, we worry only about *real* Banach spaces.

Let X be a (real) uniformly convex Banach space. For $0 \leq \varepsilon \leq 2$ let $\delta_X(\varepsilon)$ be defined by

$$\delta_X(\varepsilon) = \inf\left\{1 - \left\|\frac{x+y}{2}\right\| : x, y \in S_X, \|x - y\| = \varepsilon\right\}.$$

The function $\delta_X: [0,2] \to [0,1]$ is called the *modulus of convexity* of the space X and plainly $\delta_X(\varepsilon) > 0$ whenever $\varepsilon > 0$. Often we suppress X's role and denote the modulus by just $\delta(\varepsilon)$. Naturally, the modulus of convexity plays a key role in all that we do throughout this chapter.

Our attention throughout this chapter will be focused on the following three theorems:

Theorem 1 (S. Kakutani). *Every bounded sequence in a uniformly convex Banach space admits of a subsequence whose arithmetic means are norm convergent.*

Theorem 2 (M. Kadeč). *If $\sum_n x_n$ is an unconditionally convergent series in the uniformly convex space X, then*

$$\sum_n \delta(\|x_n\|) < \infty.$$

Theorem 3 (N. and V. Gurarii). *If the normalized Schauder basis (x_n) spans a uniformly convex space X, then there is a $p > 1$ and an $A > 0$ such that $\sum_n a_n x_n \in X$ whenever $(a_n) \in l_p$*

$$\left\|\sum_n a_n x_n\right\| \leq A\|(a_n)\|_p.$$

We start by studying δ; more precisely, we show

1. $\delta(\varepsilon) = \inf\{1 - \|(x+y)/2\| : x, y \in B_X, \|x - y\| = \varepsilon\}$,
2. $\delta(\varepsilon_1) \leq \delta(\varepsilon_2)$ whenever $0 \leq \varepsilon_1 \leq \varepsilon_2 \leq 2$,
3. $\delta(\varepsilon_1)/\varepsilon_1 \leq \delta(\varepsilon_2)/\varepsilon_2$ whenever $0 < \varepsilon_1 \leq \varepsilon_2 \leq 2$.

These facts follow from the corresponding facts about uniformly convex Banach spaces of finitely many dimensions and the following more or less obvious consequence of the definition of the modulus of convexity:

$$\delta(\varepsilon) = \inf\{\delta_Y(\varepsilon) : Y \text{ is a finite-dimensional subspace of } X\}.$$

This in hand we will prove statements 1, 2, and 3 for finite-dimensional X; as one might expect, the compactness of closed bounded sets eases the proof of each claim.

In each of the next three lemmas, X is a finite-dimensional uniformly convex space.

Lemma 4. $\delta(\varepsilon) = \inf\{1 - \|(x+y)/2\| : \|x\|, \|y\| \leq 1, \|x - y\| = \varepsilon\}$.

Proof. We begin with a remark about local maxima for linear functionals: *whenever $\varphi \in S_{X^*}$ achieves a local maximum at $x \in S_X$, then $|\varphi(x)|$ is a global maximum for $|\varphi|$ on S_X.*

Why is this so? Well, take any $\varepsilon > 0$ and find $u \in S_X$ so that $\varphi(u) > 1 - \varepsilon$. If λ is close enough to 0, then

$$\varphi\left(\frac{x + |\lambda| u}{\|x + |\lambda| u\|}\right) \leq \varphi(x);$$

so

$$\varphi(x + |\lambda| u) = \varphi(x) + |\lambda|\varphi(u) \leq \|x + |\lambda| u\| \varphi(x)$$

and

$$|\lambda|\varphi(u) \leq (\|x + |\lambda| u\| - 1)\varphi(x) \leq |\lambda| \|\varphi(x)|.$$

From this we conclude that $|\varphi(x)| \geq \varphi(u) > 1 - \varepsilon$. ε is arbitrarily chosen after the identity of x has been established; so $|\varphi(x)| = 1$.

Now let's see that $\delta(\varepsilon)$ is indeed the quantity:

$$\inf\left\{1 - \left\|\frac{x+y}{2}\right\| : x, y \in B_X, \|x - y\| = \varepsilon\right\}.$$

What we show is that the above inf is attained (in the presence of the added hypothesis that $\dim X < \infty$) when on the unit sphere.

Let $0 < \varepsilon \leq 2$ and choose $x, y \in B_X$ so that

$$\|x + y\| = \sup\{\|u + v\| : u, v \in B_X, \|u - v\| = \varepsilon\}.$$

Assume that $\|x\| \leq \|y\|$ (so $\|y\| \neq 0$).

First we show that $\|y\|$ is necessarily 1. In fact, if we let $c = (1 - \|y\|)/2$, then $0 \leq c \leq 1$. Considering the vectors

$$x_1 = \frac{(1-c)x + cy}{\|y\|} \quad \text{and} \quad y_1 = \frac{(1-c)(y) + cx}{\|y\|},$$

we find that $x_1, y_1 \in B_X$ and $\|x_1 - y_1\| = \varepsilon$. Therefore,

$$\|x_1 + y_1\| \leq \|x + y\|.$$

But

$$\|x_1 + y_1\| = \frac{1}{\|y\|}\|x + y\|.$$

Since $\|y\| \leq 1$, it follows from this last inequality and our choice of x, y that $\|y\| = 1$.

Having ascertained that $\|y\| = 1$, what about x? Of course, if $\|x\| = 1$, too, then we are done. Suppose $\|x\| < 1$. Pick $\varphi \in S_{X^*}$ so that

$$\varphi\left(\frac{x+y}{\|x+y\|}\right) = 1.$$

For any $z \in B_X$ with $\|z - y\| = \varepsilon$ we have

$$\varphi(z + y) \leq \|z + y\| \leq \|x + y\| = \varphi(x + y),$$

and so

$$\varphi(z) \leq \varphi(x).$$

φ attains its maximum value on $B_X \cap (y + \varepsilon S_X)$ at x. Suppose that we let $U = \{u \in S_X : y + \varepsilon u \in B_X \setminus S_X\}$. U is relatively open in S_X and contains $(x - y)/\varepsilon$. By what we have just done, φ attains its maximum value throughout U at $(x - y)/\varepsilon$. Our opening remark alerts us to the fact that $|\varphi|$ attains its global maximum on S_X at $(x - y)/\varepsilon$. Plainly $|\varphi((x - y)/\varepsilon)| = 1$. Recalling that $\varepsilon = \|x - y\|$, we are left with the possibilities that $\varphi(x - y) = \|x - y\|$ or $\varphi(x - y) = -\|x - y\|$. The first of these possibilities is ruled out by our supposition that $\|x\| < 1$; indeed,

$$\varphi(x) = \tfrac{1}{2}\varphi((x + y) + (x - y)) = \tfrac{1}{2}(\|x + y\| + \|y - x\|) \geq \tfrac{1}{2}\|2y\| = 1$$

makes sense if $\varphi(x-y)=\|x-y\|$ and forces $\|x\|\geq 1$. The second possibility, $\varphi(x-y)=-\|x-y\|$, is then the reality of the situation. This firmly in mind, take any $z\in B_X$ with $\|z-y\|=\varepsilon$. Then

$$|\varphi(z-y)|\leq|\varphi(x-y)|=\varepsilon;$$

so

$$-\varepsilon=\varphi(x-y)\leq\varphi(z-y)$$

and

$$\varphi(x)\leq\varphi(z).$$

But then

$$\varphi(x+y)\leq\varphi(z+y),$$

forcing

$$\|z+y\|=\|x+y\|.$$

In short, should $\|x\|<1$, then any $z\in B_X$ such that $\|z-y\|=\varepsilon$ satisfies $\|z+y\|=\|x+y\|$. Our poor first choice of x just has to be replaced by a z in S_X such that $\|z-y\|=\varepsilon$. □

An important consequence of the above is the nondecreasing nature of δ.

Lemma 5. *δ is a nondecreasing function of ε in* $[0,2]$.

PROOF. Let $0\leq\varepsilon_1<\varepsilon_2\leq 2$.

Pick $x, y\in S_X$ so that $\|x-y\|=\varepsilon_2$ and $\|x+y\|=2(1-\delta(\varepsilon_2))$. Let $c=(\varepsilon_2-\varepsilon_1)/2\varepsilon_2$. $0\leq c\leq 1$. Set

$$x_1=(1-c)x+cy \quad\text{and}\quad y_1=(1-c)y+cx.$$

Plainly $x_1, y_1\in B_X$ and it is quickly checked that $\|x_1-y_1\|=\varepsilon_1$. Furthermore, $\|x_1+y_1\|=\|x+y\|$; so by the previous lemma we see that

$$\delta(\varepsilon_1)\leq 1-\left\|\frac{x_1+y_1}{2}\right\|=1-\left\|\frac{x+y}{2}\right\|=\delta(\varepsilon_2).$$ □

Lemma 6. *Let $\delta_1(s)$ be defined for $0\leq s$ by*

$$\delta_1(s)=\inf_{u,v\in S_X}\{\max\{\|u+sv\|,\|u-sv\|\}-1\}.$$

Then $f(s)=\delta_1(s)/s$ is nondecreasing on $[0,\infty)$ and

$$\frac{\delta(\varepsilon)}{\varepsilon}=\frac{1}{2}f\left(\frac{\varepsilon}{2(1-\delta(\varepsilon))}\right).$$

PROOF. Fixing $u, v\in S_X$ momentarily and letting $g_{u,v}(s)$ be defined by

$$g_{u,v}(s)=\max\{\|u+sv\|,\|u-sv\|\}-1,$$

we see that $g_{u,v}$ is a convex function on $[0, \infty)$ that vanishes at 0. Therefore, whenever $0 \leq s \leq t$,

$$g_{u,v}(s) = g_{u,v}\left(\frac{t-s}{t}0 + \frac{s}{t}t\right)$$
$$\leq \frac{t-s}{t} g_{u,v}(0) + \frac{s}{t} g_{u,v}(t) = \frac{s}{t} g_{u,v}(t).$$

Consequently, $g_{u,v}(s)/s$ is nondecreasing for $s \in [0, \infty)$. Taking infima, we find that $f(s)/s$ is nondecreasing, too.

Now we establish the beautiful formula

$$\frac{\delta(\varepsilon)}{\varepsilon} = \frac{1}{2} f\left(\frac{\varepsilon}{2(1-\delta(\varepsilon))}\right).$$

Let $0 < \varepsilon \leq 2$ be given. Choose $x, y \in S_X$ so that $\|x - y\| = \varepsilon$ and $\|(x+y)/2\| = 1 - \delta(\varepsilon)$. Let

$$u = \frac{x+y}{\|x+y\|} \quad \text{and} \quad v = \frac{x-y}{\|x+y\|}.$$

Of course,

$$\|u\| = 1 \quad \text{and} \quad \|v\| = \frac{\varepsilon}{\|x+y\|} = \frac{\varepsilon}{2(1-\delta(\varepsilon))}.$$

We consider $s = \|v\|$. Since $\|u \pm v\| = 1/[1 - \delta(\varepsilon)]$,

$$\delta_1(s) \leq \left\|u \pm \|v\| \frac{v}{\|v\|}\right\| - 1 = \left\|u + \|v\| \frac{v}{\|v\|}\right\| - 1$$
$$= \frac{1}{1-\delta(\varepsilon)} - 1 = \frac{\delta(\varepsilon)}{1-\delta(\varepsilon)}.$$

On the other hand, we can pick u' and v' so that $\|u'\| = 1$, $\|v'\| = s$ and $\max\{\|u' + v'\|, \|u' - v'\|\} = 1 + \delta_1(s) = 1/a$. Letting

$$x' = a(u' + v') \quad \text{and} \quad y' = a(u' - v')$$

we get $x', y' \in B_X$ and $\|x' - y'\| = 2as$. It follows that

$$\delta(2as) \leq 1 - \left\|\frac{x'+y'}{2}\right\|$$
$$= 1 - a = 1 - \frac{1}{1+\delta_1(s)}$$
$$= \frac{\delta_1(s)}{1+\delta_1(s)}.$$

Since $t/(1+t)$ is increasing on $[0, \infty)$ and $\delta_1(s) \leq \delta(\varepsilon)/[1 - \delta(\varepsilon)]$, the last quantity above is

$$\leq \frac{\delta(\varepsilon)/[1-\delta(\varepsilon)]}{1 + \delta(\varepsilon)/[1-\delta(\varepsilon)]} = \delta(\varepsilon).$$

Ah ha! δ is nondecreasing. Should $\delta(2as) = \delta(\varepsilon)$, then

$$\delta(\varepsilon) \leq \frac{\delta_1(s)}{1+\delta_1(s)}.$$

So

$$\delta(\varepsilon) \leq \delta_1(s)(1-\delta(\varepsilon)),$$

or

$$\frac{\delta(\varepsilon)}{1-\delta(\varepsilon)} \leq \delta_1(s).$$

On the other hand, $\delta(2as) < \delta(\varepsilon)$ ensures that $2as < \varepsilon$ so that

$$\delta_1(s) = \frac{1}{a} - 1 \geq \frac{2s}{\varepsilon} - 1 = \frac{\delta(\varepsilon)}{1-\delta(\varepsilon)},$$

as an easy computation involving $s = \varepsilon/2(1-\delta(\varepsilon))$ shows. The upshot of all this is that thanks to δ's monotonicity,

$$\delta_1(s) = \frac{\delta(\varepsilon)}{1-\delta(\varepsilon)}.$$

It is pretty straightforward to derive the sought-after formula from this. □

Theorem 2 is now an immediate consequence of the next lemma.

Lemma 7. *Let* $x_1, x_2, \ldots, x_n \in X$ *satisfy* $\max_{\varepsilon_i = \pm 1} \|\sum_{i=1}^n \varepsilon_i x_i\| \leq 2$. *Then* $\sum_{i=1}^n \delta(\|x_i\|) \leq 1$.

PROOF. We suppose of course that the x_i are nonzero.

Let $\varepsilon_1 = 1$ and $S_1 = \varepsilon_1 x_1$.

Let ε_2 be the sign that produces the longer vector of $\varepsilon_1 x_1 + \varepsilon_2 x_2$, i.e., $\varepsilon_2 = 1$ if $\|x_1 + x_2\| \geq \|x_1 - x_2\|$ and $\varepsilon_2 = -1$ if $\|x_1 - x_2\| > \|x_1 + x_2\|$.

Let $S_2 = \varepsilon_1 x_1 + \varepsilon_2 x_2$.

Consider the vectors

$$x = \frac{S_2}{\|S_2\|} \quad \text{and} \quad y = \frac{S_2 - 2\varepsilon_2 x_2}{\|S_2\|},$$

then $x, y \in B_X$ so that

$$\left\|\frac{x+y}{2}\right\| \leq 1 - \delta(\|x-y\|).$$

If we look to the definition of x and y, then this last inequality quickly translates into

$$\frac{\|S_2 - \varepsilon_2 x_2\|}{\|S_2\|} \leq 1 - \delta\left(\frac{2\|x_2\|}{\|S_2\|}\right),$$

which in turn is the same as

$$\delta\left(\frac{2\|x_2\|}{\|S_2\|}\right) \le 1 - \frac{\|x_1\|}{\|S_2\|} = \frac{\|S_2\| - \|S_1\|}{\|S_2\|}.$$

We record this fact in the more convenient form

$$\|S_2\|\delta\left(\frac{2\|x_2\|}{\|S_2\|}\right) \le \|S_2\| - \|S_1\|. \tag{1}$$

Pursuing things a bit further, we notice that $2/\|S_2\| \ge 1$ so that

$$\frac{\delta(\|x_2\|)}{\|x_2\|} \le \frac{\delta\left(\frac{2}{\|S_2\|}\|x_2\|\right)}{\frac{2}{\|S_2\|}\|x_2\|}$$

by the monotonicity of $\delta(\varepsilon)/\varepsilon$. It follows that

$$2\delta(\|x_2\|) \le \|S_2\|\delta\left(\frac{2\|x_2\|}{\|S_2\|}\right). \tag{2}$$

Getting expressions (1) and (2) together but eliminating the middle man, we get

$$2\delta(\|x_2\|) \le \|S_2\| - \|S_1\|.$$

Let ε_3 be the sign that produces the longer vector $S_2 + \varepsilon_3 x_3$; i.e., $\varepsilon_3 = 1$ if $\|S_2 + x_3\| \ge \|S_2 - x_3\|$, but $\varepsilon_3 = -1$ if $\|S_2 + x_3\| < \|S_2 - x_3\|$.

Let $S_3 = S_2 + \varepsilon_3 x_3$. As we did above, we now are ready to set

$$x = \frac{S_3}{\|S_3\|} \quad \text{and} \quad y = \frac{S_3 - 2\varepsilon_3 x_3}{\|S_3\|}.$$

Proceed along a parallel to that followed above, and on arrival at your planned destination you ought to find

$$2\delta(\|x_3\|) \le \|S_3\| - \|S_2\|.$$

After repeating this argument a number of times and making the usual allowances for S_0 (set it $= 0$), we have in our telescopic sights the following

$$2\delta(\|x_1\|) \le \|S_1\| - \|S_0\|$$
$$2\delta(\|x_2\|) \le \|S_2\| - \|S_1\|$$
$$2\delta(\|x_3\|) \le \|S_3\| - \|S_2\|$$
$$\vdots$$
$$2\delta(\|x_n\|) \le \|S_n\| - \|S_{n-1}\|. \qquad \square$$

In making our way to the proof of Kakutani's Theorem 1 the following result of V. P. Milman and B. J. Pettis plays an important role. Its exceedingly short proof is due to J. Lindenstrauss and L. Tzafriri and serves as an excellent example of the clarity of view improving with the years.

Theorem (Milman-Pettis). *Uniformly convex Banach spaces are reflexive.*

PROOF. Let $x^{**} \in S_{X^{**}}$. Select a net $(x_d)_{d \in D}$ from B_X such that $x^{**} = \text{weak}^* \lim_d x_d$; such a net exists through the good graces of Goldstine's theorem. Since $2x^{**}$ is the weak* limit of the doubly indexed net $(x_d + x_{d'})_{(d,d') \in D \times D}$ and the norm in X^{**} is weak* lower semicontinuous, we know that $\lim_{(d,d')} \|x_d + x_{d'}\| = 2$. The uniform convexity of X allows us to conclude that $\lim_{(d,d')} \|x_d - x_{d'}\| = 0$. Since X is complete, (x_d) converges in norm to a member of X; this can only be x^{**}. ☐

PROOF OF THEOREM 1. In light of the Milman-Pettis theorem and the Eberlein-Šmulian theorem, it is enough to show that each weakly null sequence (x_n) of terms from B_X admits of a subsequence having norm-convergent arithmetic means.

Let (x_n) be such a sequence.

Let θ be the bigger of $1 - \delta(\frac{1}{2})$ and $\frac{3}{4}$.

Let $m_1 = 2$.

If $\|x_2\| \le \frac{1}{2}$, then $\|(x_2 + x_3)/2\| \le \frac{3}{4} \le \theta$; in this case we let $m_2 = 3$.

If $\|x_2\| > \frac{1}{2}$, then there is an $m > 2$ so that $\|x_2 - x_m\| > \frac{1}{2}$.

In fact, were $\|x_2 - x_m\| \le \frac{1}{2}$ for all $m > 2$, then we would have for any x^* in B_{X^*} that

$$|x^*x_2| = \lim_m |x^*x_2 - x^*x_m| \le \overline{\lim_m} \|x_2 - x_m\| \le \tfrac{1}{2}.$$

Let m_2 be the first $m > 2$ for which $\|x_2 - x_m\| \ge \frac{1}{2}$. Since $x_{m_1}, x_{m_2} \in B_X$ we have $\|(x_{m_1} + x_{m_2})/2\| \le 1 - \delta(\frac{1}{2}) \le \theta$.

In any case we can choose $m_2 > m_1$ so that

$$\left\| \frac{x_{m_1} + x_{m_2}}{2} \right\| \le \theta.$$

Let $m_3 = m_2 + 1$.

If $\|x_{m_3}\| \le \frac{1}{2}$, then $\|(x_{m_3} + x_{m_3} + 1)/2\| \le \frac{3}{4} \le \theta$. In this case let $m_4 = m_3 + 1$.

If $\|x_{m_3}\| > \frac{1}{2}$, then there is an $m_4 > m_3$ so that $\|x_{m_3} - x_{m_4}\| \ge \frac{1}{2}$. Since x_{m_3}, x_{m_4} are in B_X we have

$$\left\| \frac{x_{m_3} + x_{m_4}}{2} \right\| \le 1 - \delta(\tfrac{1}{2}) \le \theta.$$

In any case we can choose $m_4 > m_3$ so that

$$\left\| \frac{x_{m_3} + x_{m_4}}{2} \right\| \le \theta.$$

Let $m_5 = m_4 + 1$.

Continue in this vein.

We obtain a subsequence (x_{m_k}) of (x_n) for which given any k

$$\|x_{m_{2k-1}} + x_{m_{2k}}\| \le 2\theta.$$

Before proceeding to the next step, we take note of a fact about the modulus of convexity which follows by means of an easy normalization argument involving statement 1 cited in the proof of Kadeč's theorem, namely, the fact that whenever $\|x - y\| \ge \varepsilon \max(\|x\|, \|y\|)$, then $\|x + y\| \le 2(1 - \delta(\varepsilon))\max(\|x\|, \|y\|)$.

Let (x_n^1) be the sequence defined by

$$2x_n^1 = x_{m_{2n-1}} + x_{m_{2n}};$$

$\|x_n^1\| \le \theta$ for all k. Moreover, (x_n^1) is weakly null.

Let $m_1(1) = 2$.

If $\|x_2^1\| \le \theta/2$, then $\|(x_2^1 + x_3^1)/2\| \le 3\theta/4 \le \theta^2$.

If $\|x_2^1\| > \theta/2$, then there is an $m > 2$ so that $\|x_2^1 - x_m^1\| > \theta/2$. Indeed, were $\|x_2^1 - x_m^1\| \le \theta/2$ for all $m \ge 2$, then for any $x^* \in B_{X^*}$ we would have

$$|x^*x_2^1| = \lim_m |x^*x_2^1 - x^*x_m^1| \le \overline{\lim_m} \|x_2^1 - x_m^1\| \le \theta/2.$$

Now, once $\|x_2^1 - x_m^1\| \ge \theta/2$, we have that $\|x_2^1 + x_m^1\|/2 \le (1 - \delta(\frac{1}{2}))$, and so

$$\|x_2^1 + x_m^1\| \le 2(1 - \delta(\tfrac{1}{2}))\max(\|x_2^1\|, \|x_m^1\|) \le 2\theta^2.$$

In any case there is a first $m_2(1) > m_1(1) = 2$ for which

$$\left\|x_{m_1(1)}^1 = x_{m_2(1)}^1\right\| \le 2\theta^2.$$

The attentive reader can see how we now go about selecting $m_3(1), m_4(1), \ldots$, in an increasing fashion [with $m_3(1) = m_2(1)+1$] so as to ensure that

$$\left\|x_{m_{2n-1}(1)}^1 + x_{m_{2n}(1)}^1\right\| \le 2\theta^2$$

holds for all n.

Let (x_n^2) be the sequence given by

$$2x_n^2 = x_{m_{2n-1}(1)}^1 + x_{m_{2n}(1)}^1.$$

Then $\|x_n^2\| \le \theta^2$ for all n. Moreover, (x_n^2) is weakly null.

Proceeding as before, we can select an increasing sequence $(m_k(2))$ of indices with $m_1(2) = 2$ so that for every n

$$\left\|x_{m_{2n-1}(2)}^2 + x_{m_{2n}(2)}^2\right\| \le 2\theta^3.$$

Let (x_n^3) be the sequence defined by

$$2x_n^3 = x_{m_{2n-1}(2)}^2 + x_{m_{2n}(2)}^2.$$

The iteration seems clear enough: at the pth stage we have a sequence (x_n^p) of vectors each of norm $\le \theta^p$ and such that weak $\lim_n x_n^p = 0$. We select an increasing sequence $(m_n(p))$ of indices [with $m_1(p)=2$] in such a manner that

$$\left\|x^p_{m_{2n-1}(p)}+x^p_{m_{2n}(p)}\right\|\le 2\theta^{p+1}.$$

This lets us define the sequence (x_n^{p+1}) by

$$2x_n^{p+1}=x^p_{m_{2n-1}(p)}+x^p_{m_{2n}(p)}$$

and on so doing obtain a weakly null sequence each term of which has norm $\le\theta^{p+1}$.

Now to keep careful books, we tabulate

$$\begin{aligned}
x_1^1&=\tfrac12(x_{m_1}+x_{m_2}),\\
x_1^2&=\tfrac12\left(x^1_{m_1(1)}+x^1_{m_2(1)}\right)\\
&=\tfrac12\left(\tfrac12\left(x_{m_{2m_1}(1)-1}+x_{m_{2m_1}(1)}\right)+\tfrac12\left(x_{m_{2m_2}(1)-1}+x_{m_{2m_2}(1)}\right)\right)\\
&=\tfrac14\left(x_{m_{2m_1}(1)-1}+x_{m_{2m_1}(1)}+x_{m_{2m_2}(1)-1}+x_{m_{2m_2}(1)}\right),
\end{aligned}$$

where we note that the indices in this last expression are strictly increasing as one proceeds from left to right. If we continue to backtrack, we find that for any $p\ge1$ the vector x_1^p is representable in the form

$$x_1^p=2^{-p}\left(x_{l_1(p)}+x_{l_2(p)}+\cdots+x_{l_{2^p}(p)}\right)$$

where $1<l_1(1)<l_2(1)<l_1(2)<l_2(2)<l_3(2)<l_4(2)\cdots$. Further, we have arranged things so that if $q<p$ and $1\le i\le 2^{p-q}$, then the average of the ith block of 2^q members of $x_{l_1(p)},\dots,x_{l_{2^p}(p)}$,

$$2^{-q}\left(x_{l_{(i-1)2^q+1}(p)}+\cdots+x_{l_{i2^q}(p)}\right)$$

is a member of the sequence (x_n^q) and as such has norm $\le\theta^q$.

Let $n_1=1, n_2=l_1(1), n_3=l_2(1), n_4=l_1(2), n_5=l_2(2),\dots$.

Take any $k\ge1$ and suppose $r^{2^q}\le k\le(r+1)2^q$. Then

$$\begin{aligned}
\|x_{n_1}+\cdots+x_{n_k}\|&\le\left\|x_{n_1}+\cdots+x_{n_{2^q-1}}\right\|\\
&\quad+\sum_{j=2}^{r}\left\|x_{n_{(j-1)2^q}}+\cdots+x_{n_{j2^q-1}}\right\|\\
&\quad+\left\|x_{n_{r2^q}}+\cdots+x_{n_k}\right\|\\
&\le(2^q-1)+(r-1)2^q\theta^q+2^q.
\end{aligned}$$

It follows that

$$\begin{aligned}
\overline{\lim_k}\left\|\frac{x_{n_1}+\cdots+x_{n_k}}{k}\right\|&\le\overline{\lim_{k,q}}\,\frac{(2^q-1)}{k}+\frac{(r-1)2^q\theta^q}{k}+\frac{2^q}{k}\\
&=0.
\end{aligned}$$

□

Before embarking on the proof of Theorem 3, we wish to make a point about the *inclination* of a basic sequence. If (x_n) is a basic sequence, then the inclination of (x_n) is the number

$$k = \inf_n \operatorname{distance}\left(S_{[x_1,\dots,x_n]},[x_k : k > n]\right),$$

where $[A]$ denotes, as usual, the closed linear span of A. Our point is just this: *if the basic sequence (x_n) has basis constant K and inclination k, then $kK = 1$.*

In fact, we know that for any scalars $b_1, \dots, b_m, b_{m+1}, \dots, b_{m+n}$ that

$$\left\|\sum_{i=1}^{m} b_i x_i\right\| \le K \left\|\sum_{i=1}^{m+n} b_i x_i\right\|;$$

so that should $\sum_{i=1}^m b_i x_i \in S_X$, then regardless of $b_{m+1}, \dots, b_{m+n}$, we would have

$$1 = \left\|\sum_{i=1}^{m} b_i x_i\right\| \le K \left\|\sum_{i=1}^{m} b_i x_i - \sum_{i=m+1}^{m+n} b_i x_i\right\|.$$

It follows that

$$K^{-1} \le \left\|\sum_{i=1}^{m} b_i x_i - \sum_{i=m+1}^{m+n} b_i x_i\right\|,$$

or

$$K^{-1} \le k, \quad \text{that is, } 1 \le Kk.$$

On the way toward establishing equality, take any vector of the form $\sum_{i=1}^{m+n} b_i x_i$ and look at $\sum_{i=1}^m b_i x_i \in [x_1, \dots, x_m]$. Suppose $\sum_{i=1}^m b_i x_i \ne 0$. Then

$$\operatorname{distance}\left(\frac{\sum_{i=1}^{m} b_i x_i}{\left\|\sum_{i=1}^{m} b_i x_i\right\|}, [x_j : j > m]\right) \ge k.$$

Therefore, recalling an idea of Banach, we can find an x^* such that

$$x^* \text{ vanishes on } [x_j : j > m],$$

$$x^*\left(\sum_{i=1}^{m} b_i x_i\right) = \left\|\sum_{i=1}^{m} b_i x_i\right\|,$$

and

$$\|x^*\| = \frac{1}{\operatorname{distance}\left(\sum_{i=1}^{m} b_i x_i \Big/ \left\|\sum_{i=1}^{m} b_i x_i\right\|, [x_j : j > m]\right)},$$

a number $\leq 1/k$. It follows that

$$\left\|P_m \sum_{i=1}^{m+n} b_i x_i\right\| = \left\|\sum_{i=1}^{m} b_i x_i\right\|$$
$$= x^*\left(\sum_{i=1}^{m} b_i x_i\right) = x^*\left(\sum_{i=1}^{m+n} b_i x_i\right)$$
$$\leq \frac{1}{k}\left\|\sum_{i=1}^{m+n} b_i x_i\right\|.$$

From this we see that $\|P_m\| \leq 1/k$, and so, keeping in mind the fact that $K = \sup_m \|P_m\|$, we see that $k \leq 1/K$ or $Kk \leq 1$.

Now suppose we have a normalized basic sequence (x_n) that spans a uniformly convex space. Suppose (x_n) has a basic constant K and, correspondingly, an inclination of $k = 1/K$. Let p be chosen so that

$$\left(2\left[1 - \delta_{[x_n]_{n\geq 1}}(k)\right]\right)^p < 2.$$

Since given $x, y \in S_X$ for which $\|x - y\| \geq k$ we have

$$\|x + y\| \leq 2\left(1 - \delta_{[x_n]_{n\geq 1}}(k)\right),$$

it follows that the *continuous* functions $\varphi(t)$ and $\chi(t)$ given by

$$\varphi(t) = \|x + ty\|^p, \qquad \chi(t) = 1 + t^p$$

satisfy $\varphi(1) < \chi(1)$; consequently, there is an $\eta > 0$ so that

$$\|x + ty\|^p \leq 1 + t^p,$$

whenever $\|1 - t\| \leq \eta$. Of course, we can assume η is very small, say $\eta < 1$.

Claim. For any finitely nonzero sequence (a_m) of scalars we have

$$\frac{\eta}{2}\left\|\sum_m a_m x_m\right\| \leq \|(a_m)\|_p. \tag{3}$$

The proof of this claim (and, consequently, of Theorem 3) will be an induction on the number l of nonzero terms in vectors of the form $\sum_{i=1}^{l} a_i x_i$. For $l = 1$, expression (3) is plainly so. So suppose (3) holds for vectors $\sum_m a_m x_m$ that have no more than l nonzero a_i, where $l \geq 1$, and consider a vector $\sum_m b_m x_m$, where $l+1$ of the b_m are nonzero. For sanity in notation, we assume we are looking at a vector $\sum_{i=1}^{l+1} b_i x_i$, where all the b_i are nonzero. Of course, if just one of the coefficients exceeds the left side of (3) in modulus, then we are done; so we need to see what happens when all $l+1$ of the b_i satisfy

$$|b_i| < \frac{\eta}{2}\left\|\sum_{i=1}^{l+1} b_i x_i\right\|.$$

For $1 \leq j \leq l$, consider the vectors

$$y_j = \sum_{i=1}^{j} b_i x_i, \quad \sum_{i=j+1}^{l+1} b_i x_i = z_j.$$

Plainly,

$$\big|\|y_{j+1}\| - \|y_j\|\big|, \big|\|z_{j+1}\| - \|z_j\|\big| < \frac{\eta}{2}\left\|\sum_{i=1}^{l+1} b_i x_i\right\|$$

for $j = 1, 2, \ldots, l$, where $z_{l+1} = 0$ and $y_{l+1} = \Sigma_{i=1}^{l+1} b_i x_i$. It follows easily that for some special j, $1 \leq j \leq l$

$$\big|\|y_j\| - \|z_j\|\big| < \frac{\eta}{2}\left\|\sum_{i=1}^{l+1} b_i x_i\right\|. \tag{4}$$

We suppose that $\|y_j\| \geq \|z_j\|$ and for reasons of homogeneity assume $\|y_j\| = 1$. Since

$$\sum_{i=1}^{l+1} b_i x_i = y_j + z_j,$$

it follows that

$$\left\|\sum_{i=1}^{l+1} b_i x_i\right\| \leq \|y_j\| + \|z_j\| \leq 2.$$

Expression (4) assures us that

$$\big|1 - \|z_j\|\big| = \big|\|y_j\| - \|z_j\|\big| \leq \frac{\eta}{2}\left\|\sum_{i=1}^{l+1} b_i x_i\right\| \leq \eta.$$

In light of (3) this tells us that

$$\begin{aligned}
\left(\frac{\eta}{2}\left\|\sum_{i=1}^{l+1} b_i x_i\right\|\right)^p &= \left(\frac{\eta}{2}\|y_j + z_j\|\right)^p \\
&= \left(\frac{\eta}{2}\right)^p \left\|y_j + \|z_j\| \frac{z_j}{\|z_j\|}\right\|^p \\
&\leq \left(\frac{\eta}{2}\right)^p \left(\|y_j\|^p + \|z_j\|^p\right) \quad \text{[(3) enters here]} \\
&\leq \left(\frac{\eta}{2}\left\|\sum_{i=1}^{j} b_i x_i\right\|\right)^p + \left(\frac{\eta}{2}\left\|\sum_{i=j+1}^{l+1} b_i x_i\right\|\right)^p,
\end{aligned}$$

which by our inductive hypothesis,

$$\begin{aligned}
&\leq \|(b_1, b_2, \ldots, b_j, 0, 0, \ldots)\|_p^p \\
&\quad + \|(0, \ldots, 0, b_{j+1}, \ldots, b_{l+1}, 0, \ldots)\|_p^p \\
&= \|(b_i)\|_p^p.
\end{aligned}$$

Exercises

1. *No trees grow in uniformly convexifiable spaces.* A *finite tree* in the Banach space X is a set of elements of the form $\{x_1, x_2, x_3, \ldots, x_{2^n-1}\}$, where for each plausible index k we have

$$x_k = \frac{x_{2k} + x_{2k+1}}{2}.$$

The *height* of the finite tree $\{x_1, x_2, x_3, \ldots, x_{2^n-1}\}$ is the integer $n-1$. If the finite tree $\{x_1, x_2, x_3, \ldots, x_{2^n-1}\}$ also satisfies the conditions

$$\|x_k - x_{2k}\| \geq \delta, \|x_k - x_{2k+1}\| \geq \delta$$

for all plausible k, then it is called a *δ-tree of height $n-1$*. A Banach space X has the *finite tree property* if there is a $\delta > 0$ such that B_X contains δ-trees of arbitrary heights.

   (i) Uniformly convex spaces do not have the finite tree property.

   (ii) The finite tree property is an isomorphic invariant.

   (iii) If there is a constant $K > 0$ such that for each n, a one-to-one linear operator $T_n : l_1^n \to X$ can be found with $\|T_n\| \|T_n^{-1}\|_{\mathscr{L}(T_n(l_1^n); l_1^n)} \leq K$, then X has the finite tree property.

2. *An analysis of Kakutani's proof of Theorem* 1. Suppose we are given a positive integer $m \geq 2$. We say the Banach space X has property A_m if X is reflexive and there is α, $0 < \alpha < 1$, such that given a weakly null sequence (x_n) in B_X we can find $n_1 < \cdots < n_m$ such that

$$\left\| \sum_{k=1}^{m} x_{n_k} \right\| \leq \alpha m \sup_{1 \leq k \leq m} \|x_{n_k}\|.$$

We noticed in the above proof that uniformly convex spaces have A_2.

   (i) If $m_2 \geq m_1$ and X has A_{m_1}, then X has A_{m_2}.

   (ii) If X has A_m for some $m \geq 2$, then X enjoys the Banach-Saks property.

3. *Kakutani's theorem via the (Gurarii)^2 theorem.*

   (i) If (x_n) is a bounded sequence in a Banach space X and $(\sum_{k=1}^n k^{-1} x_k)_{n \geq 1}$ is norm convergent, then so, too, is the sequence $(n^{-1} \sum_{k=1}^n x_k)_{n \geq 1}$ norm convergent.

   (ii) Derive Kakutani's theorem from the (Gurarii)^2 theorem, (i), and the results of Chapter VI.

4. *Upper and lower l_p-estimates of (Gurarii)^2 type.* Suppose there are constants $A > 0$ and $p > 1$ so that given a normalized basic sequence (x_n) in the Banach space X, then

$$\|(a_n)\|_p \leq A \left\| \sum_n a_n x_n \right\| \leq A^2 \|(a_n)\|_{p'}.$$

$p^{-1} + (p')^{-1} = 1$ holds for all scalar sequences (a_n). Show that each normalized basic sequence in X is boundedly complete.

Notes and Remarks

At the instigation of J. D. Tamarkin, J. A. Clarkson (1936) introduced the class of uniformly convex spaces. His avowed purpose, admirably achieved, was to prove the following theorem.

Theorem (J. A. Clarkson). *If X is a uniformly convex Banach space and $f:[0,1]\to X$ has bounded variation, then*

$$f'(t)=\lim_{h\to 0}\frac{f(t+h)-f(t)}{h}$$

exists almost everywhere.

Furthermore, should f be absolutely continuous, then for all t,

$$f(t)=f(0)+(Bochner)\int_0^t f'(s)\,ds.$$

By way of exhibiting nontrivial examples of uniformly convex spaces, Clarkson established "Clarkson's inequalities" and, in so doing, proved that $L_p[0,1]$ is uniformly convex whenever $1<p<\infty$. It's a short trip from the uniform convexity of $L_p[0,1]$ to that of $L_p(\mu)$, for any μ and $1<p<\infty$.

Since the appearance of uniformly convexity on the scene, many important classes of function spaces have been thoroughly researched with an eye to sorting out the uniformly convex members. It has long been known, for instance, that if $1<p<\infty$ and X is a uniformly convex space, then the space $L_p(\mu, X)$ is uniformly convex for any μ; the discovery of this fact seems to be due to E. J. McShane (1950). Indeed, McShane gave a proof of the uniform convexity of $L_p(\mu, X)$ which in order to encompass the vectorial case considerably simplified the existing proofs for plain old $L_p(\mu)$.

A *complete* characterization of the uniformly convex Orlicz spaces, regardless of whether the Orlicz norm or the Luxemburg norm is used, has been obtained through the efforts of W. A. J. Luxemburg, H. W. Milnes (1957), B. A. Akimovič (1972), and A. Kamińska (1982).

The Lorentz spaces have proved to be somewhat more elusive. The spaces $L_{p,q}$ are uniformly convexifi*able* whenever they are reflexive, i.e., if $1<p$, $q<\infty$; whether these spaces are uniformly convex in certain of their naturally occurring norms remains an enigma of sorts. For the Lorentz spaces $L_{w,p}$, I. Halperin (1954) has given some criteria for the uniform convexity; in the case of the Lorentz sequence spaces $d(a,p)$, Z. Altshuler (1975) proved that their uniform convexity is equivalent to their uniform convexifiability and gives criteria in terms of the weight a for such.

The Schatten classes C_p were shown to be uniformly convex whenever $1<p<\infty$ by C. A. McCarthy (1967). J. Arazy (1981) has proved that for a separable symmetric Banach sequence space E the associated Schatten unitary matrix space C_E is uniformly convexifiable if and only if E is; Arazy leaves open the determination of whether C_E is uniformly convex when E is, however.

The reflexivity of uniformly convex spaces was established independently by D. P. Milman (1970) and B. J. Pettis (1939). For some reason, Milman's role in this matter is more widely known; in any case, the original proofs by Milman and Pettis vary greatly. Milman's proof was an early model upon which S. Kakutani (1939) made substantial improvements; both J. Dieudonné (1942) and A. F. Ruston (1949) effected further streamlining with Dieudonné's proof quite close in spirit (if not in execution) to the proof given in this text. We owe to J. Lindenstrauss and L. Tzafriri (1977, 1979) the proof found in these pages.

Pettis's approach to the Milman-Pettis theorem is often a surprise to present-day mathematicians: he calls upon finitely additive measures for help. Actually, the main idea behind Pettis's proof comes from H. H. Goldstine (1938) who used the idea in establishing "Goldstine's theorem"; since Pettis's proof is *so* different from the others, we discuss it a bit further.

Here is the setup: realize that for any Banach space X, X^* is always (isometrically isomorphic to) a closed linear subspace of $l_\infty(B_X)$. Therefore, following the directions provided us by Chapter VII, any $x^{**} \in (X^*)^*$ has a Hahn-Banach extension to a member χ of $l_\infty(B_X)^*$ which we know to be $ba(2^{B_X})$. It follows that x^{**} has the form

$$x^{**}f = \int_{B_X} f(x)\, d\chi(x),$$

for all $f \in l_\infty(B_X)$; moreover, $\|x^{**}\| = |\chi|(B_X)$. So far the fact that B_X is the closed unit ball of a Banach space has been exploited but sparingly. Look at χ^+ and χ^-

$$\chi^+ = \frac{|\chi| + \chi}{2}, \qquad \chi^- = \frac{|\chi| - \chi}{2},$$

which are both nonnegative members of $ba(2^{B_X})$. Of course, $\chi = \chi^+ - \chi^-$; for $E \leq B_X$ define $pE = -E$ and consider $\mu(E) = \chi^+(E) + \chi^-(pE)$. *$\mu$ is a* nonnegative *member of* $ba(2^{B_X})$ *for which*

$$x^{**}x^* = \int_{B_X} x^*(x)\, d\mu(x)$$

holds for all $x^* \in X^*$. *Moreover*, $\|x^{**}\| = \mu(B_X)$. Using the integration with respect to finitely additive measures that was developed in Chapter VII, it is now easy to prove Goldstine's theorem.

What about the Milman-Pettis theorem? Well, suppose X is uniformly convex and $x^{**} \in S_{X^{**}}$. There is a sequence (x_n^*) in S_{X^*} with $x^{**}(x_n^*)$ at least $1 - 1/n$ and, naturally, one locates a sequence (x_n) in S_X for which $x_n^*(x_n) = 1$; the uniform convexity of X can (and should) be used to see that each x_n is unique in S_X with respect to the condition $x_n^* x_n = 1$. All this is a rather typical warm-up for the main effort of this proof: show $x^{**} \in X$. We assert, with Pettis, that (x_n) is a Cauchy sequence with limit x^{**}; of course, we represent x^{**} by μ à la Goldstine.

Let $\varepsilon > 0$ be given and look at $B_n(\varepsilon) = \{x \in B_X : \|x - x_n\| \leq \varepsilon\}$. By uniform convexity, there is a $\delta_\varepsilon > 0$ so that for any $x^* \in X^*$ if $x \in S_X$, $y \in B_X$, $x^*x = \|x^*\|$, and $\|x - y\| \geq \varepsilon$, then $x^*y \leq (1-\delta)\|x^*\|$. Now integrating x_n^* over $B_X = B_n(\varepsilon) \cup [B_X \setminus B_n(\varepsilon)]$ ought to lead to the estimate $\mu(B_X \setminus B_n(\varepsilon)) < (n\delta_\varepsilon)^{-1}$. From this one quickly deduces that for m, n large enough, $B_m(\varepsilon)$ and $B_n(\varepsilon)$ intersect, i.e., (x_n) is a Cauchy sequence. Suppose its limit is denoted by x_0. Then $B_0(\varepsilon)$ contains $B_n(\varepsilon/2)$ for all n sufficiently large, allowing us to conclude that $B_X \setminus B_0(\varepsilon)$ has μ-measure zero. Now it is easy to see that $x_0 = x^{**}$. In fact, if $x^* \in X^*$,

$$\begin{aligned}
|x^{**}x^* - x^*x_0| &= \left| \int_{B_X} x^*x \, d\mu(x) - \int_{B_X} x^*x_0 \, d\mu(x) \right| \\
&\leq \int_{B_X} |x^*x - x^*x_0| \, d\mu(x) \\
&= \int_{B_0(\varepsilon)} + \int_{B_X \setminus B_0(\varepsilon)} = \int_{B_0(\varepsilon)} |x^*x - x^*x_0| \, d\mu(x) \\
&\leq \|x^*\| \in \mu(B_0(\varepsilon)) \leq \|x^*\|\varepsilon,
\end{aligned}$$

which completes our proof.

We have repeated Kakutani's original proof with nary a change to be found. An alternative proof, building on the (Gurarii)^2 theorem, is indicated in the exercises; it was shown to us by D. J. H. Garling in 1978. The exercise analyzing Kakutani's proof is inspired by work of J. R. Partington (1977).

T. Nishiura and D. Waterman first demonstrated that *a Banach space with the Banach-Saks property is reflexive*. Other proofs have been offered, notably by J. Diestel (1975) and D. van Dulst (1978); still another can be found in the exercises following Rosenthal's dichotomy. A. Baernstein II (1972) gave the first example of a reflexive Banach space that does not have the Banach-Saks property; C. Seifert (1978) showed that the dual of Baernstein's space has the Banach-Saks property leaving open the question of just what property is dual to the Banach-Saks property. In affairs of a Banach-Saks nature, the wise thing is to consult the works of B. Beauzamy (1976, 1979), who gives apt characterizations of the Banach-Saks property, the Banach-Saks-Rosenthal property, and the alternating-signs Banach-Saks property.

M. Kadeč (1956) first proved Theorem 2; however, we follow T. Figiel's (1976) lead in this matter with A. T. Plant's (1981) hints along the way being of obvious help. The attentive reader will, no doubt, be suspicious of possible connections between Kadeč's result and Orlicz's theorem found in Chapter VII. In fact, if $1 < p < \infty$, then Theorem 2 encompasses Exercise 20 of Chapter VII. This follows from the determination [by Hanner (1956)] of the asymptotic behavior of the modulus of convexity of $L_p(\mu)$; precisely, for any nontrivial measure μ,

$$\delta_{L_p(\mu)}(\varepsilon) \sim \begin{cases} \varepsilon^2 & \text{if } 1 < p \le 2, \\ \varepsilon^p & \text{if } 2 \le p < \infty, \end{cases}$$

where "$\sim$" indicates that we are detailing asymptotic behavior up to a constant for ε close to zero.

While Kadeč's result does not cover the case of $L_1(\mu)$ (as Orlicz's theorem does), it does give very sharp information about uniformly convex spaces once accurate estimates have been made regarding their moduli of convexity.

For Orlicz spaces, R. P. Maleev and S. L. Troyanski (1975) have given the tightest possible estimates for the moduli of convexity; moreover, their estimates involve, in a natural way, the generating Orlicz function.

Though the moduli of convexity for Lorentz sequence spaces have been worked out by Z. Altshuler (1975, 1980), the problem for Lorentz function spaces remains wantonly open.

In a striking tour de force of Rademacher know-how, N. Tomczak-Jaegermann (1974) has shown that for $1 < p < \infty$, the C_p classes have moduli that behave like the $L_p(\mu)$-spaces. Ms. Tomczak actually proves more: *C_p has cotype 2, if $1 < p < 2$, and cotype p, if $2 < p < \infty$*. Furthermore, she shows that *the dual of any C^*-algebra has cotype* 2.

Following B. Maurey and G. Pisier (1976), we say that a Banach space X has cotype p if there is a constant $K > 0$ for which

$$\sum_{i=1}^{n} \|x_i\|^p \le K^p \int_0^1 \left\| \sum_{i=1}^{n} r_i(t) x_i \right\|^p dt$$

for any finite set $\{x_1, \dots, x_n\}$ in X; here, as usual, the functions $r_1, \dots, r_n$ are the first n Rademacher functions. Thanks to J. P. Kahane (1968), we can paraphase cotype p as follows: a Banach space X has cotype p provided $\Sigma \|x_n\|^p$ converges whenever $\Sigma_n \sigma_n x_n$ is convergent for almost all sequences (σ_n) of signs $\sigma_n = \pm 1$ in $\{-1,1\}^{\mathbb{N}}$, where $\{-1,1\}^{\mathbb{N}}$ is endowed with the natural product measure whose coordinate measures assign each singleton the fair probability of $\frac{1}{2}$.

Although the precise definition of cotype did not appear on the mathematical scene until the early seventies, W. Orlicz's original results regarding unconditionally convergent series in $L_p[0,1]$ already had delved into the notion; in fact, with but a bit of doctoring Orlicz's proofs show that $L_p[0,1]$ has cotype 2 in case $1 \le p \le 2$ while it has cotype p for $p \ge 2$. What relation then, if any, does the cotype of a uniformly convex space bear to its modulus of convexity? In answering this question we pass over some of the most beautiful and richest terrain in the theory of Banach spaces; the ambitious reader would do well to study the fertile land we are treading.

Our response starts by recalling the notion of uniform smoothness: a Banach space X is said to be uniformly smooth whenever the limit

$$\lim_{t \to 0} \frac{\|x + ty\| - \|x\|}{t}$$

exists uniformly for all $x, y \in S_X$. This notion was studied extensively by V. L. Šmulian (1941), who showed that *X is uniformly smooth if and only if X^* is uniformly convex* and *X is uniformly convex if and only if X^* is uniformly smooth*. Along the way, Šmulian also showed that *if X^* is uniformly smooth* [in fact if we only ask that $\lim_{t \to 0}(\|x + tu\| - \|x\|)/t$ exist uniformly for $y \in S_X$ for each $x \in S_X$], *then X is reflexive*; thus, yet another proof of the reflexivity of uniformly convex spaces emerges. Now uniformly smooth spaces have a modulus of their own, a modulus of smoothness whose value for any $\tau > 0$ is given by

$$\rho(\tau) = \sup\left\{\frac{\|x + y\|}{2} + \frac{\|x - y\|}{2} - 1 : x \in S_X, \|y\| = \tau\right\}.$$

A surprising development relating the modulus of convexity of X^* and the modulus of smoothness of X took place in the early days of Lindenstrauss: for $0 < \varepsilon \leq 2$ and for $0 < \tau < \infty$

$$\rho_X(\tau) = \sup_{0 < \varepsilon \leq 2} \left\{\frac{\tau\varepsilon}{2} - \delta_{X^*}(\varepsilon)\right\}.$$

From this formula, Lindenstrauss was able to deduce that whenever $\Sigma_n \rho(\|x_n\|) < \infty$, then $\Sigma_n \sigma_n x_n$ converges for some sequence (σ_n) of signs $\sigma_n = \pm 1$. In passing it should be recalled that G. Nordlander had shown in 1960 that the modulus of convexity always satisfies

$$\overline{\lim_{\varepsilon \to 0}} \frac{\delta(\varepsilon)}{\varepsilon^2} < \infty;$$

consequently, Hilbert space is as convex as possible; as one might expect, Hilbert space is also as smooth as possible. Lindenstrauss showed that if a Banach space X has an unconditional basis and is as convex and smooth as Hilbert space, X is isomorphic to Hilbert space. He asked if such were so for any Banach space.

T. Figiel and G. Pisier (1974) gave much more in response to Lindenstrauss's query than was asked for. Recall that $L_2([0,1], X)$ is uniformly convex if X is—thanks to McShane—how does the modulus of convexity of $L_2([0,1], X)$ compare with that of X? Figiel and Pisier showed that $L_2([0,1], X)$ has a modulus of convexity which is asymptotically the same as that of X; i.e., there are constants c, $C > 0$ such that

$$c \leq \underline{\lim_{\varepsilon \to 0}} \frac{\delta_{L_2([0,1], X)}(\varepsilon)}{\delta_X(\varepsilon)} \leq \overline{\lim_{\varepsilon \to 0}} \frac{\delta_2([0,1], X)(\varepsilon)}{\delta_X(\varepsilon)} \leq C.$$

Therefore, if X is as convex as possible, so too is $L_1([0,1], X)$ and similarly for X^* and $L_2([0,1], X^*) = L_2([0,1], X)^*$. But now the Kadeč theorem comes into play. Using it, Figiel and Pisier conclude that for a Banach space X that is maximally smooth and convex one has the following analogue to Khintchine's inequalities: there is a $K > 0$ so that for any finite set $\{x_1, \dots, x_n\}$ in X

$$K^{-1}\left(\sum_{i=1}^{n} \|x_i\|^2\right)^{1/2} \leq \left(\int_0^1 \left\| \sum_{i=1}^{n} r_i(t) x_i \right\|^2 dt\right)^{1/2} \leq K\left(\sum_{i=1}^{n} \|x_i\|^2\right)^{1/2};$$

however, S. Kwapien (1972) had already noticed that such an inequality is enough to identify X as among the isomorphs of Hilbert space.

What Figiel and Pisier had shown however was more. In light of Kahane's discovery that $(\int_0^1 \|\sum_{i=1}^n r_i(t) x_i\|^2 \, dt)^{1/2}$ and $(\int_0^1 \|\sum_{i=1}^n r_i(t) x_i\|^p \, dt)^{1/p}$ are equivalent expressions, the fact that $L_2([0,1], X)$ and X are equally convex may be translated to the statement that if X has a modulus of convexity of power type $\delta(\varepsilon) = \varepsilon^p$ for some $p \geq 2$, then X also has cotype p. Again, Kadeč's theorem now allows one to conclude that if $\sum_n \sigma_n x_n$ is convergent even for almost all choices (σ_n) of signs, then $\sum_n \|x_n\|^p < \infty$; indeed, $\sum_n \sigma_n x_n$'s convergence for almost all choices of signs is tantamount to the unconditional convergence of $\sum_n r_n \otimes x_n$ in $L_2([0,1], X)$.

The story is not yet over. In fact, we have left the best part of this particular tale to the last. In a remarkable chain of developments, R. C. James had introduced the super reflexive Banach spaces; P. Enflo (1972) had shown them to be precisely those spaces which can be equivalently normed in a uniformly convex manner (which might as well be our definition), or precisely those spaces which can be equivalently normed in a uniformly smooth manner, or precisely those spaces which can be equivalently normed in a simultaneously uniformly convex and uniformly smooth manner; and G. Pisier had shown that every uniformly convexifiable Banach space has an equivalent norm which is uniformly convex with power type modulus of convexity.

For the case of Banach lattices there are finer notions than cotype (and type) that have allowed for a very fine gradation of the classical function spaces. For the rundown on these events the reader is referred to the monograph of W. B. Johnson et al. (1979) and the appropriate sections of the Lindenstrauss-Tzafriri books. With particular attention to the Lorentz spaces, J. Creekmore (1981) has computed the type and cotype of the $L_{p,q}$-spaces, and N. Carothers (1981) has gone on to solve the more difficult problem of the type and cotype of the $L_{p,w}$ spaces.

Theorem 3 is due to V. I. Gurarii and N. I. Gurarii (1971); our proof follows their lead all the way. As noted in the exercises, the existence of upper and lower l_p-estimates for all normalized basic sequences is a tight restriction indeed. Actually, the restriction is much tighter than one might glean from the exercises; in fact, R. C. James (1972) has shown that in order for the conclusion of Theorem 3 to apply in a Banach space X, it is necessary (and, from Theorem 3, sufficient) that X admit an equivalent uniformly convex norm.

Bibliography

Akimovič, B. A. 1972. On uniformly convex and uniformly smooth Orlicz spaces. *Teor. Funckii Funkcional Anal. & Prilozen*, **15**, 114–120 (Russian).

Altshuler, Z. 1975. Uniform convexity in Lorentz sequence spaces. *Israel J. Math.*, **20**, 260–274.

Altshuler, Z. 1980. The modulus of convexity of Lorentz and Orlicz sequence spaces, *Notes in Banach Spaces*. Austin: University of Texas Press, pp. 359–378.

Arazy, J. 1981. On the geometry of the unit ball of unitary matrix spaces. *Integral Equations and Operator Theory*, **4**, 151–171.

Baernstein, A., II. 1972. On reflexivity and summability. *Studia Math.*, **42**, 91–94.

Beauzamy, B. 1976. Operateurs uniformement convexifiants. *Studia Math.*, **57**, 103–139.

Beauzamy, B. 1979. Banach-Saks properties and spreading models. *Math. Scand.*, **44**, 357–384.

Brunel, A. and Sucheston, L. 1972. Sur quêlques conditions equivalentes à la super-reflexivite dans les espaces de Banach. *C. R. Acad. Sci., Paris*, **275**, 993–994.

Carothers, N. 1981. Rearrangement invariant subspaces of Lorentz function spaces. *Israel J. Math.*, **40**, 217–228.

Clarkson, J. A. 1936. Uniformly convex spaces. *Trans. Amer. Math. Soc.*, **40**, 396–414.

Creekmore, J. 1981. Type and cotype in Lorentz L_{pq} spaces. *Indag. Math.*, **43**, 145–152.

Diestel, J. 1975. *Geometry of Banach Spaces—Selected Topics*, Lecture Notes in Mathematics, Vol. 485. Berlin-Heidelberg-New York: Springer-Verlag.

Dieudonné, J. 1942. La dualite dans les espaces vectoriels topologiques. *Ann. École Norm.*, **59**, 107–132.

Enflo, P. 1972. Banach spaces which can be given an equivalent uniformly convex norm. *Israel J. Math.*, **13**, 281–288.

Figiel, T. 1976. On the moduli of convexity and smoothness. *Studia Math.*, **56**, 121–155.

Figiel, T. and Pisier, G. 1974. Series aleatoires dans les espaces uniformement convexes on uniformement lisses. *C. R. Acad. Sci., Paris*, **279**, 611–614.

Garling, D. J. H. 1978. Convexity, smoothness and martingale inequalities. *Israel J. Math.*, **29**, 189–198.

Goldstine, H. H. 1938. Weakly complete Banach spaces. *Duke Math. J.*, **9**, 125–131.

Gorgadze, Z. G. and Tarieladze, V. I. 1980. On geometry of Orlicz spaces. *Probability Theory on Vector Spaces*, II (Proc. Second International Conf., Blazejewku, 1979), Lecture Notes in Mathematics, Vol. 828. Berlin-Heidelberg-New York: Springer-Verlag, pp. 47–51.

Gurarii, V. I. and Gurarii, N. I. 1971. On bases in uniformly convex and uniformly smooth Banach spaces. *Izv. Akad. Nauk.*, **35**, 210–215.

Halperin, I. 1954. Uniform convexity in function spaces. *Duke Math. J.*, **21**, 195–204.

Hanner, O. 1956. On the uniform convexity of L^p and l^p. *Ark. Mat.*, **3**, 239–244.

James, R. C. 1972. Some self-dual properties of normed linear spaces. *Symposium on Infinite Dimensional Topology, Ann. Math. Stud.*, **69**, 159–175.

James, R. C. 1972. Super-reflexive Banach spaces. *Can. J. Math.*, **24**, 896–904.

James, R. C. 1972. Super reflexive spaces with bases. *Pacific J. Math.*, **41**, 409–419.

Johnson, W. B., Maurey, B., Schechtman, G., and Tzafriri, L. 1979. *Symmetric Structures in Banach Spaces*. American Mathematical Society, Vol. 217.

Kadeč, M. I. 1956. Unconditional convergence of series in uniformly convex spaces. *Uspchi Mat. Nank N. S.*, **11**, 185–190 (Russian).

Kahane, J. P. 1968. *Some Random Series of Functions*. Lexington, Mass.: Heath Mathematical Monographs.

Kakutani, S. 1938. Weak convergence in uniformly convex spaces. *Tohoku Math. J.*, **45**, 188–193.

Kakutani, S. 1939. Weak topology and regularity of Banach spaces. *Proc. Imp. Acad. Tokyo*, **15**, 169–173.

Kamińska, A. 1982. On uniform convexity of Orlicz spaces. *Indag. Math.*, **44**, 27–36.

Kwapien, S. 1972. Isomorphic characterizations of inner product spaces by orthogonal series with vector-valued coefficients. *Studia Math.*, **44**, 583–592.

Lindenstrauss, J. 1963. On the modulus of smoothness and divergent series in Banach spaces. *Mich. Math. J.*, **10**, 241–252.

Lindenstrauss, J. and Tzafriri, L. 1977. *Classical Banach Spaces I. Sequence Spaces*. Ergebnisse der Mathematik und ihrer Grenzgebiete, Vol. 92. Springer-Verlag, Berlin-Heidelberg-New York.

Lindenstrauss, J. and Tzafriri, L. 1979. *Classical Banach Spaces II. Function Spaces*. Ergebnisse der Mathematik und ihrer Grenzgebiete, Vol. 97. Berlin-Heidelberg-New York: Springer-Verlag.

Maleev, R. P. and Troyanski, S. L. 1975. On the moduli of convexity and smoothness in Orlicz spaces. *Studia Math.*, **54**, 131–141.

Maurey, B. and Pisier, G. 1973. Characterisation d'une classe d'espaces de Banach par des proprietes de series aleatoires vectorielles. *C. R. Acad. Sci., Paris*, **277**, 687–690.

Maurey, B. and Pisier, G. 1976. Series de variables aleatories vectorielles independants et proprietes geometriques des espaces de Banach. *Studia Math.*, **58**, 45–90.

McCarthy, C. A. 1967. C_p. *Israel J. Math.*, **5**, 249–271.

McShane, E. J. 1950. Linear functionals on certain Banach spaces. *Proc. Amer. Math. Soc.*, **11**, 402–408.

Milman, D. P. 1938. On some criteria for the regularity of spaces of the type (B). *Dokl. Akad. Nauk. SSSR*, **20**, 243–246.

Milman, V. D. 1970. Geometric theory of Banach spaces, I: Theory of basic and minimal systems. *Russian Math. Surv.*, **25**, 111–170.

Milman, V. D. 1971. Geometric theory of Banach spaces, II: Geometry of the unit sphere. *Russian Math. Surv.*, **26**, 79–163.

Milnes, H. W. 1957. Convexity of Orlicz spaces. *Pacific J. Math.*, **7**, 1451–1486.

Nishiura, T. and Waterman, D. 1963. Reflexivity and summability. *Studia Math.*, **23**, 53–57.

Nordlander, G. 1960. The modulus of convexity in normed linear spaces. *Ark. Mat.*, **4**, 15–17.

Partington, J. R. 1977. On the Banach-Saks property. *Math. Proc. Cambridge Phil. Soc.*, **82**, 369–374.

Pettis, B. J. 1939. A proof that every uniformly convex space is reflexive. *Duke Math. J.*, **5**, 249–253.

Pisier, G. 1973. Sur les espaces de Banach qui ne contiennent pas uniformement de l_n^1. *C. R. Acad. Sci., Paris*, **277**, 991–994.

Pisier, G. 1975. Martingales with values in uniformly convex spaces. *Israel J. Math.*, **20**, 326–350.

Plant, A. T. 1981. A note on the modulus of convexity. *Glasgow Math. J.*, **22**, 157–158.

Ruston, A. F. 1949. A note on convexity of Banach spaces. *Proc. Cambridge Phil. Soc.*, **45**, 157–159.

Seifert, C. J. 1978. The dual of Baernstein's space and the Banach-Saks property. *Bull. Acad. Polon. Sci.*, **26**, 237–239.

Šmulian, V. L. 1941. Sur la structure de la sphere unitaire dans l'espace de Banach. *Mat. Sb.*, **9**, 545–561.

Tomczak-Jaegermann, N. 1974. The moduli of smoothness and convexity and the Rademacher average of trace class S_p. *Studia Math.*, **50**, 163–182.

Van Dulst, D. 1978. *Reflexive and Superreflexive Banach Spaces*. Amsterdam: Mathematical Centre Tracts, Volume 102.

CHAPTER IX

Extremal Tests for Weak Convergence of Sequences and Series

This chapter has two theorems as foci. The first, due to the enigmatic Rainwater, states that for a bounded sequence (x_n) in a Banach space X to converge weakly to the point x, it is necessary and sufficient that $x^*x = \lim_n x^*x_n$ hold for each extreme point x^* of B_{X^*}. The second improves the Bessaga-Pelczynski criterion for detecting c_0's absence; thanks to Elton, we are able to prove that in a Banach space X without a copy of c_0 inside it, any series $\Sigma_n x_n$ for which $\Sigma_n |x^*x_n| < \infty$ for each extreme point x^* of B_{X^*} is unconditionally convergent.

The inclusion of these results provides us the opportunity to present the geometric background that allows their proof. This is an opportunity not to be missed! The Krein-Milman theorem and its converse due to Milman, Bauer's characterization of extreme points, and Choquet's integral representation theorem are all eye-opening results. Each contributes to the proof of Rainwater's theorem.

The approach to Elton's theorem requires a discussion of some of the most subtle yet enjoyable developments in geometry witnessed in the recent past. Our presentation is based on the Bourgain-Namioka "Superlemma." From it we derive another result of Bessaga and Pelczynski, this one to the effect that in separable duals, closed bounded convex sets always have extreme points. Using Choquet's theorem and the Bochner integral, we then show that dual balls with a norm separable set of extreme points are norm separable. This arsenal stockpiled, we describe the delightful arguments of Fonf that serve as a necessary but engaging prelude to the proof of Elton's theorem.

Rainwater's Theorem

Our interests in representation theory are quite mundane. We want to be able to *test* convergence in the weak topology with but a minimum of muss or fuss; more particularly, we want to be able to *test weak convergence of sequences* by the most economical means available and integral representation theory opens several avenues of approach to such possibilities.

For the remainder of this section, E will be a real locally convex Hausdorff linear topological space with topological dual E^*.

Before starting with the famous theorem of Krein and Milman, we recall that a point x of a convex set K is called *extreme* if x cannot be written as a convex combination $\lambda y + (1-\lambda)z$, $0 \le \lambda \le 1$, of two distinct points y, z of K.

The Krein-Milman Theorem. *Let K be a nonempty compact convex subset of E. Then K has extreme points and is, in fact, the closed convex hull of its extreme points.*

PROOF. We start by introducing the notion of "extremal subset." A subset A of a convex set B is extremal in B if A is a nonvoid convex subset of B with the property that should $x, y \in B$ and $\lambda x + (1-\lambda)y \in A$ for some $0 < \lambda < 1$, then $x, y \in A$. Of course, an extremal set with but one element consists of an extreme point. Naturally, we are looking for small extremal subsets of K.

Let ξ be the collection of all nonempty closed extremal subsets of K (plainly, $K \in \xi$); order ξ by $K_1 \le K_2$ whenever $K_2 \subseteq K_1$. The compactness of K along with a bit of judicious Zornication produces a maximal $K_0 \in \xi$. We claim K_0 is a singleton. Indeed, if $x, y \in K_0$ are distinct, there is a linear continuous f on E with $f(x) < f(y)$. $K_0 \cap \{z : f(z) = \max f(K_0)\}$ is then a proper closed extremal subset of K_0, a contradiction.

K has extreme points.

Let C be the closed convex hull of the set of extreme points of K. Can $K \setminus C$ have any points? Well, if $x \in K \setminus C$, then there is a linear continuous functional f on E such that $\max f(C) < f(x)$. Looking at $\{z \in K : f(z) = \max f(K)\}$, we should see a closed extremal subset of K which entirely misses C. On the other hand, each closed extremal subset of K contains an extreme point on K, doesn't it? This completes our proof. □

Suppose C is a compact subset of E and let μ be a regular Borel probability measure on C. We say a point x of E *is represented by* μ (or *is the barycenter of* μ) if for each $f \in E^*$ we have

$$f(x) = \int_C f(c)\, d\mu(c).$$

To be sure of our footing, we prove that every regular Borel probability measure on a reasonable compact set has a barycenter.

Theorem 1. *Suppose the closed convex hull K of a closed set $F(\subseteq E)$ is compact. Then each regular Borel probability measure μ on F has a unique barycenter in K.*

PROOF. The restriction $f|_F$ of any $f \in E^*$ to F is plainly μ-integrable for any regular Borel probability measure μ on F. Take any such μ. We claim that the hyperplanes

$$\left\{x \in E: f(x) = \int_F f\, d\mu\right\} = E_f,\ f \in E^*$$

intersect K in a common point.

Since K is compact, we need only show that given $f_1, \ldots, f_n \in E^*$, then

$$K \cap E_{f_1} \cap E_{f_2} \cap \cdots \cap E_{f_n} \neq \varnothing.$$

From this the existence of a barycenter (for μ) in K follows. Consider the operator $T: E \to \mathbb{R}^n$ given by $Ty = (f_1(y), \ldots, f_n(y))$; TK is compact and convex, T being linear and continuous. Should $(\int_F f_1\, d\mu, \ldots, \int_F f_n\, d\mu)$ not belong to TK, then there would be a functional $a = (a_1, \ldots, a_n) \in \mathbb{R}^{n*} = \mathbb{R}^n$ such that

$$\sup\{a \cdot Ty : y \in K\} < a \cdot \left(\int_F f_1\, d\mu, \ldots, \int_F f_n\, d\mu\right).$$

Let $g = \sum_{i=1}^n a_i f_i$. Then

$$\begin{aligned}\sup\{g(y) : y \in K\} &= \sup\left\{\sum_{i=1}^n a_i f_i(y) : y \in K\right\}\\ &= \sup\{a \cdot Ty : y \in K\}\\ &< a \cdot \left(\int_F f_1\, d\mu, \ldots, \int_F f_n\, d\mu\right)\\ &= \sum_{i=1}^n a_i \int_F f_i\, d\mu = \int_F \sum_{i=1}^n a_i f_i\, d\mu = \int_F g\, d\mu\\ &\leq \sup\{g(y) : y \in F\}\\ &\leq \sup\{g(y) : y \in K\};\end{aligned}$$

this is a contradiction, and the proof is complete. □

Uniqueness of barycenters is, or ought to be, obvious.

With an eye to Banach spaces it ought to be recalled that whether you are looking at a Banach space in its norm topology, a Banach space in its weak topology, or the dual of a Banach space in its weak* topology—in each case the fact is that the closed convex hull of a compact set is compact, too.

Being the barycenter of a regular Borel probability measure that lives on a given compact set C in E means being some sort of average of points of C. More precisely, we have the following theorem.

Theorem 2. *For any compact subset C in E, a point x of E belongs to the closed convex hull of C if and only if there exists a regular Borel probability measure μ on C whose barycenter (exists and) is x.*

Proof. If μ is a regular Borel probability measure on C and has x as its barycenter, then for any $f \in E^*$ we know

$$f(x) = \int_C f\,d\mu \leq \sup f(C) \leq \sup f(\overline{\mathrm{co}}C).$$

Were x not in $\overline{\mathrm{co}}C$, there would be an $f \in E^*$ violating $f(x) \leq \sup f(\overline{\mathrm{co}}C)$.

Conversely, suppose $x \in \overline{\mathrm{co}}(C)$. Then there is a net $(\sigma_d)_{d \in D}$ of members of $\mathrm{co}(C)$ converging to x. Each σ_d is of the form

$$\sigma_d = \sum_i a_i^d y_i^d \quad \text{(finite sum)},$$

where $a_i^d > 0$, $\Sigma_i a_i^d = 1$, and $y_i^d \in C$. Let μ_d be the regular Borel probability measure

$$\mu_d = \sum_i a_i^d \delta_{y_i^d},$$

where $\delta_c \in C(C)^*$ is the usual evaluation functional at $c \in C$. Directed as they are by the same set D as the net $(\sigma_d)_{d \in D}$, the μ_d form a net with values in the weak* compact set $B_{C(C)^*}$ and as such have a convergent subnet $(\mu_s)_{s \in S}$ with a limit μ that is quickly seen to also be a regular Borel probability measure on C. Naturally, x is the barycenter of μ; in fact, if $f \in E^*$ is given, then

$$\begin{aligned} f(x) &= \lim_s f(\sigma_s) \\ &= \lim_s \int_C f(c)\,d\mu_s(c) = \int_C f(c)\,d\mu(c). \end{aligned}$$ □

Now for a real touch of elegance we characterize the extreme points of a compact set by means of their representing measures.

Theorem 3 (Bauer's Characterization of Extreme Points). *Let K be a nonempty compact convex subset of E. A point x of K is an extreme point of K if and only if δ_x is the only regular Borel probability measure on K that represents x.*

Proof. If $x \in K$ is not an extreme point, then there are $y, z \in K$ with $y \neq z$ so that $x = \frac{1}{2}y + \frac{1}{2}z$. Plainly, $\frac{1}{2}\delta_y + \frac{1}{2}\delta_z$ is a regular Borel probability measure on K that represents x and differs from δ_x.

Suppose x is an extreme point of K and μ is a regular Borel probability measure on K that represents x. We claim that $\mu(C) = 0$ for each compact subset C of $K \setminus \{x\}$. The only alternative is that $\mu(C) > 0$ for some compact set $C \subseteq K \setminus \{x\}$. An easy compactness argument shows that there is a point y in this C for which $\mu(U \cap K) > 0$ for each neighborhood U of y in E. Letting U be a closed convex neighborhood of y for which $x \notin U \cap K$, we get a particularly interesting nonempty compact convex proper subset

$U \cap K$ of K. Why is $U \cap K$ of interest? Well, its μ-probability cannot be 1 because μ represents $x \notin U \cap K$, yet its μ-probability is not 0! $0 < \mu(U \cap K) < 1$. If we define μ_1 and μ_2 by

$$\mu_1(B) = \frac{\mu(B \cap U \cap K)}{\mu(U \cap K)}, \qquad \mu_2(B) = \frac{\mu(B \cap (U \cap K)^c)}{1 - \mu(U \cap K)},$$

we get regular Borel probability measures on K. Let x_1 be the barycenter of μ_1 and x_2 be the barycenter of μ_2. Each of x_1 and x_2 belongs to K; x_1 is in $U \cap K$ and so is not x. On the other hand,

$$\mu = \mu(U \cap K)\mu_1 + (1 - \mu(U \cap K))\mu_2,$$

forcing

$$x = \mu(U \cap K)x_1 + (1 - \mu(U \cap K))x_2,$$

which is a contradiction. □

Corollary 4 (Milman's Converse to the Krein-Milman Theorem). *Let K be a compact convex subset of E. If K is the closed convex hull of a set Z, then every extreme point of K lies in Z's closure.*

PROOF. Suppose x is an extreme point of $K = \overline{\text{co}}(\overline{Z})$. Then x is the barycenter of a regular Borel probability measure μ that lives on $\overline{Z}$ (Theorem 2). We can extend μ to all of K by making $\mu(B) = \mu(B \cap \overline{Z})$ (it is plain that this makes sense) for Borel sets $B \subseteq K$. The resulting measure still represents x. But now Bauer's theorem enters the foray to tell us that μ must be δ_x; since μ is supported by $\overline{Z}$, it follows that $x \in \overline{Z}$. □

We start now on our way to Choquet's theorem.

For a compact convex subset K of E we denote by $A(K)$ the space of all affine continuous real-valued functions defined on K; $f \in C(K)$ is *affine* if $f(tx + (1-t)y) = tf(x) + (1-t)f(y)$ for all $x, y \in K$ and all t, $0 \le t \le 1$. $A(K)$ is a closed linear subspace of $C(K)$ whose members separate the points of K. Among the members of $A(K)$ the discerning viewers will surely find the constants.

Let $f \in C(K)$ and define $\bar{f}: K \to (-\infty, \infty)$ by

$$\bar{f}(x) = \inf\{h(x) : h \in A(K), f \le h\}.$$

Lemma 5. *For $f, g \in C(K)$ we have*

1. *$\bar{f}$ is a concave, bounded, upper semicontinuous function on K; hence $\bar{f}$ is Borel measurable and universally integrable on K with respect to the class of all regular Borel measures on K.*
2. *$f \le \bar{f}$.*
3. *$f = \bar{f}$ if and only if f is concave.*
4. *$\overline{f+g} \le \bar{f} + \bar{g}$, but $\overline{f+g} = \bar{f} + g$, if $g \in A(K)$, and $\overline{rf} = r\bar{f}$, if $0 \le r < \infty$.*

PROOF. Parts 1 and 2 are plain, simple calculations and are corollaries to well-known facts.

Part 3 is not so direct. Suppose f is concave. Let $K_f = \{(x, r) \in Kx(-\infty, \infty): f(x) \geq r\}$. K_f is closed and convex since f is continuous and concave. Suppose there is an $x_0 \in K$ such that $f(x_0) < \bar{f}(x_0)$. It follows that there is a real-linear continuous functional λ on $E \times \mathbb{R}$ such that

$$\sup \lambda(K_f) < \lambda_0 < \lambda((x_0, \bar{f}(x_0)))$$

for some fixed real value λ_0 of λ. In particular,

$$\lambda((x_0, f(x_0))) < \lambda((x_0, \bar{f}(x_0))).$$

It follows that

$$0 < \lambda((0, \bar{f}(x_0) - f(x_0)))$$

and from this that for any $a > 0$,

$$0 < \lambda((0, a)).$$

Of course, from this we see that for any $x \in K$,

$$\lambda((x, t)) \to \pm\infty \text{ as } t \to \pm\infty.$$

The continuity of $\lambda((x, \cdot))$ for each $x \in K$ now tells us that given $x \in K$ there is an $r_x \in (-\infty, \infty)$ such that

$$\lambda((x, r_x)) = \lambda_0.$$

Notice that

$$\lambda((x, r)) = \lambda((x, r'))$$

if and only if

$$0 = \lambda((0, r - r')) = (r - r')\lambda((0, 1)),$$

which, in light of the fact that $\lambda((0,1)) > 0$, holds if and only if

$$r = r'.$$

It follows that the association $x \to r_x$ is a well-defined function $h: K \to R$. We claim for h the following:

a. $f < h$.
b. $h(x_0) < \bar{f}(x_0)$.
c. $h \in A(K)$, i.e., h is affine and continuous.

Of course, a, b, and c together contradict the definition of $\bar{f}$ and compel us to reject any alternative to $f = \bar{f}$.

a. Take $x \in K$. Then $(x, f(x)) \in K_f$. Thus,

$$\lambda((x, f(x)) < \lambda_0 = \lambda((x, h(x)));$$

so

$$0 < \lambda((0, h(x) - f(x))) = (h(x) - f(x))\lambda((0, 1)).$$

b. Similarly,

$$\lambda((x_0, h(x_0))) = \lambda_0 < \lambda((x_0, \bar{f}(x_0)));$$

so

$$0 < (\bar{f}(x_0) - h(x_0))\lambda((0,1)).$$

c. If $x, y \in K$ and $0 \le t \le 1$, then

$$\begin{aligned}\lambda((tx+(1-t)y, th(x)+(1-t)h(y))) &= t\lambda((x, h(x))) \\ &\quad + (1-t)\lambda((y, h(y))) \\ &= t\lambda_0 + (1-t)\lambda_0 = \lambda_0 \\ &= \lambda((tx+(1-t)y, h(tx+(1-t)y))).\end{aligned}$$

As in a and b, we can conclude that

$$0 = \lambda((0,1))(th(x)+(1-t)h(y) - [h(tx+(1-t)y)]),$$

and the affinity of h is established.

Finally, h is continuous. Let $(x_d)_D$ be a net in K converging to x. Let $r_d = h(x_d)$ and $r = h(x)$. Notice that (r_d) is a bounded net of reals. In fact, otherwise, there'd be a subsequence (r_{d_n}) such that $|r_{d_n}| \to \infty$. From this we see that

$$0 = \lim_n \frac{\lambda_0}{r_{d_n}^2} = \lim_n \lambda\left(\left(\frac{x_{d_n}}{r_{d_n}}, 1\right)\right) = \lambda((0,1)) > 0$$

The boundedness of (r_d) implies that any subset has a further subnet that converges; if (r_{d_p}) is a subnet of (r_d), then there is a subnet $(r_{d_q})_Q$ that converges to some real r_0. Now,

$$\begin{aligned}\lambda((x, r)) &= \lambda((x, h(x))) = \lambda_0 \\ &= \lambda_0 = \lambda((x_{d_q}, h(x_{d_q}))) \\ &= \lambda((x_{d_q}, r_{d_q})) \to \lambda((x, r_0)).\end{aligned}$$

$r = r_0$ and $h(x_{d_q}) \to h(x)$. The continuity of h is established.

Part 4 involves some relatively straightforward computations which are just as well left to the reader's diligence. □

Lemma 6. *Let K be a nonempty compact convex* metrizable *subset of E. Then $C(K)$ contains a strictly convex member.*

Proof. The metrizability of K ensures the separability of $C(K)$ and hence that of $A(K)$. Let (h_n) be a dense sequence in $S_{A(K)}$; define $h = \Sigma_n h_n^2/2^n$. The M-test assures us that $h \in C(K)$. Plainly, h is convex. In fact, h is strictly convex. Indeed, if $x, y \in K$ and $x \neq y$, then there is an n so that $h_n(x) \neq h_n(y)$—remember $A(K)$ separates the points of K. If we now consider $0 < t < 1$ and let $s = 1 - t$, then

$$\begin{aligned} h_n^2(tx + sy) &= t^2h_n^2(x) + s^2h_n^2(y) + 2sth_n(x)h_n(y) \\ &= th_n^2(x) + sh_n^2(y) - st\left[h_n(x) - h_n(y)\right]^2 \\ &< th_n^2(x) + sh_n^2(y). \end{aligned}$$

It follows that h too satisfies the strict inequality

$$h(tx + sy) < th(x) + sh(y),$$

i.e., h is strictly convex. □

We are ready for the real highlight of this section: Choquet's theorem. It is from this remarkable theorem that we derive the result of Rainwater.

Choquet's Integral Representation Theorem. *Let K be a nonempty compact convex metrizable subset of a locally convex Hausdorff space E. Then each point of K is the barycenter of a regular Borel probability measure that is concentrated on the extreme points of K.*

More precisely, if $x \in K$, then there is a regular Borel probability measure μ defined on K for which μ (extreme points of K) $= 1$ and for which given any $f \in A(K)$,

$$f(x) = \int_K f(k)\, d\mu(k).$$

PROOF. First things first. The set of extreme points of K is a Borel set. In fact, the complement of this set is easily seen to be

$$\bigcup_{n=1}^{\infty} \left\{ \tfrac{1}{2}y + \tfrac{1}{2}z = x : y, z \in K, d(x, y), d(x, z) \geq \frac{1}{n} \right\},$$

where d is a metric generating K's topology. The point of this remark should be well-taken: The set of extreme points of K is a $\mathscr{G}_\delta$-set, and so μ(extreme points of K) $= 1$ makes sense for any Borel measure μ.

Now on to the proof proper.

Let $x \in K$, and let $h \in C(K)$ be strictly convex. Define F_x: linear span $\{A(K), h\} \to (-\infty, \infty)$ by

$$F_x(a + th) = a(x) + t\bar{h}(x).$$

Clearly F_x is linear on its indicated domain.

Define $p : C(K) \to (-\infty, \infty)$ by

$$p(f) = \bar{f}(x).$$

p is a sublinear, positively homogeneous functional on $C(K)$.

Claim. p dominates F_x on the linear span of $A(K)$ and h.

To see the claim's basis, look at a vector $a + th$. If $t \geq 0$, then $F_x(a + th) = a(x) + t\bar{h}(x) = \overline{a + th}(x) = p(a + th)$. If $0 > t$, then $a + th$ is concave; so

$F_x(a + th) = a(x) + t\bar{h}(x) \leq a(x) + th(x) = \overline{a + th}(x) = p(a + th)$. Either way the claim is well-founded.

The Hahn-Banach theorem now lets us extend F_x to all of $C(K)$ keeping the domination by p as a control. Call the extension F_x, too, and study it for a bit. First, note that if $g \in C(K)$ and $g \geq 0$, then $-g \leq 0$ so that

$$-F_x(g) = F_x(-g) \leq p(-g) = \overline{-g}(x) \leq 0,$$

and $F_x(g) \geq 0$. F_x is a positive linear functional on $C(K)$: F_x is represented by a positive regular Borel measure μ on K. Since $F_x(1) = 1$, the measure μ is a probability measure. Of course, μ represents x. In fact, if $f \in A(K)$, then

$$f(x) = F_x(f) = \int_K f(k)\, d\mu(k).$$

It remains to be seen that μ (extreme points of K) $= 1$. This we do in two steps:

I. $\int h(k)\, d\mu(k) = \int \bar{h}(k)\, d\mu(k)$.
II. $\{x \in K : h(x) = \bar{h}(x)\}$ consists of only extreme points of K.

I. $\bar{h}(x) = F_x(h) = \int h\, d\mu \leq \int \bar{h}\, d\mu \leq \int a\, d\mu$ for all $a \in A(K)$ such that $h \leq a$. It follows that for each such a,

$$\bar{h}(x) = \int h\, d\mu \leq \int \bar{h}\, d\mu \leq \int a\, d\mu = F_x(a) = \bar{a}(x) = a(x),$$

and so by definition of $\bar{h}$ we get all the quantities to the left of $\int a\, d\mu$ the same, including $\int h\, d\mu$ and $\int \bar{h}\, d\mu$.

II. If x is a nonextreme point of K, then there are distinct points $y, z \in K$ such that $x = \frac{1}{2}y + \frac{1}{2}z$. Since h is strictly convex,

$$\begin{aligned} h(x) = h\left(\frac{y}{2} + \frac{z}{2}\right) &< \frac{h(y)}{2} + \frac{h(z)}{2} \\ &\leq \frac{\bar{h}(y)}{2} + \frac{\bar{h}(z)}{2} \leq \bar{h}\left(\frac{y+z}{2}\right) = \bar{h}(x). \end{aligned}$$

This completes the proof. □

Now as a corollary to the Choquet theorem, we present Rainwater's theorem.

Rainwater's Theorem. *Let X be a Banach space and (x_n) be a bounded sequence in X. Then in order that (x_n) converge weakly to $x \in X$, it is both necessary and sufficient that $x^*x = \lim_n x^*x_n$ holds for each extreme point x^* of B_{X^*}.*

Proof. We take two small steps before arriving at the finish. Before the first, we notice that the theorem need only be proved for real Banach spaces X.

Our first step entails proving the theorem in case X is a *separable* Banach space. For such a space, B_{X^*} is weak* compact, convex, *and* metrizable. Therefore, we are set up for Choquet's theorem should we find a way to use it—and be sure we will! Suppose (x_n) is a bounded sequence in X such that $x^*x = \lim_n x^*x_n$ holds for each extreme point x^* of B_{X^*}, where x is the hoped-for weak limit of (x_n). Take any $x^* \in B_{X^*}$. Then Choquet's theorem gives us a regular Borel probability measure μ on B_{X^*}(weak*) such that

$$a(x^*) = \int_{\text{ex } B_{X^*}} a(y^*)\, d\mu(y^*)$$

for each $a \in A(B_{X^*}(\text{weak}^*))$ where ex B_{X^*} denotes the set of extreme points of B_{X^*}. Viewing members of X as being in $A(B_{X^*}(\text{weak}^*))$, we get

$$xx^* = \int_{\text{ex } B_{X^*}} x(y^*)\, d\mu(y^*) = \int_{\text{ex } B_{X^*}} \lim_n x_n(y^*)\, d\mu(y^*)$$

$$= \lim_n \int_{\text{ex } B_{X^*}} x_n(y^*)\, d\mu(y^*) = \lim_n x_n(x^*)$$

by the boundedness of (x_n) and Lebesgue's bounded convergence theorem. It follows that x is the weak limit of (x_n).

For a general Banach space X, we suppose (x_n) is a bounded sequence in X and x is an element of X such that

$$\lim_n x^*x_n = x^*x$$

holds for every extreme point x^* of B_{X^*}. Let X_0 be the closed linear span of $\{x_n : n \geq 1\} \cup \{x\}$. Then X_0 is a separable Banach space. We claim that x is the weak limit of (x_n) in X_0; once verified, the Hahn-Banach theorem assures us that (x_n) converges to x weakly in X, too. To show that x is the weak limit of (x_n) in X_0, we will show that $y^*x = \lim_n y^*x_n$ for each extreme point y^* of $B_{X_0^*}$ and then apply the known verity of the theorem for separable spaces.

Well, take any extreme point y^* of $B_{X_0^*}$. Let $HB(y^*)$ denote the set of all $x^* \in B_{X^*}$ such that $x^*|_{X_0} = y^*$. It is easy to see that $HB(y^*)$ is a nonempty convex weak* compact subset of B_{X^*}; furthermore, since y^* is an extreme point of $B_{X_0^*}$, $HB(y^*)$ is an extremal subset of B_{X^*}. It follows that $HB(y^*)$ contains some extreme point z^* of B_{X^*}; of course, now we know that

$$y^*x = z^*x = \lim_n z^*x_n = \lim_n y^*x_n. \qquad \square$$

rollary. *A bounded sequence (x_n) in the Banach space X is weakly Cauchy and only if* $\lim_n x^*x_n$ *exists for each extreme point x^* of B_{X^*}.*

OOF. It need only be remarked that a sequence (x_n) is weakly Cauchy if d only if given increasing sequences (k_n) and (j_n) of positive integers the sequence $(x_{k_n} - x_{j_n})$ is weakly null. In light of Rainwater's theorem, this remark is enough to prove the corollary.

Elton's Theorem

Rainwater's theorem gives a strong hint of the control extreme points of a dual ball exercise on weak convergence. There is a corresponding result in the theory of series due to John Elton. It can be formulated as follows: for a Banach space X to be without c_0 subspaces it is necessary and sufficient that $\Sigma_n x_n$ converges whenever $\Sigma_n |x^* x_n| < \infty$ for each extreme point x^* of B_{X^*}. The purpose of this section is to prove this elegant result of Elton.

We start the section with a treatment of the Superlemma. Though we need only the weak* version of this stunning geometric fact, a complete exposition hurts no one. We then apply the Superlemma to derive a theorem of Bessaga and Pelczynski to the effect that in separable dual spaces, closed bounded convex sets have extreme points; here we follow Isaac Namioka's lead. This having been done, we supply a natural criterion for the dual of a Banach space to be separable, namely, that the dual ball have a norm-separable set of extreme points; Choquet's theorem plays an important role here. After all the groundwork has been laid, we pass to a proof of Fonf's theorem: whenever the dual of a Banach space has only countably many extreme points the space is c_0 rich. From here it is clear (though not easy) sailing to Elton's theorem.

We start with a lemma discovered initially in its second, or weak*, version by Isaac Namioka and sharpened by Jean Bourgain. This mild-looking lemma of Namioka and Bourgain is known to its public as "Superlemma"!

A *slice* of a set is the intersection of the set with a half-plane.

Superlemma. *Let C, C_0, and C_1 be closed bounded convex subsets of the Banach space X and let $\varepsilon > 0$. Suppose that*

1. *C_0 is a subset of C having diameter $< \varepsilon$.*
2. *C is not a subset of C_1.*
3. *C is a subset of $\overline{\text{co}}(C_0 \cup C_1)$.*

Then there is a slice of C having diameter $< \varepsilon$ that intersects C_0.

PROOF. For $0 \leq r \leq 1$ define

$$D_r = \{(1-\lambda)x_0 + \lambda x_1 : r \leq \lambda \leq 1, x_0 \in C_0, x_1 \in C_1\}.$$

Each D_r is convex, $\overline{D}_0$ contains C—this is just 3—and $D_1 = C_1$.

Notice that for $r > 0$, $\overline{D}_r$ does *not* contain C. To wit: since we've supposed $C \not\subseteq C_1$, there must be an $x^* \in X^*$ such that

$$\sup x^* C_1 < \sup x^* C;$$

were $C \subseteq \overline{D}_r$ $(r > 0)$, then we would have

$$\begin{aligned}\sup x^*C &\le \sup x^*\overline{D}_r \\ &= \sup x^*D_r \\ &\le (1-r)\sup x^*C_0 + r\sup x^*C_1 \\ &\le (1-r)\sup x^*C + r\sup x^*C_1,\end{aligned}$$

which leads to the conclusion that

$$\sup x^*C \le \sup x^*C_1.$$

Now notice that $C\setminus\overline{D}_r \subseteq \overline{D}_0\setminus\overline{D}_r$ and $D_0\setminus\overline{D}_r$ is dense in $\overline{D}_0\setminus\overline{D}_r$. Take $x \in D_0\setminus\overline{D}_r$. Then x is in D_0; so x is a convex combination $(1-\lambda)x_0 + \lambda x_1$, where $x_0 \in C_0$ and $x_1 \in C_1$. x is not in D_r; so $0 \le \lambda < r$. It follows that $\|x - x_0\| = \lambda\|x_0 - x_1\| < r\sup\{\|y - z\|: y \in C_0, z \in C_1\} = r\delta$. But now observe that any y in $C\setminus\overline{D}_r$ can be approximated by an x in $D_0\setminus\overline{D}_r$ as closely as you please; each such x is itself within $r\delta$ of a point in C_0. The upshot of this is that the diameter of $C\setminus\overline{D}_r$ is $\le 2r\delta + \operatorname{diam} C_0$.

If we choose $r > 0$ to be very small indeed, then $2r\delta + \operatorname{diam} C_0 < \varepsilon$; now the fact that $C\setminus\overline{D}_r$ is nonvoid allows us to pick a slice of C disjoint from $\overline{D}_r$. Since C_1 is a subset of D_r, $C_0\setminus\overline{D}_r$ is nonempty; so we can even pick our slice of C to contain a given point of $C_0\setminus\overline{D}_r$. Let it be done. □

Of great value in studying duals is the following:

Superlemma (Weak* Version). *Let K, K_0 and K_1 be weak* compact convex subsets of X^* and let $\varepsilon > 0$. Suppose that*

1. *K_0 is a subset of K having diameter $< \varepsilon$.*
2. *K is not a subset of K_1.*
3. *K is a subset of $\operatorname{co}(K_0 \cup K_1)$.*

Then there is a weak slice of K of diameter $< \varepsilon$ that intersects K_0.*

The proof of the weak* version of the Superlemma is virtually identical with that of the Superlemma itself; certain minor modifications need to be made. These are that the sets D_r are of the form

$$D_r = \{(1-\lambda)x_0^* + \lambda x_1^* : r \le \lambda \le 1, x_0^* \in K_0, x_1^* \in K_1\}.$$

The D_r are weak* compact and convex with $D_0 = \operatorname{co}(K_0 \cup K_1)$; (3) tells us that D_0 contains K. Now when we separate K from K_1, we can do so with a weak* continuous linear functional, if we wish. In any case, the end result is the same: for $r > 0$, K is not a subset of D_r. Next, a computation [here things are a bit quicker because $K \subseteq \operatorname{co}(K_0 \cup K_1)$]. As in the proof of the Superlemma, we see that the diameter of $K\setminus D_r$ is strictly less than $2r\sup\{\|u_0^* - u_1^*\|: u_0^* \in K_0, u_1^* \in K_1\} + \operatorname{diam} K_0$; on choosing r very small, we arrange things so that $K \setminus D_r$ has diameter $< \varepsilon$. K_1 is a subset of D_r so $K_0 \setminus D_r$ is nonempty. Taking a point x_0^* of $K_0 \setminus D_r$ and slicing in a weak* continuous fashion by the separation theorem, we obtain a weak* slice of K that contains x_0^* and is contained in $K \setminus D_r$. This is the slice we want!

Theorem 7. *Let X be a separable Banach space with separable dual X^*. Then the identity map id_K on K is weak*-norm continuous at a weak* dense $\mathcal{G}_\delta$ set of points of K whenever K is a weak* compact subset of X^*.*

Proof. For each $\varepsilon > 0$ let A_ε be the union of all $W \cap K$, where W is a weak* open set in X^* for which the norm diameter of $W \cap K$ is $\leq \varepsilon$. Plainly each A_ε is weak* open in K. Moreover, the points of weak*-norm continuity of id_K are exactly those of $\cap_n A_{1/n}$. Should we show that each A_ε is weak* dense in K, then Baire's category theorem will let us conclude that $\cap_n A_{1/n}$ is weak* dense, too, and, of course, a weak* $\mathcal{G}_\delta$ in K.

X^* is separable; so we can find a sequence $(x_n^*(\varepsilon))$ in X^* such that

$$K = \bigcup_n \left(K \cap \left(x_n^* + \frac{\varepsilon}{2} B_{X^*} \right) \right).$$

Each of the sets $K \cap (x_n^* + (\varepsilon/2) B_{X^*})$ is weak* closed; hence, Baire's category theorem assures us that if we let W_n be the relative weak* interior of $K \cap (x_n^* + (\varepsilon/2) B_{X^*})$ in K, then $\cup_n W_n$ is weak* dense in K. Since the W_n clearly have norm diameter $\leq \varepsilon$, they are among those sets that go into making A_ε what it is, which, in part, is weak* dense in K. □

Theorem 7 is due to Isaac Namioka and so is Theorem 8.

Theorem 8. *Let X be a separable Banach space with separable dual X^*. Let K be a weak* compact convex subset of X^*. Then the set of points of weak*-norm continuity of the identity map id_K of K meets the set $\mathrm{ext}\, K$ of extreme points of K in a set that is a dense $\mathcal{G}_\delta$-subset of $(\mathrm{ext}\, K, \text{weak}^*)$.*

Proof. We already know from Theorem 7 that id_K has lots of points of weak*-norm continuity in (K, weak^*)—a dense $\mathcal{G}_\delta$-set of them in fact. In proving the present result, we will follow the lead of the proof of Theorem 7 and apply the weak* version of Superlemma to pull us through any difficulties encountered.

To start with, for each $\varepsilon > 0$ let B_ε be the set

$$\{ u^* \in \mathrm{ext}\, K : u^* \text{ has a neighborhood } W^* \text{ in } (K, \text{weak}^*) \text{ with } \|\cdot\| \text{ diam} < \varepsilon \}.$$

Each B_ε is open in $(\mathrm{ext}\, K, \text{weak}^*)$; we claim that each is dense therein as well.

Let W^* be a weak* open subset of X^* that intersects $\mathrm{ext}\, K$; of course $W^* \cap \overline{\mathrm{ext}\, K}^{\text{weak}^*}$ is nonempty, too. By Theorem 7, the set of points of weak*-norm continuity of $\mathrm{id}_{\overline{\mathrm{ext}\, K}^{\text{weak}^*}}$ is a dense $\mathcal{G}_\delta$-subset of

$(\overline{\text{ext}\, K}^{\text{weak}^*}, \text{weak}^*)$. It follows that there must be a weak* open set V^* in X^* such that V^* intersects $\overline{\text{ext}\, K}^{\text{weak}^*}$ in a nonvoid subset of $W^* \cap \overline{\text{ext}\, K}^{\text{weak}^*}$ having norm diameter $\leq \varepsilon/2$, say. Define K_0 and K_1 as follows:

$$K_0 = \text{weak* closed convex hull of } V^* \cap \overline{\text{ext}\, K}^{\text{weak}^*}$$

and

$$K_1 = \text{weak* closed convex hull of } \overline{\text{ext}\, K}^{\text{weak}^*} \setminus V^*.$$

K_0 is weak* compact and convex and so is K_1. The norm diameter of K_0 is $\leq \varepsilon/2$, and K_0 is contained in K. Since the set $\overline{\text{ext}\, K}^{\text{weak}^*} \setminus V$ is weak* closed in K_1, it is weak* compact; Milman's theorem alerts us to the fact that $\text{ext}\, K_1 \subseteq \overline{\text{ext}\, K}^{\text{weak}^*} \setminus V^*$. On the other hand, K is the convex hull of $K_0 \cup K_1$, and V^* does intersect $\overline{\text{ext}\, K}^{\text{weak}^*}$. Hence K_1 does not contain K, and the stage has been set for the entrance of Superlemma. On cue Superlemma produces a weak* slice S^* of K having norm diameter $< \varepsilon$ that intersects K_0 and misses K_1. S^* contains a point u^* of $\text{ext}\, K$ in its weak* interior. Since S^* has norm diameter $< \varepsilon$, we know that $u^* \in B_\varepsilon$. Finally, notice that $\overline{\text{ext}\, K}^{\text{weak}^*} \setminus V^* \subseteq K_1$, a set disjoint from S^*; therefore, $u^* \in \overline{\text{ext}\, K}^{\text{weak}^*} \cap V^* \subseteq W^*$, and so $u^* \in W^*$, too. $u^* \in B_\varepsilon \cap W^*$ and B_ε is dense in $\overline{\text{ext}\, K}^{\text{weak}^*}$.

Naturally, the points of weak*-norm continuity of id_K inside $\text{ext}\, K$ are precisely those points that find themselves in $\bigcap_n B_{1/n}$. It suffices, therefore, to note that in the weak* topology $\text{ext}\, K$ is a Baire space. Why is this so? Well, X is separable so weak* compact subsets of X^* are weak* metrizable. Further, we saw in the proof of Choquet's theorem that in a metrizable compact convex set the complement of the set of extreme points is a countable union of closed sets; the set of extreme points is a $\mathcal{G}_\delta$. Of course, $\mathcal{G}_\delta$-subsets of completely metrizable spaces are Baire spaces, as the usual proof of the Baire category theorem so obviously indicates. The proof of Theorem 8 is complete. □

Okay, let X^* be separable and let C be a nonempty closed bounded convex subset of X^*. C's weak* closure K is weak* compact and convex, and Theorem 8 applies to K. Let Z be the set of points of weak*-norm continuity of id_K. Take a $z^* \in Z$. Since C is weak* dense in K, there is a net $(x_d^*)_D$ in C that converges to z^* in the weak* topology—actually we have a sequence in C converging to z^* because K is weak* metrizable. Of course, z^* being a point of weak*-norm continuity assures us that z^* is the norm limit of $(x_d^*)_D$, too. But C is norm closed; so $z^* \in C$.

We have just demonstrated the following.

Theorem 9 (Bessaga-Pelczynski). *In separable dual spaces, nonempty closed bounded convex sets have extreme points.*

Our next stepping stone involves recognizing a separable dual by how many extreme points its dual ball has. Here's the main result.

Theorem 10. *Let X be a separable Banach space and suppose the set* ext B_{X^*} *of extreme points of B_{X^*} is a norm separable set. Then X^* is separable.*

PROOF. Several ingredients provide just the right mix to make a proof to Theorem 10. We present them in more-or-less arbitrary order.

1. Let S be a separable metric space and Y be a Banach space. Suppose $f: S \to Y$ is a continuous bounded function and μ is a probability Borel measure defined on S. Then f is Bochner integrable with respect to μ.

We denote by $C_b(S)$ the Banach space of all continuous bounded real-valued functions defined on the separable metric space S and by $\mathscr{P}(S)$ the convex set of all probability Borel measures defined on S. Take special note of the inclusion of $\mathscr{P}(S)$ within $B_{C_b(S)^*}$, making it natural to consider $\mathscr{P}(S)$ in its weak* topology.

2. Let S be a separable metric space and Y be a Banach space. Suppose $f: S \to Y$ is a continuous bounded function. Then the map $I_f: \mathscr{P}(S) \to Y$ given by $I_f(\mu) =$ Bochner $\int f \, d\mu$ is weak*-weak continuous.

In fact to show I_f is weak*-weak continuous, it suffices to show that $y^* I_f$ is weak* continuous on $\mathscr{P}(S)$ for each $y^* \in Y^*$. Since

$$y^* I_f(\mu) = y^* \int f \, d\mu = \int y^* f \, d\mu,$$

a useful property of the Bochner integral, this weak* continuity is an immediate consequence of $y^* f$'s membership in $C_b(S)$ for each $y^* \in Y^*$.

Closer to the spirit of the theorem itself is item 3.

3. Let S be a norm-separable subset of B_{X^*}. Then the norm Borel subsets of S and the sets of the form $S \cap B$, where B is weak* Borel subset of B_{X^*} coincide.

Let $\mathscr{Y}$ denote the collection of all subsets of S having the form $S \cap B$, where B is a weak* Borel subset of B_{X^*}. $\mathscr{Y}$ contains a base for the norm topology of S, namely, sets $S \cap B$, where B is a closed ball of X^*. Take a norm open set U in S. Each $x \in U$ is contained in the interior of a closed ball B_x for which $S \cap B_x \subseteq U$. Since S is separable, a countable number of closed balls B_x are needed to cover U. Of course, U is therefore a member of $\mathscr{Y}$. It follows that all the norm Borel subsets of S belong to $\mathscr{Y}$, and 3 is proved.

We are now ready to prove Theorem 10. S will be used to denote the set ext B_{X^*} in the norm topology, and the function f encountered in 1 and 2 will be the formal identity from S into X^*. As we saw in 1 and 2, I_f is weak*-weak continuous from $\mathscr{P}(S)$ into X^*.

Take a point mass $\delta_s \in \mathscr{P}(S)$: $I_f(\delta_s) = f(s) = s \in S$.

Take a convex combination $\sum_{i=1}^n a_i \delta_{s_i}$ of point masses:

$$I_f\left(\sum_{i=1}^n a_i \delta_{s_i}\right) = \sum_{i=1}^n a_i f(s_i) \in \operatorname{co}(S).$$

Take a weak* limit $\mu \in \mathscr{P}(S)$ of convex combinations $(\mu_d)_D$ of point masses: $I_f(\mu) = I_f(\text{weak* lim}_D \mu_d) = \text{weak lim}_D I_f(\mu_d) \in \overline{\text{co}(S)}^{\text{weak}}$.

Of course, each $\mu \in \mathscr{P}(S)$ is such a weak* limit; so we get $I_f(\mathscr{P}(S))$'s containment in $\overline{\text{co}(S)}^{\text{weak}}$, the weakly closed convex hull of S. By Mazur's theorem, $I_f(\mathscr{P}(S))$ is contained in the norm closed convex hull of S and so is norm separable.

Now let us see that B_{X^*} is itself contained in $I_f(\mathscr{P}(S))$.

Take $x^* \in B_{X^*}$. X is separable making B_{X^*} weak* metrizable and ext B_{X^*} a weak* $\mathscr{G}_\delta$-subset of B_{X^*}. By Choquet's theorem there is a regular Borel probability measure μ on ext B_{X^*} with

$$x^*x = \int_{\text{ext } B_{X^*}} y^*(x)\, d\mu(y^*);$$

part 3 assures us that we need not worry about whether we are speaking about the Borel sets of S in the norm topology or the weak* topology. Of course, the Bochner integral $\int_S f\, d\mu$ is actually at work above, and the formula above just says

$$x^*x = \left(\int_S f\, d\mu\right)(x)$$

for each $x \in X$; in other words, $x^* = \int_S f\, d\mu = I_f(\mu)$.

$$B_{X^*} \subseteq I_f(\mathscr{P}(S)) \subseteq \overline{\text{co}(S)}^{\text{norm}}. \qquad \square$$

Our next lemma, due to John Elton, indicates the severe limitations on the separability of the set of extreme points. Its proof will soon make another appearance.

Lemma 11. *Let X be a separable real Banach space and suppose that the set ext B_{X^*} of extreme points of B_{X^*} can be covered by a countable union of compact sets. Then X can be renormed so that its new dual unit ball has but countably many extreme points.*

PROOF. Let (K_n) be a sequence of compact subsets of X^* (each contained in B_{X^*}) for which

$$\text{ext } B_{X^*} \subseteq \bigcup_n K_n.$$

Let (ε_n) be a sequence of positive numbers for which $1 > \varepsilon_1 > \varepsilon_2 > \cdots > \varepsilon_n > \varepsilon_{n+1} \cdots \to 0$. For each n let $\mathscr{F}_n$ be a finite $\varepsilon_n/2$ net for K_n. Define $|||x|||$ by

$$|||x||| = \sup \bigcup_{n=1}^{\infty} \{(1+\varepsilon_n)|f(x)| : f \in \mathscr{F}_n\};$$

$|||x|||$ satisfies $\|x\| \leq |||x||| \leq (1+\varepsilon_1)\|x\|$ and so is an equivalent norm on X.

Obviously, $B_{X^*} \subseteq B_{(X,|||\cdot|||)^*}$. Not so obvious is the fact that

$$B_{X^*} \subseteq \{x^* \in X^* : |||x^*||| < 1\}.$$

To see that this *is* so, suppose otherwise. Then there exists an $x_0^* \in X^*$ for which

$$|||x_0^*||| = 1 = \|x_0^*\|.$$

Pick $x_0^{**} \in X^{**}$ such that

$$|||x_0^{**}||| = 1 = x_0^{**}(x_0^*).$$

Plainly $\|x_0^{**}\| = 1$, too. Now the set $\{x^* \in X^* : \|x^*\| = 1 = x_0^{**}x^*\}$ is a nonempty closed convex subset of X^*. By Theorem 10, X^* is separable; so by Theorem 9, the set $\{x^* \in X^* : \|x^*\| = 1 = x_0^{**}x^*\}$ has an extreme point, say x_e^*. x_e^* belongs to K_m for some m; therefore, $\|x_e^* - x_m^*\| < \varepsilon_m/2$ for some $x_m^* \in \mathscr{F}_m$. Of course,

$$\begin{aligned} x_0^{**}(x_m^*) &\geq x_0^{**}(x_e^*) - x_0^{**}(x_e^* - x_m^*) \\ &> 1 - \left(\frac{\varepsilon_m}{2}\right), \end{aligned}$$

and so

$$x_0^{**}((1+\varepsilon_m)x_m^*) > (1+\varepsilon_m)\left(1 - \frac{\varepsilon_m}{2}\right) > 1.$$

x_0^{**} has committed the gravest of mathematical sins: while proclaiming that $|||x_0^{**}||| = 1$, x_0^{**} has achieved a value >1 at an element, $(1+\varepsilon_m)x_m^*$, having $|||\cdot|||$ length no more than 1.

Next we notice that for each n, $(1+\varepsilon_n)\mathscr{F}_n$ is a subset of $B_{(X,|||\cdot|||)^*}$ and that, in fact, $B_{(X,|||\cdot|||)^*}$ is the weak* closed convex hull of $\cup_n(\pm(1+\varepsilon_n)\mathscr{F}_n)$. Why is this last assertion so? Well, if there were an x^* in $B_{(X,|||\cdot|||)^*}$ absent from the weak* closed convex hull of $\cup_n(\pm(1+\varepsilon_n)\mathscr{F}_n)$, then there would exist a weak* continuous linear functional $x \in X$ of $|||\cdot|||$ length 1 and an $\varepsilon > 0$ so that $x^*x = 1$, yet $|(1+\varepsilon_n)y_n^*(x)| \leq 1 - \varepsilon$ for all n and all $y_n^* \in \mathscr{F}_n$. A look at the definition of $|||x|||$ will establish our assertion.

Since $\cup_n(\pm(1+\varepsilon_n)\mathscr{F}_n)$ generates $B_{(X,|||\cdot|||)^*}$, Milman's theorem assures us that each extreme point of $B_{(X,|||\cdot|||)^*}$ belongs to $\overline{\cup_n(\pm(1+\varepsilon_n)\mathscr{F}_n)}^{\text{weak}^*}$. Let's look and see where the weak* limit points of $\cup_n(\pm(1+\varepsilon_n)\mathscr{F}_n)$ fall. Take a weak* convergent sequence (u_k^*) the terms of which belong to $\cup_n(\pm(1+\varepsilon_n)\mathscr{F}_n)$. If (u_k^*) repeatedly returns to one of the sets $\pm(1+\varepsilon_n)\mathscr{F}_n$, then it is clear from the finiteness of $\mathscr{F}_n$ that the weak* limit of (u_k^*) is also in $\pm(1+\varepsilon_n)\mathscr{F}_n$. Otherwise, there is an increasing sequence (n_k) of positive integers and a subsequence (v_k^*) of (u_k^*) for which $v_k^* \in \pm(1+\varepsilon_{n_k})\mathscr{F}_{n_k}$. By our judiciously placed constraints on (ε_n) we see that $\|\text{weak}^* \lim v_k^*\| \leq 1$; therefore, $|||\text{weak}^* \lim v_k^*||| < 1$ and $\text{weak}^* \lim v_k^*$ is *not* an extreme point of $B_{(X,|||\cdot|||)^*}$! In other words, *all extreme points of $B_{(X,|||\cdot|||)^*}$ are in the countable* set

$$\bigcup_n (\pm(1+\varepsilon_n)\mathscr{F}_n). \qquad \square$$

We are rapidly closing in on the finis of Elton's theorem. One giant step is contained in the next beautiful result of V. Fonf.

Theorem 12. *Let X be a separable real Banach space of infinite dimension whose dual unit ball has but countably many extreme points. Then X contains an isomorph of c_0.*

PROOF. Suppose $\operatorname{ext} B_{X^*} = \{\pm x_n^*\}_{n\geq 1}$ and let $1 > \varepsilon_1 > \varepsilon_2 > \cdots > \varepsilon_n \downarrow 0$. Define a new norm on X by

$$|||x||| = \sup\left\{|x^*x| : x^* \in \overline{\operatorname{co}\{\pm(1+\varepsilon_n)x_n^* : n\geq 1\}}^{\text{weak}^*}\right\}.$$

Then $|||\cdot|||$ is a norm on X satisfying $\|x\| \leq |||x||| \leq (1+\varepsilon_1)\|x\|$ for all x. Correspondingly, $B_{X^*} \subseteq B_{(X,|||\cdot|||)^*} \subseteq (1+\varepsilon_1)B_{X^*}$. As in Lemma 11, we claim that $B_{X^*} \subseteq \{x^* : |||x^*||| < 1\}$. (Were this not so, there would be an $x_0^* \in X^*$ for which $\|x_0^*\| = 1 = |||x_0^*|||$. Take $x_0^{**} \in X^{**}$ such that $x_0^{**}x_0^* = 1 = |||x_0^{**}|||$. Plainly, $\|x_0^{**}\| = 1$, too. B_{X^*} has but a countable number of extreme points; so Theorem 10 ensures the separability of X^*; Theorem 9 now assures us of the presence of an extreme point x_e^* in the norm closed bounded convex set $\{x^* : \|x^*\| = 1 = x_0^{**}x^*\}$. Since this set is extremal in B_{X^*}, x_e^* is in the list $\{\pm x_n^*\}_{n\geq 1}$. Therefore, $x_e^* = \pm x_{n_e}^*$, and so $(1+\varepsilon_{n_e})x_{n_e}^*$ has $|||\cdot|||$-length 1. But this gives $|x_0^{**}(x_{n_e}^*)| = 1 + \varepsilon_{n_e} > 1$, a contradiction.

An easy separation argument shows that $B_{(X,|||\cdot|||)^*}$ is the weak* closed convex hull of $\{\pm(1+\varepsilon_n)x_n^* : n\geq 1\}$; so $\operatorname{ext} B_{(X,|||\cdot|||)^*}$ is contained in $\overline{\{\pm(1+\varepsilon_n)x_n^* : n\geq 1\}}^{\text{weak}^*}$ thanks to Milman's theorem. A weak* convergent sequence taken from the set $\{\pm(1+\varepsilon_n)x_n^* : n\geq 1\}$ converges either to a point of the set or to a point of B_{X^*} (which cannot be an extreme point of $B_{(X,|||\cdot|||)^*}$); it follows that all the extreme points of $B_{(X,|||\cdot|||)^*}$ find themselves in $\{\pm(1+\varepsilon_n)x_n^* : n\geq 1\}$. In other words, there is a subsequence (y_n^*) of (x_n^*) and a subsequence (δ_n) of (ε_n) such that

$$\operatorname{ext} B_{(X,|||\cdot|||)^*} \subseteq \{\pm(1+\delta_n)y_n^* : n\geq 1\}.$$

A key consequence of the above development is this: given a finite-dimensional subspace F of $(X, |||\cdot|||)$ there is an n_F such that

$$\operatorname{ext} B_{(F,|||\cdot|||)^*} \subseteq \{\pm(1+\delta_n)y_n^*|_F\}_{1\leq n\leq n_F}.$$

Since every member of $\operatorname{ext} B_{(F,|||\cdot|||)^*}$ has an extension that's extreme in $B_{(X,|||\cdot|||)^*}$, it is clear that each extreme point of $B_{(F,|||\cdot|||)^*}$ is of the form $\sigma_n(1+\delta_n)y_n^*|_F$ for some n and sign σ_n. This, in tandem with the nondecreasing nature of the linear subspaces of F^* spanned by $\{\pm(1+\delta_m)y_m^* : 1\leq m \leq n\}$, will soon produce the required n_F.

We now build a normalized sequence (x_n) in $(X, |||\cdot|||)$.

Take any $x_1 \in X$ such that $|||x_1|||=1$. The collection $\{x^*: |||x^*|||=1=x^*x_1\}$ is a nonempty extremal weak* compact convex subset of $B_{(X,|||\cdot|||)^*}$; as such it contains an extreme point of $B_{(X,|||\cdot|||)^*}$—say $\sigma_{n_1}(1+\delta_{n_1})y^*_{n_1}$. Of course, $|(1+\delta_{n_1})y^*_{n_1}(x_1)|=1$.

Take $x_2 \in X$ such that $|||x_2|||=1$ and $y_1^*(x_2)=\cdots=y^*_{n_1}(x_2)=0$. Let F be the linear span of x_1 and x_2. Pick $n_2 > n_1$ so that

$$\operatorname{ext} B_{(F,|||\cdot|||)^*} \subseteq \{\pm(1+\delta_n)y_n^*|_F\}_{1\le n\le n_2}.$$

Take $x_3 \in X$ such that $|||x_3|||=1$ and $y_1^*(x_3)=\cdots=y^*_{n_2}(x_3)=0$. Let F be the linear span of x_1, x_2, and x_3. Pick $n_3 > n_2$ so that

$$\operatorname{ext} B_{(F,|||\cdot|||)^*} \subseteq \{\pm(1+\delta_n)y_n^*|_F\}_{1\le n\le n_3}.$$

Et cetera.

It is easy to see that (x_n) is a monotone basic sequence, i.e., for any $j,k \ge 1$ we have

$$\left|\left|\left|\sum_{i=1}^{j} a_i x_i\right|\right|\right| \le \left|\left|\left|\sum_{i=1}^{j+k} a_i x_k\right|\right|\right|.$$

Let $Z=[x_n]_{n\ge1}$ be the closed linear span of the x_n. Then $B_{(Z,|||\cdot|||)^*}$ has but countably many extreme points, each a restriction of some extreme point of $B_{(X,|||\cdot|||)^*}$ to Z. List the extreme points of $B_{(Z,|||\cdot|||)^*}$ as $\{\pm z_n^*\}$. Keeping in mind the origins of the extreme points of $B_{(X,|||\cdot|||)^*}$, two key properties of (z_n^*) come to the fore:

First, any weak* limit point of $\{\pm z_n^*\}$ that does not belong to $\{\pm z_n^*\}$ has $|||\cdot|||$ length <1.

Second, given any n there is a $k(n)$ such that $z_n^* x_m = 0$ for all $m \ge k(n)$.

Now we take dead aim on finding a c_0 in Z.

Set $\eta_1 = 1$.

Suppose coefficients $\eta_1,\ldots,\eta_n$ have been chosen so carefully that

$$\left|\left|\left|\sum_{i=1}^{n} \sigma_i\eta_i x_i\right|\right|\right| < 2$$

for any signs $\sigma_1,\sigma_2,\ldots,\sigma_n$, yet for some signs $\sigma_1^n,\ldots,\sigma_n^n$ we have

$$\left|\left|\left|\sum_{i=1}^{n-1} \sigma_i^n\eta_i x_i\right|\right|\right| < \left|\left|\left|\sum_{i=1}^{n} \sigma_i^n\eta_i x_i\right|\right|\right|.$$

We show how to pick η_{n+1}.

Set

$$\beta_{n+1} = \min\left\{\max\left\{h\ge 0: \left|\left|\left|\sum_{i=1}^{n}\sigma_i\eta_i x_i + \sigma_{n+1}hx_{n+1}\right|\right|\right| = \left|\left|\left|\sum_{i=1}^{n}\sigma_i\eta_i x_i\right|\right|\right|\right\}\right\},$$

where the minimum is taken over all 2^{n+1} $(n+1)$-tuples $(\sigma_1, \sigma_2, \ldots, \sigma_{n+1})$ of signs. Just what does β_{n+1} signify? Well, given signs $\sigma_1, \ldots, \sigma_n, \sigma_{n+1}$ the function $\varphi : [0, \infty) \to [0, \infty)$ defined by

$$\varphi(h) = \left|\left|\left| \sum_{i=1}^{n} \sigma_i \eta_i x_i + \sigma_{n+1} h x_{n+1} \right|\right|\right|$$

is continuous and nondecreasing [since (x_n) is monotone] and has value $|||\sum_{i=1}^{n} \sigma_i \eta_i x_i|||$ at 0. Therefore, the number β_{n+1} has the property that any number h bigger than β_{n+1} opts for some $(n+1)$-tuple of signs $(\bar{\sigma}_1, \ldots, \bar{\sigma}_{n+1})$ such that

$$\left|\left|\left| \sum_{i=1}^{n} \bar{\sigma}_i \eta_i x_i \right|\right|\right| < \left|\left|\left| \sum_{i=1}^{n} \bar{\sigma}_i \eta_i x_i + \bar{\sigma}_{n+1} h x_{n+1} \right|\right|\right|.$$

This is mind (along side our hopes for η_{n+1}), we choose $\lambda_{n+1} > 0$ so that for any $(n+1)$-tuple $(\sigma_1, \ldots, \sigma_{n+1})$ of signs

$$\left|\left|\left| \sum_{i=1}^{n} \sigma_i \eta_i x_i + \sigma_{n+1}(\beta_{n+1} + \lambda_{n+1}) x_{n+1} \right|\right|\right| < 2,$$

yet for some $(n+1)$ tuple of signs $(\bar{\sigma}_1, \ldots, \bar{\sigma}_{n+1})$ we have

$$\left|\left|\left| \sum_{i=1}^{n} \bar{\sigma}_n \eta_i x_i \right|\right|\right| < \left|\left|\left| \sum_{i=1}^{n} \bar{\sigma} \cdot \eta_i x_i + \bar{\sigma}_{n+1}(\beta_{n+1} + \lambda_{n+1}) x_{n+1} \right|\right|\right|.$$

Set $\eta_{n+1} = \beta_{n+1} + \lambda_{n+1}$.

Plainly, we have built the series $\sum_n \eta_n x_n$ to be a wuC. Can it converge unconditionally? If so, then the set $\{\sum_n \sigma_n \eta_n x_n : (\sigma_n)$ is a sequence of signs$\}$ would be a relatively norm compact set in Z; of course, our choice of $\eta_1 = 1$ and the monotonicity of (x_n) assures us that for any sequence (σ_n) of signs we have $|||\sum_n \sigma_n \eta_n x_n||| \geq 1$.

Here is the hitch: if K is a compact subset of $\{z \in Z : |||z||| \geq 1\}$, then there is an N so that each $z = \sum_n t_n x_n \in K$ is actually of the form $z = \sum_{n=1}^{N} t_n x_n$. The contradiction attendant to this fact for the set $\{\sum_n \sigma_n \eta_n x_n : (\sigma_n)$ is a sequence of signs$\}$ will prove that Z contains a divergent wuC and so a copy of c_0 by the Bessaga-Pelczynski theorem of Chapter V.

Let's establish the aforementioned striking feature of norm-compact subsets K of $\{z \in Z : |||z||| \geq 1\}$ by supposing it did not hold and deriving a suitable contradiction. Were K a norm compact subset of $\{z \in Z : |||z||| \geq 1\}$ that does not depend on but a finite number of the x_n, then there would be a sequence (u_n) in K and a sequence (u_n^*) among the extreme points of $B_{(Z, |||\cdot|||)^*}$, each u_n^* of the form $\sigma_{k_n}^* z_{k_n}^* (\sigma_{k_n} = \pm 1)$ for some subsequence $(z_{k_n}^*)$ of (z_n^*) such that

$$u_n^* u_n = |||u_n|||.$$

The compactness of K and the compact metric nature of $(B_{(Z, |||\cdot|||)^*}, \text{weak}^*)$

allows us to assume that

$$u_0 = \text{norm}\lim_n u_n$$

and

$$u_0^* = \text{weak}^*\lim_n u_n^*$$

both exist. Of course, $|||u_0^*||| < 1$; so

$$|||u_0||| = \lim_n |||u_n||| = \lim_n u_n^* u_n,$$

yet

$$\lim_n u_n^* u_n = u_0^*(u_0) \le |||u_0^*|||\,|||u_0||| < |||u_0|||. \qquad \square$$

Lemma 13. *Let (x_n) be a normalized basis for the Banach space X and suppose that $\Sigma_n |x^* x_n| < \infty$ for each $x^* \in \operatorname{ext} B_{X^*}$. Then the sequence (x_n^*) of coefficient functionals is a basis for X^*.*

Proof. First, we take special note of the following: if (u_j) is a normalized block basis built on (x_n), then (u_j) is weakly null. In fact, since $\|u_j\| = 1$, the u_j have uniformly bounded coefficients in their expansions according to the basis (x_n). Because we have assumed that $\Sigma_n |x^* x_n| < \infty$ for each extreme point x^* of B_{X^*}, we can conclude that $(x^* u_j)$ is null for each extreme point x^* of B_{X^*}. Now we need only apply Rainwater's theorem.

Now we show that $\lim_m \|x^* P_m - x^*\| = 0$ for each $x^* \in X^*$, where $P_m : X \to X$ is the mth expansion operator with respect to the basis (x_n), $P_m(\Sigma_n a_n x_n) = \Sigma_{n=1}^m a_n x_n$. But x^* is always the weak* limit of the sequence $(x^* P_m)$; so the only thing that can go wrong with $\lim_m \|x^* P_m - x^*\| = 0$ for each $x^* \in X^*$ would have to be the existence of an x_0^* such that the sequence $(x_0^* P_m)$ is not even Cauchy. For such an x_0^* we could find an increasing sequence (m_n) of positive integers such that

$$\left\| x_0^* P_{m_{n+1}} - x_0^* P_{m_n} \right\| > \varepsilon$$

for all n and some $\varepsilon > 0$. Correspondingly, there is for each n a $v_n \in B_X$ such that

$$\left| \left(x_0^* P_{m_{n+1}} - x_0^* P_{m_n} \right)(v_n) \right| = \left| x_0^* \left(\left(P_{m_{n+1}} - P_{m_n} \right) v_n \right) \right| > \varepsilon.$$

Look at $u_n = (P_{m_{n+1}} - P_{m_n})(v_n)$. The sequence (u_n) is a block basic sequence built from (x_n), $\|u_n\| > \varepsilon / \|x_0^*\|$ for all n, and $\|u_n\| \le 2 \sup_m \|P_m\|$. In light of our opening remarks, (u_n) must be weakly null yet $|x_0^* u_n| > \varepsilon$ for all n, which is a contradiction. It follows that $\lim_m \|x^* P_m - x^*\| = 0$ for each $x^* \in X^*$.

As a point of fact, we are done. The expansion operators P_m have adjoints $P_m^*: X^* \to X^*$ whose form is given by

$$(P_m^* x^*)(x) = (x^* P_m)(x) = \sum_{i=1}^{m} x^*(x_i) x_i^*.$$

Since $\|P_m^*\| = \|P_m\| \leq \sup_m \|P_m\| < \infty$, the sequence (x_n^*) satisfies the criterion for basic sequences; (x_n^*) is a basis for its closed linear span. In light of the previous paragraph the closed linear span of (x_n^*) is all of X^*—remember $P_m^* x^* = x^* P_m$. □

One last step:

Theorem 14. *Suppose that the Banach space X has a normalized basis (x_n) for which $\sum_n |x^* x_n| < \infty$ for each extreme point x^* of B_{X^*}. Then X contains a copy of c_0.*

PROOF. Lemma 13 assures us that the sequence (x_n^*) of coefficient functionals is itself a basis for X^*. Consequently, the operator $T: l_1 \to X^*$ given by $T(t_n) = \sum_n t_n x_n^*$ is a well-defined bounded linear one-to-one operator from l_1 into X^*. Denote by (e_n) the usual sequence of unit coordinate vectors in l_1.

If there is an N such that $T|_{[e_n]_{n \geq N}}$ is an isomorphism, then $(x_n^*)_{n \geq N}$ is equivalent to the unit vector basis of l_1. It is easy to deduce from this that $(x_n)_{n \geq N}$ is equivalent to c_0's unit vector basis.

If there is no N for which $T|_{[e_n]_{n \geq N}}$ is an isomorphism, then it's easy to manufacture a normalized block basis (u_n) with respect to (e_n) in l_1 such that $\|Tu_n\| < 2^{-n}$; these u_n are of the form

$$u_n = \sum_{i=p_n}^{q_n} s_i e_i,$$

where $1 \leq p_1 < q_1 < p_2 < q_2 < \cdots$ and $\sum_{i=p_n}^{q_n} |s_i| = 1$. Of course, (u_n) is equivalent to the unit vector basis of l_1 and the closed linear span U of the u_n is itself an isomorphic copy of l_1. Further $T(B_U)$ is a relatively compact subset of X^*.

Let $y_n \in X$ be the vector $y_n = \sum_{i=p_n}^{q_n} \operatorname{sign}(s_i) x_i$ and consider the closed linear span Y of the y_n; we are going to find a c_0 inside Y. (y_n) is a basis for Y. Since $\sum_n |x^* y_n| < \infty$ for each extreme point x^* of B_{X^*} and since each extreme point of B_{Y^*} admits of an extreme extension in B_{X^*}, $\sum_n |y^* y_n| < \infty$ for each extreme point y^* of B_{Y^*}. Notice that the sequence (y_n^*), where $y_n^* = Tu_n|_Y$, is biorthogonal to (y_n). Normalizing (y_n) we can apply Lemma 13; in any case, (y_n^*) is a basis for Y^*.

Let

$$K_m = \left\{ x^* \in X^* : \sum_n |x^* x_n| \leq m \right\}.$$

Our hypotheses assure us that

$$\operatorname{ext} B_{X^*} \subseteq \bigcup_{m=1}^{\infty} K_m.$$

Notice that if $x^* \in K_m$, then

$$\sum_n |x^* y_n| \leq \sum_n |x^* x_n| \leq m;$$

from this it follows that for $x^* \in K_m$, if we let $y^* = x^*|_Y$, then y^* is of the form $\Sigma_n b_n y_n^*$, where

$$\sum_n |b_n| = \sum_n |y^* y_n| = \sum_n |x^* y_n| \leq m.$$

Thus, any vector $\Sigma_n b_n y_n^*$ for which $\Sigma_n |b_n| \leq m$ belongs to the relatively compact set $T(mB_U)|_Y$. What we have then is

$$\begin{aligned} \operatorname{ext} B_{Y^*} &\subseteq \{x^*|_Y : x^* \in \operatorname{ext} B_{X^*}\} \\ &\subseteq \bigcup_{m=1}^{\infty} \{x^*|_Y : x^* \in K_m\} \\ &\subseteq \bigcup_{m=1}^{\infty} \overline{T(mB_U)|_Y}. \end{aligned}$$

Lemma 11 and Theorem 12 now combine to locate a c_0 inside Y. □

Finally, we are ready for John Elton's extremal test for unconditional convergence. It follows from Theorem 14 and the Bessaga-Pelczynski selection principle.

Theorem 15. *A Banach space X contains a copy of c_0 if and only if there is a divergent series $\Sigma_n x_n$ in X for which $\Sigma_n |x^* x_n| < \infty$ for each extreme point x^* of B_{X^*}.*

Exercises

1. *Dentable sets.* A bounded subset B of a Banach space X is called *dentable* if it has slices of arbitrarily small diameter.
   (i) A set B is dentable if and only if its closed convex hull is dentable.
   (ii) A set B is dentable if and only if given $\varepsilon > 0$ there is a point $x_\varepsilon \in B$ such that $x_\varepsilon \notin \overline{\operatorname{co}}(B \setminus \{y : \|x_\varepsilon - y\| < \varepsilon\})$.
   (iii) A set B is dentable if each of its countable subsets is dentable.
   (iv) Compact sets are dentable.
   (v) Closed bounded convex subsets of uniformly convex spaces are dentable.
   (vi) Weakly compact sets are dentable.

2. *Extremal scalar integrability.* Let (Ω, Σ, μ) be a probability measure space and X be a Banach space. Suppose $f: \Omega \to X$ is (strongly) μ-measurable. If X contains no copy of c_0 and if

$$\int_\Omega |x^* f(\omega)| \, d\mu(\omega) < \infty$$

for each extreme point x^* of B_{X^*}, then f is Pettis integrable.

3. *Nondual spaces.* Neither c_0 nor $L_1[0,1]$ are isomorphic to a subspace of any separable dual space.

Notes and Remarks

The original proof of the Krein-Milman theorem took place in a weak*-compact convex subset of the dual of a normed linear space; J. L. Kelley (1951) is responsible for the proof presented in the text.

Our treatment of the simplest elements of the barycentric calculus owe an obvious debt to R. R. Phelps's lectures (1966). We have added a few details, but in the main we have followed his wise leadership. The observation that extreme points have extremal extensions is due to I. Singer and is often referred to as "Singer's theorem"; it allows us to prove Rainwater's theorem by direct appeal to Choquet's theorem without recourse to the more delicate Choquet-Bishop-deLeeuw setup.

Only by exercising a will power all too rarely displayed has our discussion of Choquet theory been curtailed. This beautiful corner of abstract analysis has been the object of several excellent monographs making what we would say redundant at best. It behooves the student to study these basic texts: for a quick fix on the subject, Phelps's "Lectures on Choquet's theorem" (1966) can not be beat; a more extensive treatment is found in E. M. Alfsen's "Compact Convex Sets and Boundary Integrals" (1971), and a reading of Choquet's "Lectures on Analysis" (1969) affords the student the rare opportunity to learn from the master himself.

Some surprising advances in the theory of integral representations, closely related to the material of the section titled Elton's Theorem, have appeared since the publication of the aforementioned monographs. The most spectacular turn of events has been G. A. Edgar's proof (1975) of a Choquet theorem for certain noncompact sets. We state Edgar's original theorem.

Theorem. *Let K be a closed bounded convex subset of a separable Banach space. Suppose K has the Radon-Nikodym property. Then every point of K can be realized as the barycenter of a regular Borel probability measure on K supported by the extreme points of K.*

Naturally the "catch" in the above theorem is the assumption that the set K has the Radon-Nikodym property. A subset K of the Banach space X has the Radon-Nikodym property whenever given a probability measure space (Ω, Σ, μ) and a countably additive μ-continuous $F: \Sigma \to X$ for which $\{F(E)/\mu(E): E \in \Sigma, \mu(E) \neq 0\} \subseteq K$, there is a Bochner integrable $f: \Omega \to X$ such that

$$F(E) = \int_E f(\omega)\, d\mu(\omega)$$

for each $E \in \Sigma$. It is not yet known if the Radon-Nikodym property needs to be assumed in Edgar's theorem, but there is considerable evidence that it must.

Edgar's theorem has a nonseparable version, also due to Edgar, that is not possessed of as elegant a formulation. P. Mankiewicz has been able to prove the more general representation theorem of Edgar by reductions to the separable version. In all instances, some assumption of the Radon-Nikodym property is present.

Superlemma is due to I. Namioka in its weak* version and J. Bourgain in general. There is no better way to see the Superlemma in action than to read the Rainwater seminar notes from the University of Washington, where the raw power of this lemma is harnessed; the result is a masterful demonstration of the equivalence of the Radon-Nikodym property with a number of its sharpest geometric variants. We've used the weak* version of Superlemma much as Namioka did in his derivation of the Bessaga-Pelczynski theorem.

Incidentally the main concern of Exercise 1, dentable sets, has its roots again in matters related to the Radon-Nikodym property. The notion of dentability originated in M. A. Rieffel's study of Radon-Nikodym theorems for the Bochner integral. As a consequence of the combined efforts of Rieffel, H. Maynard, R. E. Huff, W. J. Davis, and R. R. Phelps, we can state the basic geometric characterization of the Radon-Nikodym property as follows:

Theorem. *A nonempty closed bounded convex subset K of a Banach space has the Radon-Nikodym property if and only if each nonempty subset of K is dentable.*

A finished product by summer of 1973, the dentability theorem (as it's come to be known) signaled the start of a period of excitement in the geometric affairs of the Radon-Nikodym property. Not to stray too far afield, we mention just one result that evolved during this "gold rush" and refer the student to the Diestel-Uhl AMS Surveys volume for a more complete story of the early happenings and to Bourgin's Springer Lecture Notes volume for recent developments.

Theorem (N. Dunford, B. J. Pettis, J. Lindenstrauss, C. Stegall, R. E. Huff, P. D. Morris). *Let X be a Banach space. Then the following are equivalent*:

1. *Each separable subspace of X has a separable dual.*
2. *Each nonempty closed bounded convex subset of X^* is dentable.*
3. *Each nonempty closed bounded convex subset of X^* has an extreme point.*

Theorem 10 was established by R. Haydon (1976), K. Musial (1978), and V. I. Rybakov (1977); our proof was inspired by Haydon's, but its execution differs at several crucial junctures. Subsequent to stumbling onto this variation in approach, E. Saab (1977) pointed out that he had used the same tactics to much greater advantage in deriving several generalizations of Haydon's result.

The ideas behind the proofs of Lemma 11 and Theorem 12 are due to V. Fonf (1979). They were most enjoyable to encounter, to lecture on, and to write about. Plainly speaking, they are too clever by half. Of course, Theorems 14 and 15 are due to J. Elton (1981).

Bibliography

Alfsen, E. M. 1971. *Compact Convex Sets and Boundary Integrals*. Ergebnisse der Mathematik und ihrer Grenzgebiete, Volume 57. Berlin: Springer-Verlag.

Bourgin, R. 1983. Geometric Aspects of Convex Sets with the Radon-Nikodym Property, Volume 993, Springer Lecture Notes in Mathematics. Berlin: Springer-Verlag.

Choquet, G. 1969. *Lectures on Analysis*, Lecture Notes in Mathematics. New York: W. A. Benjamin.

Edgar, G. A. 1975. A noncompact Choquet theorem. *Proc. Amer. Math. Soc.*, **49**, 354–358.

Elton, J. 1981. Extremely weakly unconditionally convergent series. *Israel J. Math.* **40**, 255–258.

Fonf, V. 1979. One property of Lindenstrauss-Phelps spaces. *Funct. Anal. Appl.* (English Trans.), **13**, 66–67.

Haydon, R. 1976. An extreme point criterion for separability of a dual Banach space and a new proof of a theorem of Carson. *Quart. J. Math. Oxford*, **27**, 379–385.

Kadec, M. I. and Fonf, V. 1976. Some properties of the set of extreme points of the unit ball of a Banach space. *Mat. Zametki*, **20**, 315–319.

Kelley, J. L. 1951. Note on a theorem of Krein and Milman. *J. Osaka Inst. Sci. Tech.*, **3**, 1–2.

Krein, M. and Milman, D. 1940. On extreme points of regularly convex sets. *Studia Math.*, **9**, 133–138.

Mankiewicz, P. 1978. A remark on Edgar's extremal integral representation theorem. *Studia Math.*, **63**, 259–265.

Musial, K. 1978. The weak Radon-Nikodym property in Banach spaces. *Studia Math.*, **64**, 151–174.

Phelps, R. R. 1966. *Lectures on Choquet's theorem*. Van Nostrand Math. Studies No. 7. Princeton: Van Nostrand.

Rainwater, J. 1963. Weak convergence of bounded sequences. *Proc. Amer. Math. Soc.*, **14**, 999.

Rybakov, V. 1977. Some properties of measures on a normed space that has the RN property. *Mat. Zametki*, **21**, 81–92.

Saab, E. 1977. Points extrémaux, séparabilité et faible K-analyticité dans les duaux d'espaces de Banach. *C. R. Acad. Sci. Paris*, **285**, 1057–1060.

CHAPTER X

Grothendieck's Inequality and the Grothendieck-Lindenstrauss-Pelczynski Cycle of Ideas

In this section we prove a profound inequality due, as the section title indicates, to Grothendieck. This inequality has played a fundamental role in the recent progress in the study of Banach spaces. It was discovered in the 1950s, but its full power was not generally realized until the late 1960s when Lindenstrauss and Pelczynski, in their seminal paper "Absolutely summing operators in $\mathscr{L}_p$ spaces and their applications," brutally reminded functional analysts of the existence and importance of the powerful ideas and work of Grothendieck. Since the Lindenstrauss-Pelczynski paper, the Grothendieck inequality has seen many proofs; in this, it shares a common feature of most deep and beautiful results in mathematics. The proof we present is an elaboration of one presented by R. Rietz. It is very elementary.

Some notational conventions are in order.

For a vector x in $\mathbf{R}^n$ the euclidean norm of x will be denoted by $|x|$. If x, y are in $\mathbf{R}^n$, then their inner product will be denoted by $x \cdot y$.

By dz we mean Lebesgue measure on $\mathbf{R}^n$, by $dG(z)$ we mean normalized Gaussian measure of mean zero and variance 1. *Don't be discouraged by the fancy description.* $dG(z)$ is given by

$$dG(z) = (2\pi)^{-n/2} e^{-|z|^2/2}\, dz;$$

in other words, for any Lebesgue-measurable real-valued function f on $\mathbf{R}^n$

$$\int_{R^n} f(z)\, dG(z) = \frac{1}{\sqrt{(2\pi)^n}} \int_{R^n} f(z) e^{-|z|^2/2}\, dz.$$

Particularly noteworthy (and crucial for our purposes) is the fact that the Gaussian measure is a product measure of smaller Gaussian measures. In particular, if $k + m = n$, then the product of k-dimensional Gaussian measure and m-dimensional Gaussian measure is n-dimensional Gaussian measure.

We denote by L_2 the L_2-space $L_2(\mathbf{R}^n, dG)$ of n-dimensional Gaussian measure. If $f \in L_2$, then the norm of f will be denoted by $\|f\|_2$ and the inner product of f with a $g \in L_2$ will be denoted by (f, g).

Finally, some functions of special interest: if $x \in \mathbf{R}^n$, then we define $\varphi_x : \mathbf{R}^n \to \mathbf{R}$ by $\varphi_x(y) = x \cdot y$ and $s_x : \mathbf{R}^n \to \mathbf{R}$ by $s_x(y) = s(x \cdot y)$, where s denotes the signum function.

Lemma 1. *If* $|x| = 1 = |y|$, *then* $(\varphi_x, \varphi_y) = x \cdot y$.

PROOF. Since $|x| = 1 = |y|$, there is a vector y' orthogonal to y such that

$$x = (x \cdot y) y + y'.$$

For this y' and all z we have

$$x \cdot z = (x \cdot y)(y \cdot z) + y' \cdot z,$$

and so

$$(x \cdot z)(y \cdot z) = (x \cdot y)(y \cdot z)^2 + (y' \cdot z)(y \cdot z).$$

Integrating with respect to z (and changing to the φ_x notation), we have

$$\begin{aligned}(\varphi_x, \varphi_y) &= \int (x \cdot z)(y \cdot z)\, dG(z) \\ &= (x \cdot y) \int |y \cdot z|^2\, dG(z) + \int (y' \cdot z)(y \cdot z)\, dG(z) \\ &= (x \cdot y)(\varphi_y, \varphi_y) + (\varphi_{y'}, \varphi_y). \qquad (1)\end{aligned}$$

Let's compute.

$$\begin{aligned}(\varphi_y, \varphi_y) &= \frac{1}{\sqrt{(2\pi)^n}} \int_{-\infty}^{\infty} \cdots \int_{-\infty}^{\infty} \int_{-\infty}^{\infty} (y_1 z_1 + y_2 z_2 + \cdots + y_n z_n)^2 \\ &\quad \times e^{-(z_1^2 + z_2^2 + \cdots + z_n^2)/2}\, dz_1\, dz_2 \cdots dz_n \\ &= \frac{1}{\sqrt{(2\pi)^n}} \int_{-\infty}^{\infty} \cdots \int_{-\infty}^{\infty} \int_{-\infty}^{\infty} (y_1 z_1 + I(z_1))^2 \\ &\quad \times e^{-z_1^2/2}\, dz_1\, e^{-(z_2^2 + \cdots + z_n^2)/2}\, dz_2 \cdots dz_n,\end{aligned}$$

where $I(z_1) = y_2 z_2 + \cdots + y_n z_n$ is independent of the coordinate z_1,

$$\begin{aligned}&= \frac{1}{\sqrt{(2\pi)^n}} \int_{-\infty}^{\infty} \cdots \int_{-\infty}^{\infty} \left[\int_{-\infty}^{\infty} (y_1^2 z_1^2 + 2 y_1 z_1 I(z_1) + I^2(z_1)) e^{-z_1^2/2}\, dz_1 \right] \\ &\quad \times e^{-(z_2^2 + \cdots + z_n^2)/2}\, dz_2 \cdots dz_n \\ &= \frac{1}{\sqrt{(2\pi)^n}} \int_{-\infty}^{\infty} \cdots \int_{-\infty}^{\infty} \sqrt{2\pi}\, (y_1^2 + I^2(z_1)) \\ &\quad \times e^{-(z_2^2 + \cdots + z_n^2)/2}\, dz_2 \cdots dz_n,\end{aligned}$$

where $\int_{-\infty}^{\infty} 2 y_1 z_1 I(z_1) e^{-z_1^2/2}\, dz_1 = 0$ since the integrand is an odd function of z_1,

$$= \frac{1}{\sqrt{(2\pi)^{n-1}}}\left\{y_1^2 + \int_{-\infty}^{\infty}\cdots\int_{-\infty}^{\infty}(y_2z_2 + \cdots + y_nz_n)^2\right.$$

$$\times e^{-(z_2^2+\cdots+z_n^2)2}\,dz_2\cdots dz_n,$$

which after $n-1$ more such computations will eventually be reduced to being

$$= y_1^2 + y_2^2 + \cdots + y_n^2 = |y|^2 = 1.$$

On the other hand,

$$(\varphi_{y'}, \varphi_y) = \frac{1}{\sqrt{(2\pi)^n}}\int_{-\infty}^{\infty}\cdots\int_{-\infty}^{\infty}\int_{-\infty}^{\infty}(y_1'z_1 + y_2'z_2 + \cdots + y_n'z_n)$$

$$\times(y_1z_1 + y_2z_2 + \cdots + y_nz_n)\,dG(z)$$

$$= \frac{1}{\sqrt{(2\pi)^n}}\int_{-\infty}^{\infty}\cdots\int_{-\infty}^{\infty}\int_{-\infty}^{\infty}(y_1'z_1 + I'(z_1))(y_1z_1 + I(z_1))$$

$$\times e^{-z_1^2/2}\,dz_1 e^{-(z_2^2+\cdots+z_n^2)/2}\,dz_2\cdots dz_n,$$

where $I(z_1) = y_2z_2 + \cdots + y_nz_n$ and $I'(z_1) = y_2'z_2 + \cdots + y_n'z_n$ are independent of the coordinate z_1,

$$= \frac{1}{\sqrt{(2\pi)^n}}\int_{-\infty}^{\infty}\cdots\int_{-\infty}^{\infty}\left[y_1'y_1z_1^2 + y_1I'(z_1)z_1 + y_1'I(z_1)z_1 + I'(z_1)I(z_1)\right]$$

$$\times e^{-z_1^2/2}\,dz_1 e^{-(z_2^2+\cdots+z_n^2)/2}\,dz_2\cdots dz_n$$

which again taking into account the oddness of certain integrands,

$$= \frac{1}{\sqrt{(2\pi)^n}}\int_{-\infty}^{\infty}\cdots\int_{-\infty}^{\infty}\sqrt{2\pi}\,(y_1'y_1 + I'(z_1)I(z_1))e^{-(z_2^2+\cdots+z_n^2)/2}\,dz_2\cdots dz_n$$

$$= \frac{1}{\sqrt{(2\pi)^{n-1}}}\left\{y_1'y_1 + \int_{-\infty}^{\infty}\cdots\int_{-\infty}^{\infty}(y_2'z_2 + \cdots + y_n'z_n)(y_2z_2 + \cdots + y_nz_n)\right.$$

$$\times e^{-(z_2^2+\cdots+z_n^2)/2}\,dz_2\cdots dz_n,$$

which after another $n-1$ such computations is seen to be

$$= y_1'z_1 + \cdots + y_n'y_n = y'\cdot y = 0.$$

Now a look at Eq. (1) will provide the finishing touches to the proof. □

The proof of the previous lemma, viewed from the proper perspective, illuminates the special role of the Gaussian distribution in the present setup. Consider for a moment what is going on when you try to compute the expected value of a real-valued random variable with respect to the Gaussian distribution. How can one cut back on the amount of actual computation to be done?

First, note that the measure one is integrating with respect to has two pieces so to say: The Lebesgue measure dz and the Gaussian weight $(2\pi)^{-n/2}e^{-|z|^2/2}$.

A well-known and important property enjoyed by Lebesgue measure is the fact that it is invariant under isometries. So, should f and g be different real-valued random variables on $\mathbb{R}^n$ such that, for some isometry τ of $\mathbb{R}^n$, $f(z)=g(\tau z)$ holds for all $z\in\mathbb{R}^n$, then $\int f(z)\,dz=\int g(z)\,dz$ automatically obtains.

Now take into account the second piece of our amusing little puzzle: the Gaussian weight $(2\pi)^{-n/2}e^{-|z|^2/2}$. The values of this weight agree at any $x, y\in\mathbb{R}^n$ having the same lengths. Therefore, should f and g be real-valued random variables such that whenever f attains the value $f(x)$ at a vector x, then there is one and only one vector y_x of the same length as x such that $g(y_x)$ attains the same value $f(x)$, then the Gaussian weight of itself cannot tell the random variables apart:

$$f(x)(2\pi)^{-n/2}e^{-|x|^2}=g(y_x)(2\pi)^{-n/2}e^{-|y_x|^2/2}.$$

Therefore, taking into account the Lebesgue measure's disregard for isometries and the Gaussian weight's laissez faire attitude toward vectors of the same length we see that: *should f and g be real-valued random variables such that there is a linear isometry τ of $\mathbb{R}^n$ onto itself for which $f(z)=g(\tau z)$ for all $z\in\mathbb{R}^n$, then $\int f(z)\,dG(z)=\int g(\tau z)\,dG(z)=\int g(z)\,dG(z)$ holds.*

Lemma 2. *Suppose $|x|=1=|y|$. Then*

1. $(\varphi_x, s_y)=(x\cdot x)\sqrt{2/\pi}$.
2. $\|\varphi_x-s_x\|^2=2-2\sqrt{2/\pi}$.

PROOF. As in Lemma 1, we let y' be a vector perpendicular to y so that $x=(x\cdot y)y+y'$. Now we note that

$$\begin{aligned}(\varphi_x, s_y)&=\int(x\cdot z)s(y\cdot z)\,dG(z)\\&=(x\cdot y)\int(y\cdot z)s(y\cdot z)\,dG(z)+\int(y'\cdot z)s(y\cdot z)\,dG(z)\\&=(x\cdot y)\int|y\cdot z|\,dG(z)+\int(y'\cdot z)s(y\cdot z)\,dG(z).\end{aligned}$$

Now it is clear that $\int|y\cdot z|\,dG(z)=\sqrt{2/\pi}\int_0^\infty te^{-t^2/2}\,dt$. Indeed, by our remarks above, the integral $\int|y\cdot z|\,dG(z)$ has the same value as $\int g(z)\,dG(z)$ for any real-valued random variable g on $\mathbb{R}^n$ obtained from $|y\cdot z|$ by means of an isometry of $\mathbb{R}^n$. Which g to choose? Well, the function $g(z)=|z_1|$ is easily seen to be obtainable from $|y\cdot z|$ by composition with a suitable isometry of $\mathbb{R}^n$. Therefore,

$$\int|y\cdot z|\,dG(z)=\int|z_1|\,dG(z)$$

and this latter integral is quickly seen to be equal to $\sqrt{2/\pi}$.

What of $\int (y' \cdot z) s(y \cdot z)\, dG(z)$? Of course, its value is 0. Since y' and y are perpendicular, we can move y' to $(|y'|, 0, 0, \ldots, 0)$ and y to $(0, 1, 0, \ldots, 0)$ by means of a linear isometry of $\mathbf{R}^n$ onto itself. Therefore,

$$\int (y' \cdot z) s(y \cdot z)\, dG(z) = \int z_1 \operatorname{sign} z_2\, dG(z) = 0$$

by direct and simple computation.

If we take all these comments to account, we get

$$\begin{aligned}(\varphi_x, s_y) &= (x \cdot y) \int |y \cdot z|\, dG(z) + \int (y' \cdot z) s(y \cdot z)\, dG(z)\\ &= (x \cdot y)\sqrt{\frac{2}{\pi}}.\end{aligned}$$

Part 2 is now easy to derive. In fact,

$$\begin{aligned}\|\varphi_x - s_x\|_2^2 &= (\varphi_x - s_x, \varphi_x - s_x)\\ &= (\varphi_x, \varphi_x) - 2(\varphi_x, s_x) + (s_x, s_x),\end{aligned}$$

which be Lemma 1 and part 1 of this lemma is

$$\begin{aligned}&= x \cdot x - 2x \cdot x\sqrt{\frac{2}{\pi}} + \int |x \cdot z|^2\, dG(z)\\ &= 1 - 2\sqrt{\frac{2}{\pi}} + 1 = 2 - 2\sqrt{\frac{2}{\pi}}\end{aligned}$$

since $\int |x \cdot z|^2\, dG(z) = 1$ (something we saw in Lemma 1). □

We now can state and prove the "fundamental theorem of the metric theory of tensor products."

Grothendieck's Inequality. *There is a universal constant $K_G > 0$ such that for any n and any $n \times n$ real matrix (a_{ij}) we have*

$$|(a_{ij})|_{\mathscr{H}} = \sup\left\{\left|\sum_{i=1}^n \sum_{j=1}^n a_{ij}(x_i, y_j)\right| : x_i, y_j \in \textit{some common Hilbert space},\ \|x_i\|, \|y_j\| \le 1\right\}$$

$$\left(= \sup\left\{\sum_{i=1}^n \left\|\sum_{j=1}^n a_{ij} x_j\right\| : x \in \textit{Hilbert space},\ \|x_j\| \le 1\right\}\right)$$

$$\le K_G \sup\left\{\left|\sum_{i=1}^n \sum_{j=1}^n a_{ij} s_i t_j\right| : |s_i|, |t_j| \le 1\right\}.$$

Proof. For the sake of normal relations we assume (a_{ij}) satisfies

$$\sup\left\{\left|\sum_{i=1}^{n}\sum_{j=1}^{n} a_{ij}s_i t_j\right| : |s_i|, |t_j| \leq 1\right\} = 1.$$

To start, we pick unit vectors $y_1, \ldots, y_n$ in some Hilbert space that closely approximate the quantity $|(a_{ij})|_{\mathscr{H}} = \sup\{\sum_{i=1}^{n}\|\sum_{j=1}^{n} a_{ij}x_j\| : x_j$ an element of a Hilbert space, $\|x_j\| \leq 1\}$. Let's say that $y_1, \ldots, y_n$ are selected to satisfy

$$|(a_{ij})|_{\mathscr{H}} = (1+\varepsilon)\sum_{i=1}^{n}\left\|\sum_{j=1}^{n} a_{ij}y_j\right\|.$$

The span of the vectors $y_1, \ldots, y_n$ is at most n dimensional; so we can put them in $\mathbb{R}^n$ quite comfortably. Now choose unit vectors $x_1, \ldots, x_n$ in $\mathbb{R}^n$ so that

$$\left\|\sum_{j=1}^{n} a_{ij}y_j\right\| = \sum_{j=1}^{n} a_{ij}y_j \cdot x_i,$$

this we can do for each $i = 1, \ldots, n$ through the use of the Riesz representation theorem. Of course,

$$\frac{|(a_{ij})|_{\mathscr{H}}}{1+\varepsilon} = \sum_{i=1}^{n}\sum_{j=1}^{n} a_{ij}(x_i \cdot y_j) = \sum_{i,j} a_{ij}x_i \cdot y_j.$$

Preparations are completed; let's calculate.

Notice that for any $x, y \in \mathbb{R}^n$,

$$(s_x, s_y) = (\varphi_x, s_y) + (s_x, \varphi_y) - (\varphi_x, \varphi_y) - (\varphi_x - s_x, s_y - \varphi_y).$$

Therefore,

$$\begin{aligned}
\sum_{i,j} a_{ij}(s_{x_i}, s_{y_j}) &= \sum_{i,j} a_{ij}(\varphi_{x_i}, s_{y_j}) + \sum_{i,j} a_{ij}(s_{x_i}, \varphi_{y_j}) \\
&\quad - \sum_{i,j} a_{ij}(\varphi_{x_i}, \varphi_{y_j}) - \sum_{i,j} a_{ij}(\varphi_{x_i} - s_{x_i}, s_{y_j} - \varphi_{y_j}), \\
&= \sum_{i,j} a_{ij}x_i \cdot y_j\sqrt{\frac{2}{\pi}} + \sum_{i,j} a_{ij}x_i \cdot y_j\sqrt{\frac{2}{\pi}} \\
&\quad - \sum_{i,j} a_{ij}x_i \cdot y_j - \sum_{i,j} a_{ij}(\varphi_{x_i} - s_{x_i}, s_{y_j} - \varphi_{y_j}) \\
&= \left(2\sqrt{\frac{2}{\pi}} - 1\right)\sum_{i,j} a_{ij}x_i \cdot y_j - \sum_{i,j} a_{ij}(\varphi_{x_i} - s_{x_i}, s_{y_j} - \varphi_{y_j}).
\end{aligned}$$

Let's rewrite the pertinent parts of the above identity:

$$\sum_{i,j} a_{ij}(s_{x_i}, s_{y_j}) = \left(2 - \sqrt{\frac{2}{\pi}} - 1\right)\sum_{i,j} a_{ij} x_i \cdot y_j - \sum_{i,j} a_{ij}(\varphi_{x_i} - s_{x_i}, s_{y_j} - \varphi_{y_j}).$$

Observe that each of the terms on the right involves in some way quantities related to $|(a_{ij})|_{\mathscr{H}}$. The first does so because of our choice of the x_i and y_j, whereas the second is dominated by a constant multiple of $|(a_{ij})|_{\mathscr{H}}$; indeed, a simple normalization argument shows that

$$\left|\sum_{i,j} a_{ij}(\varphi_{x_i} - s_{x_i}, s_{y_j} - \varphi_{y_j})\right| \leq |(a_{ij})|_{\mathscr{H}} \|\varphi_{x_i} - s_{x_i}\|_2 \|s_{y_j} - \varphi_{y_j}\|_2$$

which by part 2 of Lemma 2 is

$$\leq |(a_{ij})|_{\mathscr{H}} \left(2 - 2\sqrt{\frac{2}{\pi}}\right)^{1/2} \left(2 - 2\sqrt{\frac{2}{\pi}}\right)^{1/2}$$

$$= |(a_{ij})|_{\mathscr{H}} \left(2 - 2\sqrt{\frac{2}{\pi}}\right).$$

It now follows that

$$\left|\sum_{i,j} a_{ij}(s_{x_i}, s_{y_j})\right| \geq \left\{\left(\frac{2\sqrt{2/\pi} - 1)}{1+\varepsilon} - \left(2 - 2\sqrt{\frac{2}{\pi}}\right)\right\} |(a_{ij})|_{\mathscr{H}}.$$

But our assumption on (a_{ij}) that

$$\sup\left\{\left|\sum_{i,j} a_{ij} s_i t_j\right| : |s_i|, |t_j| \leq 1\right\} = 1$$

assures us of a bound on $|\sum_{i,j} a_{ij}(s_{x_i}, s_{y_j})|$:

$$\left|\sum_{i,j} a_{ij}(s_{x_i}, s_{y_j})\right| = \left|\int \sum_{i,j} a_{ij} s(x_i \cdot z) s(y_j \cdot z)\, dG(z)\right|$$

$$\leq \int \left|\sum_{i,j} a_{ij} s(x_i \cdot z) s(y_j \cdot z)\right| dG(z)$$

which, since $|s(x_i \cdot z)|, |s(y_i \cdot z)| \leq 1$ for any z, is

$$\leq \int 1\, dG(z) = 1.$$

It follows that for a fixed (but arbitrary) $\varepsilon > 0$,

$$1 \geq \left\{\left(2\sqrt{\frac{2}{\pi}} - 1\right)(1+\varepsilon)^{-1} - \left(2 - 2\sqrt{\frac{2}{\pi}}\right)\right\} |(a_{ij})|_{\mathscr{H}}.$$

On reflection one sees that this implies

$$1 \geq \left\{ \left(2\sqrt{\frac{2}{\pi}} - 1 \right) - \left(2 - 2\sqrt{\frac{2}{\pi}} \right) \right\} \|(a_{ij})\|_{\mathscr{H}}$$

$$\geq \left(4\sqrt{\frac{2}{\pi}} - 3 \right) \|(a_{ij})\|_{\mathscr{H}}$$

an estimate good enough to prove Grothendieck's inequality. □

One quick corollary to Grothendieck's inequality follows.

Generalized Littlewood Inequality. *Let $(a_{ij})_{i,j=1}^{\infty}$ be an infinite real matrix and suppose that for each N*

$$\left| \sum_{i,j=1}^{N} a_{ij} t_i s_j \right| \leq M$$

whenever all the t_i and s_j have absolute value ≤ 1. If $(x_{ki})_{k,i=1}^{\infty}$ is any real matrix such that

$$\left(\sum_k x_{ki}^2 \right)^{1/2} \leq C$$

for each i, then

$$\sum_j \left(\sum_k \left(\sum_i x_{ki} a_{ij} \right)^2 \right)^{1/2} \leq K_G C M.$$

Proof. It is clear that our hypothesis about (a_{ij}) implies that

$$\sum_i |a_{ij}| \leq M$$

for each $j \geq 1$. Further, the class of (x_{ki}) under consideration are plainly subject to the restrictions

$$|x_{ki}| \leq C$$

for each k, i. Hence, all the series $\sum_i x_{ki} a_{ij}$ are absolutely convergent (regardless of k, j); this allows us to assume each sum arising has but N summands with full confidence that the usual limiting arguments will be available to carry the argument to its natural conclusion.

Let $\vec{x}_i = (x_{1i}, x_{2i}, \ldots, x_{Ni})$ denote the ith column of $(x_{ki})_{k,i=1}^{N}$. View $\vec{x}_i$ as a vector in l_2^N; $\|\vec{x}_i\| \leq C$ for each i—this is precisely what the hypothesis on the admissible class of (x_{ki}) means. Grothendieck's inequality now tells us that

$$\sum_{j=1}^{N} \left\| \sum_{i=1}^{N} a_{ij} \vec{x}_i \right\| \leq K_G M \sup_{1 \leq i \leq N} \|\vec{x}_i\|$$

$$\leq K_G M C;$$

from here to the desired conclusion is just an easy exercise in interpreting the norms involved. □

The Grothendieck-Lindenstrauss-Pelczynski Cycle

We turn now to some of the early applications of Grothendieck's inequality. More precisely, we exhibit several instances of pairs (X, Y) of Banach spaces for which every bounded linear operator from X to Y is absolutely 2-summing. Of interest here is the pleasant fact that the X and Y arising include spaces that are truly classical Banach spaces.

First, isolate the $\mathscr{L}_p$-spaces of Lindenstrauss and Pelczynski.

Let $1 < \lambda < \infty$ and $1 \leq p \leq \infty$. A Banach space X is called a $\mathscr{L}_{p,\lambda}$*-space* if given a finite-dimensional subspace B of X there is a finite-dimensional subspace E of X containing B and an invertible linear map $T: E \to l_p^{\dim E}$ such that $\|T\|\,\|T^{-1}\| \leq \lambda$.

Every $L_p(\mu)$-space is a $\mathscr{L}_{p,\lambda}$-space for all $\lambda > 1$; c_0 and all $C(K)$-spaces are $\mathscr{L}_{\infty,\lambda}$-spaces for each $\lambda > 1$. Once a space is an $\mathscr{L}_{p,\lambda}$-space for some $\lambda > 1$, it is called an $\mathscr{L}_p$-space. The next two results were proved (more or less) for the classical infinite-dimensional models of $\mathscr{L}_p$-spaces by Grothendieck and clarified by Lindenstrauss and Pelczynski, who also recognized the finite-dimensional character of their statements.

Theorem (Grothendieck-Lindenstrauss-Pelczynski). *If X is a $\mathscr{L}_1$-space and Y is a Hilbert space, then every bounded linear operator $T: X \to Y$ is absolutely 1-summing.*

PROOF. Let $x_1, \ldots, x_n \in X$ be given and suppose $\sum_{i=1}^n |x^* x_i| \leq C\|x^*\|$ holds for every $x^* \in X^*$. Suppose X is a $\mathscr{L}_{1,\lambda}$-space; there is an integer $m > 0$ and an invertible operator $G: l_1^m \overset{\text{(into)}}{\to} X$ whose range contains $x_1, \ldots, x_n$ with $\|G\| = 1$ and $\|G^{-1}\| \leq \lambda$. Suppose $y_1, \ldots, y_n \in l_1^m$ are chosen so that $Gy_i = x_i$ for $i = 1, 2, \ldots, n$. Put

$$y_1 = a_{11}e_1 + a_{12}e_2 + \cdots + a_{1m}e_m = \sum_{j=1}^m a_{1j}e_j$$

$$y_2 = a_{21}e_1 + a_{22}e_2 + \cdots + a_{2m}e_m = \sum_{j=1}^m a_{2j}e_j$$

$$\vdots$$

$$y_n = a_{n1}e_1 + a_{n2}e_2 + \cdots + a_{nm}e_m = \sum_{j=1}^m a_{nj}e_j.$$

Let $y^* = (s_1, \ldots, s_m) \in l_\infty^m = (l_1^m)^*$ have norm ≤ 1 and take any real numbers

$t_1, \ldots, t_n$ of absolute value ≤ 1. Then

$$\left|\sum_{ij} a_{ij} t_i s_j\right| \leq \sum_i |t_i| \left(\left|\sum_j a_{ij} s_j\right|\right)$$

$$\leq \sum_i \left|\sum_j a_{ij} s_j\right|$$

$$= \sum_i |y^* y_i|$$

$$= \sum_i |y^* G^{-1} x_i|$$

$$= \sum_i |(G^{-1*} y^*) x_i|$$

$$\leq C\|G^{-1*} y^*\| \leq C\|G^{-1}\| \leq C\lambda.$$

Therefore, by Grothendieck's inequality,

$$\sum_{i=1}^n \|Tx_i\| = \sum_{i=1}^n \|TGy_i\| = \sum_{i=1}^n \left\|T \sum_{j=1}^m a_{ij} Ge_j\right\|$$

$$= \sum_{i=1}^n \left\|\sum_{j=1}^m a_{ij} TGe_j\right\| \leq K_G C\lambda \sup_{1 \leq j \leq m} \|TGe_j\|$$

$$\leq K_G C\lambda \|T\| < \infty.$$

That T is absolutely 1-summing follows from this. □

Theorem (Grothendieck-Lindenstrauss-Pelczynski). *Every operator from a $\mathscr{L}_\infty$-space to a $\mathscr{L}_1$-space is 2-summing.*

Proof. Let X be a $\mathscr{L}_{\infty,\lambda}$-space, Y be a $\mathscr{L}_{1,\rho}$-space and $S: X \to Y$ be a bounded linear operator.

Take any $x_1, \ldots, x_N \in X$. There is a $c > 0$ such that

$$\sum_{k=1}^N |x^* x_k|^2 \leq c^2 \|x^*\|^2$$

for any $x^* \in X^*$. There is an $m \geq 1$ and an invertible operator T from l_∞^m into X such that $\{x_1, \ldots, x_N\}$ is contained in Tl_∞^m, $\|T\| = 1$ and $\|T^{-1}\| \leq \lambda$. Let $z_1, \ldots, z_N \in l_\infty^m$ be chosen so that $Tz_h = x_h$. Again, there's a finite-dimensional subspace E of Y containing STl_∞^m and an invertible operator $R: E \to l_1^k$ ($k = \dim E$) with $\|R\| = 1$ and $\|R^{-1}\| \leq \rho$. Together, we conclude to the existence of an operator $S_0 = RST: l_\infty^m \to l_1^k$ and elements $z_1, \ldots, z_N$ of l_∞^m

such that given any $z^* \in l_1^m = (l_\infty^m)^*$,

$$\begin{aligned}\sum_{h=1}^{N} |z^* z_h|^2 &= \sum_{h=1}^{N} |z^* T^{-1} x_h|^2 \\ &= \sum_{h=1}^{N} \left|T^{-1*} z^*(x_h)\right|^2 \\ &\leq c^2 \|T^{-1*} z^*\|^2 \\ &\leq c^2 \lambda^2 \|z^*\|^2. \end{aligned} \tag{2}$$

Our special purpose is to show that $\sum_{h=1}^{N} \|S_0 z_h\|^2$ is bounded by some constant dependent only on c, λ, and $\|S_0\|$. Once this is done, it will follow that

$$\begin{aligned}\sum_{h=1}^{N} \|Sx_h\|^2 &= \sum_{h=1}^{N} \|R^{-1} S_0 z_h\|^2 \\ &\leq \|R^{-1}\|^2 \sum_{h=1}^{N} \|S_0 z_h\|^2 \end{aligned}$$

is bounded by a constant dependent only on c (a scaling factor), λ, $\|S_0\| \leq \|S\|$, and ρ; i.e., S is 2-summing.

Okay, let's bound $\sum_{h=1}^{N} \|S_0 z_h\|^2$.

Let $e_1, \ldots, e_m$ be the usual unit vector basis of l_∞^m and $f_1, \ldots, f_k$ be the usual unit vector basis of l_1^k. Define the matrix (a_{ij}) by

$$S_0 e_i = \sum_{j=1}^{k} a_{ij} f_j.$$

Notice that for any $u^* = (u_1, \ldots, u_k) \in S_{(l_1^k)^*} = S_{l_\infty^k}$ and any reals $t_1, \ldots, t_m, s_1, \ldots, s_k$ satisfying $|t_i|, |s_j| \leq 1$ we have

$$\begin{aligned}\left|\sum_{i,j} a_{ij} t_i s_j u_j\right| &= \left|(u_1 s_1, \ldots, u_k s_k)\left(S_0 \sum_{i=1}^{m} t_i e_i\right)\right| \\ &\leq \|(u_1 s_1, \ldots, u_k s_k)\|_{l_\infty^k} \|S_0\| \left\|\sum_{i=1}^{m} t_i e_i\right\|_{l_\infty^m} \\ &\leq \|S_0\|. \end{aligned} \tag{3}$$

Look at $z_1, \ldots, z_N$ and their representation in terms of $e_1, \ldots, e_m$:

$$z_h = \sum_{i=1}^{m} z_{hi} e_i.$$

From (2) (letting z^* take turns being each of the m different coordinate functionals on l_∞^m) we get

$$\sum_{h=1}^{N} |z_{hi}|^2 \leq c^2 \lambda^2 \tag{4}$$

for each $i=1,\dots,m$. In tandem with (3) and the generalized Littlewood inequality, (4) gives us

$$\sum_{j=1}^{k} u_j\left(\sum_{h=1}^{N}\left(\sum_{i=1}^{m} z_{hi}a_{ij}\right)^2\right)^{1/2} \leq c\lambda K_G\|S_0\|,$$

and this is so for any $(u_1,\dots,u_k)\in S_{(l_1^k)^*}$. It follows that the k-tuple

$$\left(\left(\sum_h\left(\sum_i z_{hi}a_{i1}\right)^2\right)^{1/2},\dots,\left(\sum_h\left(\sum_i z_{hi}a_{ik}\right)^2\right)^{1/2}\right)$$

has l_1^k-norm $\leq c\lambda K_G\|S_0\|$. Equivalently,

$$\sum_{j=1}^{k}\left|\sum_{h=1}^{N}\left(\left|\sum_{i=1}^{m} z_{hi}a_{ij}\right|\right)^2\right|^{1/2} \leq c\lambda K_G\|S_0\|.$$

Looking carefully at what is involved in this last inequality, we see that it just states that if the vectors $v_1,\dots,v_k$ in l_2^N are given by

$$v_j=\left(\sum_{i=1}^{m} z_{1i}a_{ij},\dots,\sum_{i=1}^{m} z_{Ni}a_{ij}\right),$$

then

$$\sum_{j=1}^{k}\|v_j\|_{l_2^N}\leq c\lambda K_G\|S_0\|.$$

The triangle inequality to the rescue:

$$\left\|\sum_{j=1}^{N} v_j\right\|_{l_2^N}\leq c\lambda K_G\|S_0\|,$$

or

$$\sum_{h=1}^{N}\left(\sum_{j=1}^{k}\left|\sum_{i=1}^{m} z_{hi}a_{ij}\right|\right)^2 \leq c^2\lambda^2K_G^2\|S_0\|^2.$$

Since

$$\begin{aligned}\|S_0 z_h\|_{l_1^k} &= \left\|\sum_{i=1}^{m} z_{hi}S_0e_i\right\|_{l_1^k}\\ &=\left\|\sum_{j=1}^{k}\left(\sum_{i=1}^{m} z_{hi}a_{ij}\right)f_j\right\|_{l_1^k}=\sum_{j=1}^{k}\left|\sum_{i=1}^{m} z_{hi}a_{ij}\right|\end{aligned}$$

we see that

$$\begin{aligned}\sum_{h=1}^{N}\|S_0z_h\|^2 &= \sum_{h=1}^{N}\left(\sum_{j=1}^{k}\left|\sum_{i=1}^{m} z_{hi}a_{ij}\right|\right)^2\\ &\leq c^2\lambda^2K_G^2\|S_0\|^2.\end{aligned}$$

This ends the proof. □

We rush to point out that l_2 is isomorphic to a subspace of $L_1[0,1]$, thanks to Khintchine's inequality. Consequently, every operator from a $\mathscr{L}_\infty$-space to l_2 is absolutely 2-summing.

Banach Spaces Having Unique Unconditional Bases

Theorem 3 (Lindenstrauss-Pelczynski). *Let* (x_n) *be a normalized unconditional basis for* l_1. *Then* (x_n) *is equivalent to the unit vector basis.*

PROOF. Suppose $K>0$ is chosen so that

$$\left\|\sum_n b_n a_n x_n\right\| \le K \left\|\sum_n a_n x_n\right\|$$

holds for any $(b_n) \in B_{l_\infty}$ and any sequence of scalars (a_n) for which $\Sigma_n a_n x_n \in l_1$. Since l_1 imbeds isometrically into $L_1[0,1]$ and $\Sigma_n a_n x_n$ is unconditionally convergent,

$$\sum_n |a_n|^2 = \sum_n \|a_n x_n\|^2 < \infty$$

by Orlicz's theorem. It follows that the operator $T: l_1 \to l_2$ defined by $T(\Sigma_n a_n x_n) = (a_n)$ is well-defined. T is linear, one to one, and bounded, too; the boundedness of T follows from the proof of Orlicz's theorem or from a closed-graph argument, if you please. By Grothendieck's inequality we know that T is absolutely summing. Consequently, if (a_n) is a scalar sequence for which $\Sigma_n a_n x_n \in l_1$, then

$$\begin{aligned}
\sum_n |a_n| &= \sum_n \|T a_n x_n\| \\
&\le \pi_1(T) \sup_{\varepsilon_n = \pm 1} \left\|\sum_n \varepsilon_n a_n x_n\right\| \\
&\le \pi_1(T) K \left\|\sum_n a_n x_n\right\| \\
&\le \pi_1(T) K \sum_n |a_n|.
\end{aligned}$$

It follows from this that (x_n) is equivalent to l_1's unit vector basis. □

Remark: If one feels like bypassing Orlicz's theorem for another application of Grothendieck's inequality, then accommodations can be made. What is needed, of course, is the assurance that $\Sigma_n |a_n|^2 < \infty$ for any $\Sigma_n a_n x_n \in l_1$. To achieve this without recourse to Orlicz, one can define the operator $S: c_0 \to l_1$ by $S(\lambda_n) = \Sigma_n \lambda_n a_n x_n$; Grothendieck's inequality leads us to believe S is absolutely 2-summing from which the square summability of (a_n) is an easy consequence.

A result parallel to that expressed about l_1 in Theorem 3 holds for c_0 as well. In fact, suppose (x_n) is a normalized unconditional basis for c_0 and let (x_n^*) be the sequence of (x_n)'s coefficient functionals. It is plain that $\Sigma_n|x^*x_n|<\infty$ for each extreme point x^* of $B_{c_0^*}=B_{l_1}$; so (x_n^*) is a basis for l_1, this because of Lemma 13 in Chapter IX. It is easy to see that the pleasure derived from (x_n)'s unconditionality is shared by (x_n^*); the sequence (x_n^*) is an unconditional basis for l_1. The fact that $\|x_n\|=1$ for all n tells us that $1\leq\|x_n^*\|\leq 2M$, where M is the basis constant of (x_n). It follows from Theorem 3 that the normalized unconditional basis $(x_n^*/\|x_n^*\|)$ of l_1 is equivalent to the unit vector basis of l_1; from this it is easy to conclude that (x_n^*) is equivalent to the unit vector basis of l_1, too. This, though, is tantamount to (x_n)'s equivalence with the unit vector basis of c_0. For recording purposes, we summarize the above discussion.

Theorem 4 (Lindenstrauss-Pelczynski). *Let (x_n) be a normalized unconditional basis of c_0. Then (x_n) is equivalent to the unit vector basis.*

What spaces other than c_0 and l_1 have unique unconditional bases? Here is one: l_2. In fact, if $x_1,\ldots,x_n\in l_2$, then it is an easy consequence of the parallelogram law to show that given $y_1,\ldots,y_n\in l_2$,

$$\sum_{(\theta_i)_{i=1}^n\in\{\pm1\}^n}\left\|\sum_{i=1}^n\theta_i y_i\right\|^2=2^n\sum_{i=1}^n\|y_i\|^2.$$

From this it follows easily that if (x_n) is a normalized unconditional basis for l_2, then $\Sigma_i a_i x_i\in l_2$ if and only if $\Sigma_i|a_i|^2<\infty$.

c_0, l_1, and l_2 all have unique normalized unconditional bases. Any others? The startling answer is No! This result, due to Lindenstrauss and Zippin, is one of the real treasures in the theory of Banach spaces. It is only with the greatest reluctance that we do not pursue the proof of this result here.

Exercises

1. *$L_p[0,1]$ is a $\mathcal{L}_p$-space.* $L_p[0,1]$ is a $\mathcal{L}_{p,1+\varepsilon}$-space for every $\varepsilon>0$.
2. *$C(K)$ is a $\mathcal{L}_\infty$-space.* If K is a compact Hausdorff space, then $C(K)$ is a $\mathcal{L}_{\infty,1+\varepsilon}$-space for each $\varepsilon>0$. (*Hint*: You might find that partitions of unity serve as a substitute for measurable partitions of $[0,1]$.)
3. *Lattice bounded operators into $L_2[0,1]$.* Let $T: X\to L_2[0,1]$ be a bounded linear operator. Suppose there is a $g\in L_2[0,1]$ such that
$$|Tx|\leq g \quad \text{almost everywhere}$$
for each $x\in B_X$. Show that T is absolutely 2-summing.
4. *Hilbert-Schmidt operators on $L_2[0,1]$.* Let $T: L_2[0,1]\to L_2[0,1]$ be a bounded linear operator for which $T(L_2[0,1])\subseteq L_\infty[0,1]$ setwise. Then T is a Hilbert-Schmidt operator.

Notes and Remarks

The importance of the Lindenstrauss-Pelczynski paper to the revival of Banach space theory cannot be exaggerated. On the one hand, the challenge of Grothendieck's visionary program was reissued and a call to arms among abstract analysts made; on the other hand, Lindenstrauss and Pelczynski provided leadership by crystalizing many notions, some perhaps only implicitly present in Grothendieck's writings, central to the development of a real structure theory. They solved long-standing problems. They added converts to the Banach space faith with enticing problems. Their work led to meaningful relationships with other important areas of mathematical endeavor.

No doubt the leading role in the Lindenstrauss-Pelczynski presentation was played by Grothendieck's inequality. They followed Grothendieck's original scheme of proof, an averaging argument pursued on the n-sphere of Euclidean space with rotation invariant Haar measure gauging size, though they did provide, as one might expect, a few more details than Grothendieck did.

Interestingly enough, many of the other proofs of Grothendieck's inequality have come about in applications of Banach space ideas to other areas of analysis.

B. Maurey (1973) proved a form of Grothendieck's inequality while looking for the general character of his now-famous factorization scheme. He borrowed some ideas from H. P. Rosenthal's work (1973) on subspaces of L_p, improved on them and, with G. Pisier, molded them into the notions of type and cotype.

G. Pisier settled a problem of J. Ringrose in operator theory by proving the following stunning C^* version of Grothendieck's inequality.

Theorem. *Let $\mathscr{A}$ be a C^*-algebra and E be a Banach space of cotype 2; suppose either $\mathscr{A}$ or E satisfies the bounded approximation property. Then every operator from $\mathscr{A}$ to E factors through a Hilbert space.*

The result itself generalizes Grothendieck's inequality but more to the point, Pisier's proof suggested (to him) a different approach to the original inequality through the use of interpolation theory.

J. L. Krivine (1973) in studying Banach lattices proved the following lattice form of Grothendieck's inequality.

Theorem. *Let X and Y be Banach lattices and $T: X \to Y$ be a bounded linear operator. Then for any $x_1, \ldots, x_n \in X$ we have*

$$\left\| \left(\sum_{i=1}^{n} |Tx_i|^2 \right)^{1/2} \right\| \leq K_G \|T\| \left\| \left(\sum_{i=1}^{n} |x_i|^2 \right)^{1/2} \right\|,$$

where K_G is the universal Grothendieck constant.

Of course, sense must be made of the square of a member of a Banach lattice, but this causes no difficulty for a Krivine; he made sense of it and derived the above inequality, thereby clearing the way for some remarkably sharp theorems in the finer structure theory of Banach lattices.

To cite but one such advance, we need to introduce the Orlicz property: a Banach space X has the *Orlicz property* if given an unconditionally convergent series $\Sigma_n x_n$ in X, then $\Sigma_n \|x_n\|^2 < \infty$. Orlicz showed that $L_p[0,1]$ has the Orlicz property whenever $1 \leq p \leq 2$. As we mentioned, Orlicz's proof can be easily adapted to show the somewhat stronger feature of the spaces $L_p[0,1]$ for $1 \leq p \leq 2$, namely, they have cotype 2. With Krivine's version of Grothendieck's inequality in hand, B. Maurey was able to establish the following improvement of a result of Dubinsky, Pelczynski, and Rosenthal (1972).

Theorem. *If X is a Banach lattice, then X has cotype 2 if and only if X has the Orlicz property.*

Generally, it is so that spaces having cotype 2 have the Orlicz property; however, *it is not known if every Banach space with the Orlicz property has cotype* 2.

A. Pelczynski and P. Wojtaszczyk were studying absolutely summing operators from the disk algebra to l_2 when they discovered their proof of what is essentially Grothendieck's inequality. They observed that an old chestnut of R. E. A. C. Paley (1933) could, with some work, be reinterpreted as saying that there is an absolutely summing operator from the disk algebra onto l_2. Using this and the lifting property of l_1, they were able to deduce that every operator from l_1 to l_2 is absolutely summing. Incidentally, they also noted that the existence of an absolutely summing operator from the disk algebra onto l_2 serves as a point of distinction between the disk algebra and any space of continuous functions. Any absolutely summing operator from a $\mathscr{L}_\infty$-space to l_2 is compact; so the existence of a quotient map from the disk algebra onto l_2 implies that the disk algebra is not isomorphic as a Banach space to any $C(K)$-space.

It is of more than passing interest that the Pelczynski-Wojtaszczyk proof that the disk algebra is not isomorphic to any $\mathscr{L}_\infty$-space had already been employed by S. V. Kisliakov, at least in spirit. Kisliakov (1976) showed that for $n \geq 2$ the spaces $C^k(I^n)$ of k-times continuously differentiable functions on the n-cube are not $\mathscr{L}_\infty$-spaces by exhibiting operators from their duals to Hilbert space that fail to be absolutely 1-summing.

As is usual in such matters, the precise determination of the best constant that works in Grothendieck's inequality has aroused considerable curiosity. Despite the optimistic hopes of a number of mathematicians, this constant appears on the surface to be unrelated to any of the old-time favorite constants; J. L. Krivine has provided a scheme that hints at the best value of the Grothendieck constant and probably sheds considerable light (for those who will see) on the exact nature of Grothendieck's inequality.

We have presented the results of the section entitled The Grothendieck-Lindenstrauss-Pelczynski Cycle much as Lindenstrauss and Pelczynski did without confronting some small technical difficulties which arise when one pursues the full strength of Theorem 4, namely, *if* $1 \le p \le 2$, *then every operator from a* $\mathscr{L}_\infty$ *to a* $\mathscr{L}_p$ *is absolutely 2-summing.*

The development of the structure of $\mathscr{L}_p$-spaces has been one of the crowning successes of Banach space theory following the Lindenstrauss-Pelczynski breakthrough. This is not the place for one to read of the many nuts cracked in the subject's development; rather, we preach patience while awaiting volume III of the Lindenstrauss-Tzafriri books wherein the complete story of the $\mathscr{L}_p$-spaces is to be told.

A Banach space Y is said to have the *Grothendieck property* if every operator from Y to l_2 is absolutely 1-summing. Theorem 3 shows that $\mathscr{L}_1$-spaces have the Grothendieck property. S. V. Kisliakov (1976) and G. Pisier (1978) have each shown that if R is a reflexive subspace of $L_1[0,1]$, then $L_1[0,1]/\mathrm{R}$ has the Grothendieck property; so $\mathscr{L}_1$-spaces are not alone in the enjoyment of the Grothendieck property. S. Kaijser (0000) has given another view of the Kisliakov-Pisier theorem with an eye toward broader applications. More recently, J. Bourgain (0000) has shown that L_1/H^1 has the Grothendieck property. *For which subspaces* X *of* $L_1[0,1]$ *does* L_1/X *have the Grothendieck property*? *If* X *is a subspace of* $L_1[0,1]$, *is isomorphic to a dual space, then does* L_1/X *have the Grothendieck property*?

Returning again to Banach spaces of cotype 2, we ought to mention that the $\mathscr{L}_p$-spaces have cotype 2 whenever $1 \le p \le 2$. Again they are not alone in this situation. N. Tomczak-Jaegermann (1974) has shown that the dual of any C^*-algebra as well as the Schatten classes C_p for $1 \le p \le 2$ have cotype 2. Gorgadze and Tarieladze (1980) have found criteria for Orlicz spaces to have cotype 2, and J. Creekmore (1981) has determined which of the Lorentz spaces L_{pq} have cotype 2. Again, G. Pisier and S. V. Kisliakov found that $L_1[0,1]/X$ has cotype 2 whenever X is a reflexive subspace of $L_1[0,1]$ and, in an awesome display of analytical power, J. Bourgain has shown that L_1/H^1 has cotype 2. Pisier builds on Bourgain's result to settle in the negative one form of an early conjecture of Grothendieck in tensor products; on the other hand, Pisier uses the considerable machinery available in spaces with cotype to give an alternative solution to the same Grothendieck conjecture in the presence of some approximation property. We cite Pisier's factorization theorem.

Theorem. *Let* X *and* Y *be Banach spaces with both* X^* *and* Y *having cotype* 2. *Then every approximable operator from* X *to* Y *factors through a Hilbert space.*

This generalizes Pisier's C^* analogue of Grothendieck inequality and the original Grothendieck inequality.

There have been many applications of absolutely summing operators and Grothendieck's inequality that bear close study. Instead of going into an encyclopedic account, we state a coupie of our favorites and hope the student will bear in mind the rapidly growing body of splendid examples that solidify the position of importance held by the theory of absolutely p-summing operators. The following results depend on the theory of absolutely summing operators and are but two of our favorites.

Theorem (G. Bennett, B. Maurey, A. Nahoum). *If* $\Sigma_n f_n$ *is an unconditionally convergent series in* $L_1[0,1]$, *then* $\Sigma_n f_n/\ln(n+1)$ *converges almost everywhere.*

An elegant proof of this is was uncovered by P. Orno (1976).

Theorem (A. Tonge, N. Varapoulos). *If a Banach algebra is a* $\mathscr{L}_\infty$*-space, then it is an algebra of operators.*

A. Tonge and his co-workers have developed the ideas essential to the proof of the above theorem to give a number of striking criteria for the representation of a Banach algebra as an algebra of operators or, even, as a uniform algebra.

As we noted in the text, the fact that c_0, l_1, and l_2 have a unique normalized unconditional basis characterizes these spaces. To be fair, the realization by Lindenstrauss and Pelczynski that c_0, l_1, and l_2 have unique unconditional bases was quite startling. Earlier, A. Pelczynski and I. Singer (1964) had shown that once an infinite-dimensional Banach space has a basis, it has infinitely many nonequivalent bases; so the Lindenstrauss-Pelczynski discovery was bound to be a surprise of sorts. The proof by Lindenstrauss and M. Zippin (1969) that only the spaces c_0, l_1, and l_2 have unconditional bases was based in large part on Zippin's earlier isolation of c_0 and the separable l_p as the only spaces with perfectly homogeneous bases.

Bibliography

Bourgain, J. 1984. Absolutely summing operators on H^∞ and the cotype property.
Bourgain, J. 1984. New Banach space properties of the disc algebra and H^∞.
Both of the above, *Acta Math., to appear.*

Creekmore, J. 1981. Type and cotype in Lorentz L_{pq} spaces. *Indag. Math.*, **43**, 145–152.

Dubinsky, E., Pelczynski, A., and Rosenthal, H. P. 1972. On Banach spaces for which $\pi_2(\alpha_\infty, X) = B(\alpha_\infty, X)$. *Studia Math.*, **44**, 617–634.

Gorgadze, Z. G. and Tarieladze, V. I. 1980. On geometry of Orlicz spaces, *Probability Theory on Vector Spaces, II* (Proc. Second International Conf., Blazejewko, 1979), Lecture Notes in Mathematics, Vol. 828. Berlin-Heidelberg-New York: Springer-Verlag, pp. 47–51.

Grothendieck, A. 1956. Resume de la theorie metrique des produits tensoriels topologiques. *Bol. Soc. Mat. Sao. Paulo*, **8**, 1–79.

Kaijser, S. 0000. A new proof on the existence of non-L^1 spaces satisfying Grothendieck's theorem.

Kisliakov, S. V. 1976. Sobolev imbedding operators and non-isomorphism of certain Banach spaces. *Funct. Anal. Appl.*, **9**, 290–294.

Kisliakov, S. V. 1976. On spaces with a "small" annihilator. In *Studies in Linear Operators and Theory of Functions*, Seminar Leningrad Math. Inst., Vol. 7, 192–195.

Krivine, J. L. 1973–1974. Theoremes de factorization dans les espaces reticules. Seminaire Maurey-Schwartz.

Krivine, J. L. 1977–1978. Constantes de Grothendieck et fonctions de type positif sur les spheres. Seminaire sur la Geometrie des espaces de Banach.

Lindenstrauss, J. and Pelczynski, A. 1968. Absolutely summing operators in $\mathscr{L}_p$ spaces and their applications. *Studia Math.*, **29**, 275–326.

Lindenstrauss, J. and Zippin, M. 1969. Banach spaces with a unique unconditional basis. *J. Funct. Anal.*, **3**, 115–125.

Maurey, B. 1972–1973. Une normelle demonstration d'un theoreme de Grothendieck. Seminaire Maurey-Schwartz.

Maurey, B. 1973–1974. Type et cotype dan les espaces munis de structures locales inconditionnalles. Seminaire Maurey-Schwartz.

Maurey, B. and Pisier, G. 1973. Caracterisation d'une classe d'espaces de Banach par des proprietes de series aleatoires vectorielles. *C. R. Acad. Sci., Paris*, **277**, 687–690.

Maurey, B. and Pisier, G. 1976. Series de variables aleatories vectorielles independants et proprietes geometriques des espaces de Banach. *Studia Math.*, **58**, 45–90.

Orno, P. 1976. A note on unconditionally converging series in L_p. *Proc. Amer. Math. Soc.*, **59**, 252–254.

Paley, R. E. A. C. 1933. On the lacunary coefficients of power series. *Ann. Math.*, **34**, 615–616.

Pelczynski, A. 1977. *Banach Spaces of Analytic Functions and Absolutely Summing Operators*. CBMS regional conference series, Vol. 30.

Pelczynski, A. and Singer, I. 1964. On non-equivalent bases and conditional bases in Banach spaces. *Studia Math.*, **25**, 5–25.

Pisier, G. 1978. Une normelle classe d'espaces de Banach verificant le theoreme de Grothendieck. *Ann. l'Inst. Fourier (Grenoble)*, **28**, 69–90.

Pisier, G. 1978. Grothendieck's theorem for noncommutative C^* algebras, with an appendix on Grothendieck's constants. *J. Funct. anal.*, **29**, 397–415.

Pisier, G. 1980. Un theoreme sur les operateurs lineaires entre espaces de Banach qui se factorisent par un espace de Hilbert. *Ann. Sci. Ec. Norm. Sup.*, **13**, 23–43.

Pisier, G. 1981. A counterexample to a conjecture of Grothendieck. *C. R. Acad. Sci. Paris*, **293**, 681–684.

Rietz, R. 1974. A proof of the Grothendieck inequality. *Israel J. Math.*, **19**, 271–276.

Rosenthal, H. P. 1975. On subspaces of L_p. *Ann. Math.*, **97**, 344–373.

Tomczak-Jaegermann, N. 1974. The moduli of smoothness and convexity and the Rademacher averages of trace class S_p. *Studia Math.*, **50**, 163–182.

Zippin, M. 1966. On perfectly homogeneous bases in Banach spaces. *Israel J. Math.*, **4**, 265–272.

An Intermission: Ramsey's Theorem

Some notation, special to the present discussion, ought to be introduced. If A and B are subsets of the set $\mathbf{N}$ of natural numbers, then we write $A < B$ whenever $a < b$ holds for each $a \in A$ and $b \in B$. The collection of finite subsets of A is denoted by $\mathscr{P}_{<\infty}(A)$ and the collection of infinite subsets of A by $\mathscr{P}_{\infty}(A)$. More generally for $A, B \subseteq \mathbf{N}$ we denote by $\mathscr{P}_{<\infty}(A, B)$ the collection

$$\{X \in \mathscr{P}_{<\infty}(\mathbf{N}) : A \subseteq X \subseteq A \cup B, A < X \setminus A\}$$

and by $\mathscr{P}_{\infty}(A, B)$ the collection

$$\{X \in \mathscr{P}_{\infty}(\mathbf{N}) : A \subseteq X \subseteq A \cup B, A < X \setminus A\}.$$

It might be useful to think of $\mathscr{P}_{<\infty}(A, B)$ as the collection of finite subsets of $A \cup B$ that "start with A" and similarly of $\mathscr{P}_{\infty}(A, B)$ as the collection of infinite subsets of $A \cup B$ that "start with A." Of course, $\mathscr{P}_{<\infty}(\varnothing, A)$ and $\mathscr{P}_{\infty}(\varnothing, A)$ are just $\mathscr{P}_{<\infty}(A)$ and $\mathscr{P}_{\infty}(A)$, respectively; we use $\mathscr{P}_{<\infty}(A)$ and $\mathscr{P}_{\infty}(A)$ in such cases—it's shorter.

The notation settled, we introduce a topology on $\mathscr{P}_{\infty}(\mathbf{N})$ by taking for a basis, sets of the form $\mathscr{P}_{\infty}(A, B)$, where $A \in \mathscr{P}_{<\infty}(\mathbf{N})$ and $B \in \mathscr{P}_{\infty}(\mathbf{N})$. It is easy to show that the collection of sets $\mathscr{P}_{\infty}(A, B)$ of the prescribed form do indeed form a base for a topology τ on $\mathscr{P}_{\infty}(\mathbf{N})$ which on a bit of reflection is seen to be stronger (has more open sets) than the relative product topology (henceforth called the *classical topology*) on $\mathscr{P}_{\infty}(\mathbf{N})$. We find in what follows that the topology τ is particularly well suited for proving results with a combinatorial bent.

Before studying τ, we investigate a bit of mathematical sociology.

Let $\mathscr{S} \subseteq \mathscr{P}_{\infty}(\mathbf{N})$, $A \in \mathscr{P}_{<\infty}(\mathbf{N})$ and $B \in \mathscr{P}_{\infty}(\mathbf{N})$. We say that B *accepts* A (into $\mathscr{S}$) if $\mathscr{P}_{\infty}(A, B) \subseteq \mathscr{S}$. Should there be no infinite subsets C of B that accept A into $\mathscr{S}$, then we say that B *rejects* A (from $\mathscr{S}$).

For the time being, keep $\mathscr{S} \subseteq \mathscr{P}_{\infty}(\mathbf{N})$ fixed and address all acceptances and rejections as relative to $\mathscr{S}$.

Lemma 1. *Let $A \in \mathcal{P}_{<\infty}(\mathbb{N})$ and $B \in \mathcal{P}_{\infty}(\mathbb{N})$.*

1. *Suppose B accepts A. Then each $C \in \mathcal{P}_{\infty}(B)$ also accepts A.*
2. *Suppose B rejects A. Then each $C \in \mathcal{P}_{\infty}(B)$ also rejects A.*

PROOF. Part 1 is immediate from the fact that whenever $C \in \mathcal{P}_{\infty}(B)$, $\mathcal{P}_{\infty}(A, C) \subseteq \mathcal{P}_{\infty}(A, B)$.

Part 2 is clear, too, since were there a $C \in \mathcal{P}_{\infty}(B)$ that did not reject A it would be because there is a $D \in \mathcal{P}_{\infty}(C)$ that accepted A. Plainly, this is unheard of for members D of $\mathcal{P}_{\infty}(B)$ once B rejects A. □

Lemma 2. *There is a $Z \in \mathcal{P}_{\infty}(\mathbb{N})$ which accepts or rejects each of its finite subsets.*

PROOF. (By diagonalization.) To start with we observe a more-or-less obvious consequence of the notions of acceptance and rejection: *given $A \in \mathcal{P}_{<\infty}(\mathbb{N})$ and $B \in \mathcal{P}_{\infty}(\mathbb{N})$ there is a $C \in \mathcal{P}_{\infty}(B)$ that either accepts or rejects A*; in other words, B is never *entirely* ambivalent towards A. Why? Well, if you consider the possibilities, either there is a $C \in \mathcal{P}_{\infty}(B)$ that accepts A or there isn't. If no $C \in \mathcal{P}_{\infty}(B)$ accepts A, it's tantamount to the rejection of A by B.

Now let $B \in \mathcal{P}_{\infty}(\mathbb{N})$.

Choose $X_0 \in \mathcal{P}_{\infty}(B)$ such that X_0 either accepts or rejects $\varnothing$; such an X_0 can be found by our initial comments.

Let $z_0 = \min X_0$.

Again by our opening marks we see that there is an $X_1 \in \mathcal{P}_{\infty}(X_0 \setminus \{z_0\})$ such that X_1 accepts or rejects $\{z_0\}$. *Take particular note*: X_1 is an infinite subset of X_0 and as such must accept or reject $\varnothing$ according to X_0's whimsy; this is in accordance with Lemma 1.

Let $z_1 = \min X_1$ and notice that $z_0 < z_1$.

Once again observe that there is an $X_2' \in \mathcal{P}_{\infty}(X_1 \setminus \{z_1\})$ such that X_2' accepts or rejects $\{z_1\}$. Taking particular notice of what went on before, we mark down the fact that X_2' accepts or rejects $\{z_0\}$ and $\varnothing$ in accordance with X_1's treatment of them. Now observe that there must be an $X_2 \in \mathcal{P}_{\infty}(X_2')$ that either accepts or rejects $\{z_0, z_1\}$. Again, Lemma 1 ensures that X_2 accepts some of the sets $\varnothing$, $\{z_0\}$, $\{z_1\}$, and $\{z_0, z_1\}$ and rejects the rest.

Briefly our next step finds us letting $z_2 = \min X_2$, taking $X_3''' \in \mathcal{P}_{\infty}(X_2 \setminus \{z_2\})$ to accept or reject $\{z_2\}$ with us mindful of the fact that X_3''' a fortiori treats $\varnothing$, $\{z_0\}$, $\{z_1\}$, and $\{z_0, z_1\}$ as X_2 does. Pick $X_3'' \in \mathcal{P}_{\infty}(X_3''')$ to accept or reject $\{z_0, z_2\}$, $X_3' \in \mathcal{P}_{\infty}(X_3'')$ to accept or reject $\{z_1, z_2\}$, and $X_3 \in \mathcal{P}_{\infty}(X_3')$ to accept or reject $\{z_0, z_1, z_2\}$. The last pick of the litter, X_3, accepts some of the subsets of $\{z_0, z_1, z_2\}$ and rejects the rest.

Our procedure is clear. $Z = \{z_0, z_1, z_2, \ldots\}$ is our set. □

We have proved more than advertised; we have actually shown the following.

Lemma 2′. *Given $Y \in \mathscr{P}_\infty(\mathbb{N})$, there is a $Z \in \mathscr{P}_\infty(Y)$ that accepts some members of $\mathscr{P}_{<\infty}(Z)$ and rejects the rest.*

Lemma 3. *Suppose $Z \in \mathscr{P}_\infty(\mathbb{N})$ accepts or rejects each of its finite subsets (on an individual basis).*

1. *If $A \in \mathscr{P}_{<\infty}(Z)$ and Z rejects A, then Z rejects $A \cup \{n\}$ for all but finitely many $n \in Z$.*
2. *If Z rejects $\varnothing$, then there is a $C \in \mathscr{P}_\infty(Z)$ that rejects each of its finite subsets.*

Proof. 1. By hypothesis, Z accepts any $A \cup \{n\}$ $(n \in Z)$ it does not reject; hence, were part 1 to fail,

$$B = \{n \in Z : Z \text{ accepts } A \cup \{n\}\} \in \mathscr{P}_\infty(Z).$$

Consider $\mathscr{P}_\infty(A, B)$. If $X \in \mathscr{P}_\infty(A, B)$, then $A \subseteq X \subseteq A \cup B$, $A < X \setminus A$ and X is infinite. Let $n = \min X \setminus A$; $n \in B$, and so Z accepts $A \cup \{n\}$. It follows that

$$\mathscr{P}_\infty(A \cup \{n\}, Z) \subseteq \mathscr{S}.$$

Furthermore,

$$X \in \mathscr{P}_\infty(A \cup \{n\}, Z)$$

by choice of n. In sum we have

$$\mathscr{P}_\infty(A, B) \subseteq \bigcup_{n \in B} \mathscr{P}_\infty(A \cup \{n\}, B) \subseteq \mathscr{S},$$

which is a contradiction. Z rejects A; so B has to, too.

Part 2 follows from part 1 in much the same way as Lemma 2 was deduced. □

It is now time for a rare treat—an application of sociology to mathematics. We say a collection $\mathscr{S}$ of members of $\mathscr{P}_\infty(\mathbb{N})$ is a *Ramsey collection* if there exists an $S \in \mathscr{P}_\infty(\mathbb{N})$ such that either $\mathscr{P}_\infty(S) \subseteq \mathscr{S}$ or $\mathscr{P}_\infty(S) \subseteq \mathscr{P}_\infty(\mathbb{N}) \setminus \mathscr{S}$. $\mathscr{S}$ is called a *completely Ramsey collection* if for each $A \in \mathscr{P}_{<\infty}(\mathbb{N})$ and each $B \in \mathscr{P}_\infty(\mathbb{N})$ there is an $S \in \mathscr{P}_\infty(B)$ such that either $\mathscr{P}_\infty(A, S) \subseteq \mathscr{S}$ or $\mathscr{P}_\infty(A, S) \subseteq \mathscr{P}_\infty(\mathbb{N}) \setminus \mathscr{S}$.

Lemma 4. *Every τ-open set in $\mathscr{P}_\infty(\mathbb{N})$ is a Ramsey collection.*

Proof. Let $\mathscr{S}$ be a τ-open subset of $\mathscr{P}_\infty(\mathbb{N})$. By Lemma 2, there is a $Z \in \mathscr{P}_\infty(\mathbb{N})$ that accepts or rejects each of its own finite subsets on an individual basis, acceptance (and rejection) being relative to $\mathscr{S}$. Of course, if

Z accepts $\varnothing$, then $\mathscr{P}_\infty(Z) \subseteq \mathscr{S}$. If Z rejects $\varnothing$, then Lemma 3 tells us there is a $Y \in \mathscr{P}_\infty(Z)$ that rejects each of its finite subsets. *We claim that* $\mathscr{P}_\infty(Y) \cap \mathscr{S} = \varnothing$. In fact, were $X \in \mathscr{P}_\infty(Y) \cap \mathscr{S}$, then there would be an $A \in \mathscr{P}_{<\infty}(\mathbb{N})$ and a $B \in \mathscr{P}_\infty(\mathbb{N})$ such that $X \in \mathscr{P}_\infty(A, B) \subseteq \mathscr{S}$ since $\mathscr{S}$ is τ-open. But $\mathscr{P}_\infty(A, B)$'s containment in $\mathscr{S}$ entails B accepting A, whereas $X \in \mathscr{P}_\infty(A, B)$ implies that X also accepts A by Lemma 1. Since $A \subseteq X$ is part and parcel of $X \in \mathscr{P}_\infty(A, B)$, we have X accepting one of its own finite subsets. Since $X \in \mathscr{P}_\infty(Y)$, Y cannot reject this same finite subset of X, by Lemma 1—a contradiction. Our claim follows and with it Lemma 4. □

We are ready for a major step.

Theorem (Nash-Williams). *Every τ-open set is completely Ramsey. Consequently every τ-closed set is completely Ramsey.*

PROOF. Let $\mathscr{S}$ be τ-open, $A \in \mathscr{P}_{<\infty}(\mathbb{N})$ and $B \in \mathscr{P}_\infty(\mathbb{N})$. Suppose $\beta : \mathbb{N} \to B$ is a one-to-one increasing map of $\mathbb{N}$ onto B: β induces a τ-continuous function f from $\mathscr{P}_\infty(\mathbb{N})$ into itself. Define $g : \mathscr{P}_\infty(\mathbb{N}) \to \mathscr{P}_\infty(\mathbb{N})$ by

$$g(Y) = Y \cup A;$$

g is also τ-continuous. Since $\mathscr{S}$ is τ-open, $(g \circ f)^{-1}(\mathscr{S})$ is too; by Lemma 4, $(g \circ f)^{-1}(\mathscr{S})$ is Ramsey. Hence, there is an $X \in \mathscr{P}_\infty(\mathbb{N})$ such that either

$$\mathscr{P}_\infty(X) \subseteq (g \circ f)^{-1}(\mathscr{S})$$

or

$$\mathscr{P}_\infty(X) \subseteq \mathscr{P}_\infty(\mathbb{N}) \backslash (g \circ f)^{-1}(\mathscr{S}).$$

Looking at $f(X) = Y$, we get a member Y of $\mathscr{P}_\infty(B)$ which satisfies either

$$\mathscr{P}_\infty(Y) \subseteq g^{-1}(\mathscr{S})$$

or

$$\mathscr{P}_\infty(Y) \subseteq \mathscr{P}_\infty(\mathbb{N}) \backslash g^{-1}(\mathscr{S}).$$

Now notice that $\mathscr{P}_\infty(A, Y)$ is a subcollection of $\{D \cup A : A \subseteq D \cup A \subseteq Y \cup A, D \in \mathscr{P}_\infty(\mathbb{N})\} \subseteq g(\mathscr{P}_\infty(Y))$. Reflecting on this we see that either

$$\mathscr{P}_\infty(A, Y) \subseteq g(\mathscr{P}_\infty(Y)) \subseteq \mathscr{S}$$

or

$$\begin{aligned}\mathscr{P}_\infty(A, Y) &\subseteq g(\mathscr{P}_\infty(\mathbb{N}) \backslash g^{-1}(\mathscr{S})) \\ &= \{E \cup A : E \in \mathscr{P}_\infty(\mathbb{N}) \backslash g^{-1}(\mathscr{S})\} \\ &= \{E \cup A : E \in \mathscr{P}_\infty(\mathbb{N}), E \cup A \notin \mathscr{S}\} \\ &\subseteq \mathscr{P}_\infty(\mathbb{N}) \backslash \mathscr{S}.\end{aligned}$$

This is as it should be and proves that τ-open sets are completely Ramsey.

That τ-closed sets are completely Ramsey is an easy consequence of the completely Ramsey nature of τ-open sets and the observation: $\mathscr{S} \subseteq \mathscr{P}_\infty(\mathbb{N})$ *is completely Ramsey if and only if* $\mathscr{P}_\infty(\mathbb{N}) \setminus \mathscr{S}$ *is.* □

Corollary. *Every classical open set and every classical closed set in* $\mathscr{P}_\infty(\mathbb{N})$ *is completely Ramsey. Consequently, given such a set* $\mathscr{S}$, *if* $M \in \mathscr{P}_\infty(\mathbb{N})$ *then there is an* $L \in \mathscr{P}_\infty(M)$ *such that either* $\mathscr{P}_\infty(L) \subseteq \mathscr{S}$ *or* $\mathscr{P}_\infty(L) \subseteq \mathscr{P}_\infty(\mathbb{N}) \setminus \mathscr{S}$.

For many applications the above corollary is enough; however, as awareness of the power inherent in such combinatorial results has spread, more sophisticated constructions have been (successfully) attempted exploiting a remarkable feature of completely Ramsey collections: every classical Borel subset of $\mathscr{P}_\infty(\mathbb{N})$ is completely Ramsey. The proof that this is so will proceed in two steps: first, we show that subsets of $(\mathscr{P}_\infty(\mathbb{N}), \tau)$ enjoying the Baire property are completely Ramsey; then we demonstrate that the Baire property is shared by a σ-algebra of subsets of $(\mathscr{P}_\infty(\mathbb{N}), \tau)$ that contains the open sets.

If T is a topological space and $S \subseteq T$, we say S *has the Baire property* if there is an open set U whose symmetric difference with S is meager (of the first category, is the countable union of nowhere dense sets).

Lemma 5. *Every subset* $\mathscr{S}$ *of* $(\mathscr{P}_\infty(\mathbb{N}), \tau)$ *having the Baire property is completely Ramsey.*

Proof. First notice that if $\mathscr{S}$ is nowhere dense, then given $A \in \mathscr{P}_{<\infty}(\mathbb{N})$ and $B \in \mathscr{P}_\infty(\mathbb{N})$ there is a $C \in \mathscr{P}_\infty(B)$ such that $\mathscr{P}_\infty(A, C) \subseteq \mathscr{P}_\infty(\mathbb{N}) \setminus \mathscr{S}$. In fact, $\bar{\mathscr{S}}$ is completely Ramsey; so there is a $C \in \mathscr{P}_\infty(B)$ such that $\mathscr{P}_\infty(A, C) \subseteq \mathscr{P}_\infty(\mathbb{N}) \setminus \bar{\mathscr{S}} \subseteq \mathscr{P}_\infty(\mathbb{N}) \setminus \mathscr{S}$ or $\mathscr{P}_\infty(A, C) \subseteq \bar{\mathscr{S}}$. But $\bar{\mathscr{S}}$'s interior is empty and so $\bar{\mathscr{S}}$ cannot contain any $\mathscr{P}_\infty(A, C)$.

Next, we show that if $\mathscr{S}$ is meager, then given A, $A \in \mathscr{P}_{<\infty}(\mathbb{N})$ and $B \in \mathscr{P}_\infty(\mathbb{N})$ there is a $C \in \mathscr{P}_\infty(B)$ such that $\mathscr{P}_\infty(A, C) \subseteq \mathscr{P}_\infty(\mathbb{N}) \setminus \mathscr{S}$. Suppose $\mathscr{S}$ if of the form $\cup_{n=0}^{\infty} \mathscr{S}_n$, where each $\mathscr{S}_n$ is nowhere dense. Let $A_0 = A$ and pick $B_0 \in \mathscr{P}_\infty(B)$ such that $\mathscr{P}_\infty(A_0, B_0)$ is a subset of $\mathscr{P}_\infty(\mathbb{N}) \setminus \mathscr{S}_0$ and $A_0 < B_0$; this is the gist of our opening observation. Let a_0 be the first member of B_0 and set $A_1 = A_0 \cup \{a_0\}$. Pick $B_1 \in \mathscr{P}_\infty(B_0 \setminus \{a_0\})$ such that $\mathscr{P}_\infty(A_1, B_1) \subseteq \mathscr{P}_\infty(\mathbb{N}) \setminus B_1 \subseteq B_0$; so $\mathscr{P}_\infty(A_0, B_1) \subseteq \mathscr{P}_\infty(\mathbb{N}) \setminus \mathscr{S}_1$ and $A_1 < B_1$. Suppose we've defined A_n and B_n with $A_n < B_n$ and $\mathscr{P}_\infty(\tilde{A}, B_n) \subseteq \mathscr{P}_\infty(\mathbb{N}) \setminus \mathscr{S}_n$ for any $A_0 \subseteq \tilde{A} \subseteq A_n$. Set $A_{n+1} = A_n \cup \{a_n\}$, where a_n is the least element of B_n; choose $B_{n+1} \in \mathscr{P}_\infty(B_n \setminus \{a_n\})$ so that for each $A_0 \subseteq \tilde{A} \subseteq A_{n+1}$ we have $\mathscr{P}_\infty(\tilde{A}, B_{n+1}) \subseteq \mathscr{P}_\infty(\mathbb{N}) \setminus \mathscr{S}_{n+1}$. Let $C = \cup_{n=0}^{\infty} A_n$. Then $\mathscr{P}_\infty(A, C)$ is disjoint from $\mathscr{S}_n$ for all n; indeed, $\mathscr{P}_\infty(A, C) \subseteq \cup_{A_0 \subseteq \tilde{A} \subseteq A_n} \mathscr{P}_\infty(\tilde{A}, B_n)$ for each n, itself a set disjoint from $\mathscr{S}_n$.

It is worth pointing out at this juncture that we have shown that a meager set $\mathscr{S}$ in $(\mathscr{P}_\infty(\mathbb{N}), \tau)$, in addition to being completely Ramsey, is actually nowhere dense. Remember that the sets $\mathscr{P}_\infty(A,C)$ form a neighborhood basis for τ.

Now let $\mathscr{S}$ be a subset of $(\mathscr{P}_\infty(\mathbb{N}),\tau)$ having the Baire property; represent $\mathscr{S}$ in the form $\mathscr{S}_0\Delta\mathscr{S}_1$, where $\mathscr{S}_0$ is open and $\mathscr{S}_1$ is meager. Take any $A\in\mathscr{P}_{<\infty}(\mathbb{N})$ and any $B\in\mathscr{P}_\infty(\mathbb{N})$. We know that there is a $\tilde{C}\in\mathscr{P}_\infty(B)$ such that $\mathscr{P}_\infty(A,\tilde{C})$ is contained in $\mathscr{P}_\infty(\mathbb{N})\backslash\mathscr{S}_1$ from the preceeding paragraphs; since each τ-open set is completely Ramsey, we can also find a $C\in\mathscr{P}_\infty(\tilde{C})$ such that $\mathscr{P}_\infty(A,C)\subseteq\mathscr{P}_\infty(\mathbb{N})\backslash\mathscr{S}_0$ or $\mathscr{P}_\infty(A,C)\subseteq\mathscr{S}_0$. Let's check $\mathscr{P}_\infty(A,C)$ for containment in $\mathscr{P}_\infty(\mathbb{N})\backslash\mathscr{S}$ or $\mathscr{S}$. If $\mathscr{P}_\infty(A,C)\subseteq\mathscr{S}_0$, then $\mathscr{P}_\infty(A,C)\subseteq\mathscr{S}$ since $\mathscr{P}_\infty(A,C)\subseteq\mathscr{P}_\infty(A,\tilde{C})$ a set disjoint from $\mathscr{S}_1$. On the other hand, if $\mathscr{P}_\infty(A,C)\subseteq\mathscr{P}_\infty(\mathbb{N})\backslash\mathscr{S}_0$, then $\mathscr{P}_\infty(A,C)\subseteq\mathscr{P}_\infty(A,\tilde{C})$ and so is disjoint from $\mathscr{S}_1$; $\mathscr{P}_\infty(A,C)$ is disjoint from $\mathscr{S}_0$ and $\mathscr{S}_1$, hence from $\mathscr{S}-\mathscr{P}_\infty(A,C)\subseteq\mathscr{P}_\infty(\mathbb{N})\backslash\mathscr{S}$. Regardless of the case in hand, $\mathscr{P}_\infty(A,C)$ is contained in either $\mathscr{S}$ or $\mathscr{P}_\infty(\mathbb{N})\backslash\mathscr{S}$. □

Lemma 6. *Let T be a topological space. Then the collection of subsets of T having the Baire property forms a σ-algebra containing the open sets of T.*

PROOF. First of all, if a set has the Baire property, so does its complement. To see this, notice that closed sets have the Baire property differing as they do from their interior by their nowhere dense boundary. Notice too that any set with a meager symmetric difference from another having the Baire property has the Baire property. Consequently, if A is a Baire set and U is an open set for which $A\Delta U$ is meager, then $A^c\Delta U^c$ is meager, too; hence A^c enjoys the Baire property as often as A does.

Next, if (B_n) is a sequence of sets with the Baire property, then $\cup_n B_n$ has the Baire property. In fact, each B_n differs from an open set U_n by a meager set A_n, so that $\cup_n B_n$ differs from the open set $\cup_n U_n$ in a (meager) subset of the meager set $\cup_n A_n$. □

Theorem (Galvin-Prikry). *Every Borel subset of $(\mathscr{P}_\infty(\mathbb{N}),\tau)$ is completely Ramsey.*

Corollary (Galvin-Prikry). *Every classical Borel subset of $\mathscr{P}_\infty(\mathbb{N})$ is completely Ramsey. Consequently if $\mathscr{B}$ is a classical Borel subset of $\mathscr{P}_\infty(\mathbb{N})$ and $M\in\mathscr{P}_\infty(\mathbb{N})$, then there is an $L\in\mathscr{P}_\infty(M)$ such that either $\mathscr{P}_\infty(L)\subseteq\mathscr{B}$ or $\mathscr{P}_\infty(L)\subseteq\mathscr{P}_\infty(\mathbb{N})\backslash\mathscr{B}$.*

Notes and Remarks

The elegant combinatorial principle known as Ramsey's theorem has had a strong impact on the theory of Banach spaces. We have recourse to use this principle in our treatment of Rosenthal's l_1 theorem (following the lead of J.

Farachat) in Chapter XI and call on it frequently in recounting the proof of the Elton-Odell separation theorem in Chapter XIV. Its effects on these deliberations are so basic that a reasonably self-contained treatment seemed in order.

The roots of the combinatorial theory reach back to the classical formulation of the Ramsey theorem due to the remarkable F. P. Ramsey, himself. The original foundation:

Theorem (Ramsey). *Let $\mathscr{A}$ be a family of doubletons from the set $\mathbb{N}$ of positive integers. Then there exists an $M \in \mathscr{P}_\infty(\mathbb{N})$ so that either $\mathscr{A}$ contains* all *the doubletons from M or $\mathscr{A}$ contains none of the doubletons from M.*

Though the proof of the above is short and sweet, we prefer to send the student to E. Odell's survey of applications of Ramsey theorems to Banach space theory to find the proof, feeling sure that once started on that survey the rewards of continuing will be too obvious to leave it unstudied.

Our discussion of the infinite versions of Ramsey's theorem were influenced greatly by a seminar lead by G. Stanek and D. Weintraub on applications of Ramsey's theorem and descriptive set theory in functional analysis. In turn they were following E. E. Ellentuck's proof (1974) of the completely Ramsey nature of analytic sets. Earlier, C. St. J. A. Nash-Williams (1965) had shown that closed subsets of $\mathscr{P}_\infty(\mathbb{N})$ are completely Ramsey (we use this in Chapter XI), F. Galvin and K. Prikry (1973) showed that Borel sets are completely Ramsey, and J. Silver (1970), using metamathematical arguments, extended the Galvin-Prikry search to find analytic sets among the completely Ramsey family.

Though Ellentuck's approach bypasses the need to know even the basics about mathematical logic, we would be remiss if we did not suggest that Banach space theory is enjoying the fruits of the logician's labors. Be it under the guise of nonstandard analysis or ultraproduct arguments, modern model theory has seen too many victories in the investigation of Banach space questions to be dismissed as being of tangential interest. Rather than survey the contributions of these all-too-alien disciplines, we recommend the student make a careful study of the surveys cited in our bibliography, as well as pertinent references contained in those surveys.

Bibliography

Ellentuck, E. E. 1974. A new proof that analytic sets are Ramsey. *J. Symbolic Logic*, **39**, 163–165.

Galvin, F. and Prikry, K. 1973. Borel sets and Ramsey's theorem. *J. Symbolic Logic*, **38**, 193–198.

Heinrich, S. 1980. Ultraproducts in Banach space theory. *J. Reine Angew. Math.*, **313**, 72–104.

Henson, C. W. and Moore, L. C., Jr. 1983. Nonstandard analysis and the theory of Banach spaces. Preprint.

Luxemburg, W. A. J. 1969. A general theory of monads. In *Applications of Model Theory to Algebra, Analysis and Probability*. New York: Holt, Rinehart and Winston.

Nash-Williams, C. St. J. A. 1965. On well quasi-ordering transfinite sequences. *Proc. Cambridge Phil. Soc.*, **61**, 33–39.

Odell, E. 1981. *Applications of Ramsey Theorems to Banach Space Theory*. Austin: University of Texas Press.

Ramsey, F. P. 1929. On a problem of formal logic. *Proc. London Math. Soc.*, **30**, 264–286.

Silver, J. 1970. Every analytic set is Ramsey. *J. Symbolic Logic*, **35**, 60–64.

Sims, B. 1982. *"Ultra"-techniques in Banach Space Theory*. Queen's Papers in Pure and Applied Mathematics, Vol. 60.

CHAPTER XI
Rosenthal's l_1 Theorem

The Eberlein-Šmulian theorem tells us that in order to be able to extract from each bounded sequence in X a weakly convergent subsequence it is both necessary and sufficient that X be reflexive. Suppose we ask less. Suppose we ask only that each bounded sequence in X have a weakly Cauchy subsequence. [Recall that a sequence (x_n) in a Banach space X is weakly Cauchy if for each $x^* \in X^*$ the scalar sequence $(x^* x_n)$ is convergent.] When can one extract from each bounded sequence in X a weakly Cauchy subsequence?

Of course, a quick sufficient condition (reflexivity) is provided by the Eberlein-Šmulian theorem. But, can one extract weakly Cauchy subsequences from arbitrary bounded sequences in nonreflexive Banach spaces? The answer is "it depends on the space—sometimes yes, sometimes no."

Sometimes you can. In fact (and this was known to Banach), if X is a separable Banach space with X^* also separable, then bounded sequences in X have weakly Cauchy subsequences. Let's quickly recall the proof. Let (d_n) be a dense sequence in S_{X^*} and let (x_k) be a bounded sequence in X. The sequence $(d_1 x_k)$ is a bounded sequence of scalars and, therefore, has a convergent subsequence, say $(d_1 x_k^1)$. Now look at $(d_2 x_k^1)$; it is a bounded sequence of scalars and so has a convergent subsequence $(d_2 x_k^2)$. Of course, $(d_1 x_k^2)$ is also convergent. The coast is clear. Follow your nose down the diagonal.

Sometimes you cannot. If e_n denotes the nth unit vector in l_1, then (e_n) has no weakly Cauchy subsequence. In fact, if (n_k) is any strictly increasing sequence of positive integers, and if we consider $\lambda \in l_\infty = l_1^*$ defined by

$$\lambda_j = \begin{cases} 1 & \text{if } j = n_{2k} \text{ for some } k, \\ -1 & \text{otherwise,} \end{cases}$$

then (λe_{n_k}) is not a convergent sequence of scalars.

The purpose of this chapter is to present a startling discovery of Haskell P. Rosenthal which says that the above counterexample is, in a sense, the only one. Precisely, we show the following.

Rosenthal l_1 Theorem. *In order that each bounded sequence in the Banach space X have a weakly Cauchy subsequence, it is both necessary and sufficient that X contain no isomorphic copy of l_1.*

In fact, we do a bit better than as claimed above; we show that *if (x_n) is a bounded sequence that has no weakly Cauchy subsequences, then (x_n) admits of a subsequence (x_n') that is the unit vector basis of l_1.* [Here, when we say that (x_n') is the unit vector basis of l_1, we mean that there are constants $a, b > 0$ so that

$$a\sum_{i=1}^{n}|c_i| \le \left\|\sum_{i=1}^{n} c_i x_i'\right\| \le b\sum_{i=1}^{n}|c_i|$$

for any scalars $c_1, \ldots, c_n$ and any n.] This finer result is due to Rosenthal in the case of real scalars and to Leonard Dor (1975) in the complex case.

How to Imbed l_1 in a Banach Space

To find a copy of l_1 in a Banach space, the obvious thing to look for is l_1's unit vector basis. If one can find a sequence (x_n) in a Banach space X such that for some $a, b > 0$.

$$a\sum_{i=1}^{n}|c_i| \le \left\|\sum_{i=1}^{n} c_i x_i\right\| \le b\sum_{i=1}^{n}|c_i| \tag{1}$$

holds for any scalars $c_1, c_2, \ldots, c_n$ and any n, then one has an isomorphic copy of l_1 inside of X. We will continue to say that such a sequence *is* the unit vector basis of l_1.

An example might help illustrate how such a sequence might look. Suppose we let $X = L_1[0,1]$, the space of Lebesgue-integrable functions on $[0,1]$ with the usual norm $\|x\|_1 = \int_0^1 |x(t)|\,dt$. If we pick (x_n) to be a sequence of members of $L_1[0,1]$ for which $\|x_n\|_1 = 1$ and $x_n \cdot x_m = 0$ for $n \ne m$ (that is, the x_n are disjointly supported), then

$$\left\|\sum_{i=1}^{n} c_i x_i\right\|_1 = \int_0^1 |c_1x_1(t) + \cdots + c_nx_n(t)|\,dt$$

$$= \int_0^1 |c_1x_1(t)| + \cdots + |c_nx_n(t)|\,dt$$

(since the x_n have disjoint supports)

$$= \sum_{i=1}^{n}|c_i|\,\|x_i\|_1 = \sum_{i=1}^{n}|c_i|.$$

So such a selection of x_n gives an exact replica of the unit vector basis of l_1.

It does not take much imagination to see that if we let the x_n stay disjointly supported in $L_1[0,1]$ but let their norms wander between a and b, where $0 < a < b$, then we would still have a unit vector basis of l_1, but now (1) would express the degree of similarity present.

In fact, if one wants to get just a bit fancier, then one can even dispense with the disjointness of supports. One must take care that on the set of common support the relative contribution to the norms of the x_n is not too great. One still can get the unit vector basis of l_1 in this way.

There are at least two drawbacks to this method of producing an l_1. The first, and most obvious, is the special nature of the space X. Isn't $X = L_1[0,1]$, a very special space? Our construction is definitely tied to the structure of this particular space. Should we be looking for a way of adapting this construction to a much broader class of spaces, however, we would soon come to grips with the second drawback in the construction. Proceeding as above the resulting copy of l_1 does not sit just anywhere in $L_1[0,1] = X$, it is a *complemented* subspace; i.e., there is a continuous linear projection from X onto the constructed copy of l_1. It would be too much to expect that anytime we find a copy of l_1 in a space, one can find a complemented copy of l_1 in the space. We shall see later *why* this is too much to expect; for now let us mention that despite the more-or-less simpleminded manner in which we have built our l_1 in $L_1[0,1]$, until late 1979, no other method of building an l_1 in $L_1[0,1]$ had been found.

If the above approach is not to be generally followed, how then to proceed? By its very statement, the Rosenthal l_1 theorem indicates the need to use duality. This suggests that we look at X as a space of functions on X^*; more precisely, we view X as a subspace of the continuous functions on B_{X^*} in its weak* topology. In this way, the question of whether or not a given sequence (x_n) in X has a weak Cauchy subsequence is reduced (enlarged?) to the question of whether or not the corresponding sequence of functions on B_{X^*} has a pointwise convergent subsequence. It is in this setting that Rosenthal's l_1 theorem will be treated.

Our setup: We have a set Ω and a uniformly bounded sequence (f_n) of scalar-valued functions which is without a pointwise convergent subsequence. We want to extract a subsequence which in the $l_\infty(\Omega)$-norm is the unit vector basis of l_1.

The first task is to guess more or less what such a subsequence has to look like. Then we see if things that look right are right. These will be the chores of this section. The pruning work will be the work of the next section; the first harvest will be gathered in the last section.

What does the unit vector basis of l_1 look like when it appears in a space of bounded functions? To get a hint, we look at a bit easier problem: what does one look like in a space $C(\Omega)$ of continuous functions on a compact Hausdorff space Ω; in fact, we consider an extra replica of the unit vector basis, and to get things under way, we worry only about *real* Banach spaces.

Let (f_n) be a sequence in $C(\Omega)$ for which

$$\left\|\sum_n c_n f_n\right\|_\infty = \sum_n |c_n|$$

for any $(c_n) \in l_1$. Then for any k we know that

$$\left\|\sum_{n=1}^{k} \pm f_n\right\|_\infty = k.$$

Of course, this means in particular that f_1 has norm 1; so there is a nonempty closed subset $\Omega_1 \subseteq \Omega$ such that

$$\Omega_1 = \{\omega \in \Omega : |f_1(\omega)| = 1\}.$$

But both $f_1 + f_2$ and $f_1 - f_2$ have norm 2; so there are nonempty closed sets $\Omega_2, \Omega_3 \subseteq \Omega_1$ such that

$$\Omega_2 = \{\omega \in \Omega : |f_1(\omega) + f_2(\omega)| = 2\}$$

and

$$\Omega_3 = \{\omega \in \Omega : |f_1(\omega) - f_2(\omega)| = 2\}.$$

Since f_2 also has norm 1 and since f_2 has the same sign as does f_1 on Ω_2 but opposite sign on Ω_3, Ω_2 and Ω_3 must be disjoint. Again, $f_1 + f_2 + f_3$, $f_1 + f_2 - f_3$, $f_1 - f_2 + f_3$, and $f_1 - f_2 - f_3$ all have norm 3; so there are nonempty closed sets $\Omega_4, \Omega_5 \subseteq \Omega_2$ and $\Omega_6, \Omega_7 \subseteq \Omega_3$ such that

$$\Omega_4 = \{\omega \in \Omega : |f_1(\omega) + f_2(\omega) + f_3(\omega)| = 3\},$$

$$\Omega_5 = \{\omega \in \Omega : |f_1(\omega) + f_2(\omega) - f_3(\omega)| = 3\},$$

$$\Omega_6 = \{\omega \in \Omega : |f_1(\omega) - f_2(\omega) + f_3(\omega)| = 3\},$$

and

$$\Omega_7 = \{\omega \in \Omega : |f_1(\omega) - f_2(\omega) - f_3(\omega)| = 3\}.$$

As before Ω_4, Ω_5, Ω_6, and Ω_7 are disjoint; this time Ω_4 being the set where f_1, f_2, and f_3 have the same signs, Ω_5 being the set where f_1 and f_2 agree in sign while f_3 disagrees, Ω_6 being the set where f_1 and f_3 agree while f_2 disagrees, and Ω_7 being the set where f_1 is the disagreeable one. The procedure is set. The point is that if you have l_1's unit vector basis in a $C(\Omega)$, then there has to exist some sort of "dyadic splitting" of subsets of Ω along with a sequence of functions that agree to change signs on the successive parts of the splitting.

The idea now is search for this kind of a sequence of functions in $l_\infty(\Omega)$. A handy model already exists—the Rademacher functions.

A basic fact: In real $l_\infty([0,1])$, the Rademacher functions are the unit vector basis of l_1. In fact, suppose $t_1, \ldots, t_n$ are real numbers; let's calculate

$\|\Sigma_{i=1}^n t_i r_i\|_\infty$. Of course, the most we can expect is $\Sigma_{i=1}^n |t_i|$; our claim is that we actually get this much. Since

$$\left\| \sum_{i=1}^n t_i r_i \right\|_\infty = \left\| \sum_{i=1}^n (-t_i) r_i \right\|_\infty ,$$

we can assume $t_1 \geq 0$, and therefore, $t_1 r_1$ is $|t_1|$ on all of $[0,1]$. Looking at $t_2 r_2$, we see that on half of $[0,1]$, $t_2 r_2$ is $|t_2|$; so on this half, $t_1 r_1 + t_2 r_2$ is $|t_1| + |t_2|$. Again, on the half where $t_1 r_1 + t_2 r_2$ is $|t_1| + |t_2|$, $t_3 r_3$ achieves both values $|t_3|$ and $-|t_3|$ throughout subintervals of length $\frac{1}{4}$. The idea is (or should be) clear by now. On some interval of length 2^{1-n}, the function $t_1 r_1 + \cdots + t_n r_n$ achieves the value $|t_1| + \cdots + |t_n|$.

The purpose of the rest of the present section is to discuss just how Rademacher-like a sequence must be in order to identify it with l_1's unit vector basis.

Let Ω be a set. A sequence (Ω_n) of nonempty subsets of Ω is called a *tree of subsets* of Ω if for each n, Ω_{2n} and Ω_{2n+1} are disjoint subsets of Ω_n. Pictorially, we have

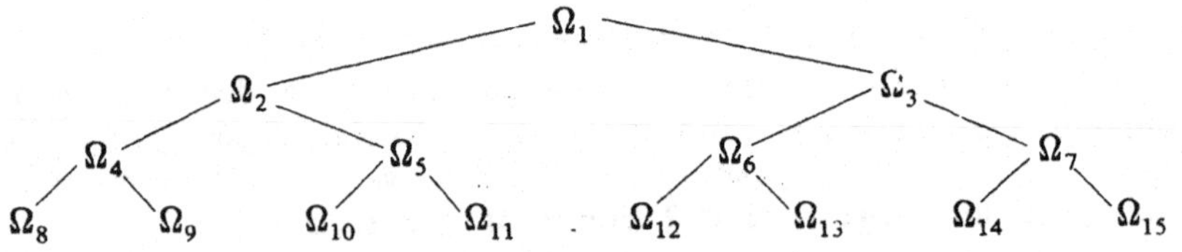

where across the rows the sets are disjoint and connecting lines indicate that lower ends are subsets of the upper.

The purpose of introducing trees of sets is obviously to mimic the dyadic splittings of $[0,1)$ so basic to the nature of the Rademacher functions. The next fact we note is a special case of a much more general scheme due to A. Pelczynski.

Proposition 1. *Let Ω be a set, (Ω_n) be a tree of subsets of Ω, B be a bounded subset of $l_\infty(\Omega)$ and $\delta > 0$. Suppose we have a (Rademacher-like) sequence (b_n) in B such that whenever $2^{n-1} \leq k < 2^n$, $(-1)^k b_n(\omega) \geq \delta$ for all $\omega \in \Omega_k$. Then (b_n) is equivalent to the unit vector basis of l_1.*

PROOF. Let $t_1, \ldots, t_n$ be real numbers. We look at $\|\Sigma_{j=1}^n t_j b_j\|_\infty$; its biggest of possible values is $(\sup_B \|b\|)\ \Sigma_{j=1}^n |t_j|$. In the generality we're dealing with, there's no hope of attaining this value; we settle for a goodly portion of this maximum possible. Similar to the case of the Rademacher functions, no harm will come our way in supposing that $t_1 < 0$. It follows that $t_1 b_1(\omega) \geq |t_1|\delta$ for all $\omega \in \Omega_1$.

The values $t_2 b_2$ attains on Ω_2 and those it attains on Ω_3 are opposite in sign and in each case all have modulus $\geq |t_2|\delta$; so on one of these "halves" all of the values of $(t_1 b_1 + t_2 b_2)$ are $\geq \delta(|t_1| + |t_2|)$.

Again, inside whichever half it is on which all of the values of $(t_1b_1 + t_2b_2)$ are $\geq \delta(|t_1| + |t_2|)$, we have a pair of disjoint Ω_i ("quarters") on one of which t_3b_3 has positive values, on the other of which t_3b_3 has negative values; regardless of which quarter we are on, the moduli of t_3b_3's values exceed $|t_3|\delta$. On the positive quarter, $t_1b_1 + t_2b_2 + t_3b_3$ stays $\geq (|t_1| + |t_2| + |t_3|)$.

It should be clear how the argument continues. Our conclusion is that the sequence (b_n) satisfies

$$\delta\left(\sum_{j=1}^{n} |t_j|\right) \leq \left\| \sum_{j=1}^{n} t_j b_j \right\|_\infty \leq \left(\sup_{b \in B} \|b\|\right) \cdot \left(\sum_{j=1}^{n} |t_j|\right)$$

for any sequence (t_n) of real numbers and all n. Of course this just says that (b_n) is equivalent to the unit vector basis of l_1.

Proposition 1 is oftentimes quite useful as is. It can, however, be significantly improved upon. Such improvements are crucial to our discussion of Rosenthal's theorem and are due to Rosenthal himself.

The first improvement:

Proposition 2. *Let Ω be a set, (Ω_n) be a tree of subsets of Ω, r be a real number, (b_n) a bounded sequence in $l_\infty(\Omega)$ and $\delta > 0$. Suppose that for n if $2^n \leq k < 2^{n+1}$ and k is even, then $b_n(\omega) \geq r + \delta$ for all $\omega \in \Omega_k$, whereas if $2^n \leq k < 2^{n+1}$ and k is odd, then $b_n(\omega) \leq r$ for all $\omega \in \Omega_k$.*

Then $(b_n)_{n \geq 2}$ is equivalent to the unit vector basis of l_1.

Our attentions are restricted to $n \geq 2$ for the sake of cleaner details only.

Proof. Some initial footwork will ease the pain of proof. First notice that we can assume that $r + \delta \neq 0$. In fact, if $r + \delta = 0$, notice that $r + \delta/2 < 0$ and the hypotheses of the proposition are satisfied with δ replaced by $\delta/2$. Next, we assume that $r + \delta > 0$.

Should this state of affairs not be in effect, we multiply all the b_n by -1 to achieve it; if $(-b_n)$ is equivalent to the unit vector basis of l_1, so is (b_n). Finally, since $r + \delta > 0$, we might as well assume $r \geq 0$ too since otherwise, we are back in the situation (more or less) of (our special case of) Pelczynski's proposition. So we prove the proposition under the added hypotheses that $r + \delta > r \geq 0$.

What we claim is true is that for any sequence (t_n) of real numbers we have (independent of n)

$$\frac{\delta}{2} \sum_{k=2}^{n} |t_k| \leq \left\| \sum_{k=2}^{n} t_k b_k \right\|_\infty; \tag{2}$$

the boundedness of the sequence (b_n) gives us the upper estimate and consequently the equivalence of (b_n) with the unit vector basis of l_1. Of course, actually we need only establish (2) for finitely nonzero sequences (t_n), and for these we can assume that their l_1 norm is 1; normalization gives (2) in general.

So we have $t_2, \ldots, t_n$ real with $\Sigma_{k=2}^n |t_k| = 1$, and we want to show that for some $\omega \in \Omega$, $|\Sigma_{k=1}^n t_k b_k(\omega)|$ exceeds $\delta/2$. To this end, suppose $1 \le m \le n$ and let

$$E_m = \bigcup_{\substack{k=2^m \\ k \text{ even}}}^{2^{m+1}-1} \Omega_k, \qquad O_m = \bigcup_{\substack{k=2^m \\ k \text{ odd}}}^{2^{m+1}-1} \Omega_k.$$

On E_m, $b_m \ge r + \delta$, whereas on O_m, $b_m \le r$. The important point about the E_m and O_m is the following property acquired from the treelike nature of the sequence (Ω_n): *if you intersect E and O being careful not to pick two with the same subscript, the result is nonempty.* This allows us to estimate the size of $\|\Sigma_{k=2}^n t_k b_k\|_\infty$.

Let $P = \{k : t_k > 0\}$ and $N = \{k : t_k < 0\}$. Then there are points $\omega_1, \omega_2 \in \Omega$ such that

$$\omega_1 \in \bigcap_{k \in P} E_k \cap \bigcap_{k \in N} O_k \quad \text{and} \quad \omega_2 \in \bigcap_{k \in N} E_k \cap \bigcap_{k \in P} O_k.$$

For ω_1 and ω_2 this means that if $t_k > 0$, then $b_k(\omega_1) \ge r + \delta$ and $b_k(\omega_2) \le r$, whereas if $t_k < 0$, then $b_k(\omega_1) \le r$ while $b_k(\omega_2) \ge r + \delta$.

At ω_1 we get

$$\begin{aligned}
\sum_{k=2}^n t_k b_k(\omega_1) &= \sum_{k \in P} t_k b_k(\omega_1) + \sum_{k \in N} t_k b_k(\omega_1) \\
&\ge \sum_{k \in P} t_k (r + \delta) + \sum_{\substack{k \in N \\ b_k(\omega_1) > 0}} t_k b_k(\omega_1) + \sum_{\substack{k \in N \\ b_k(\omega_1) \le 0}} t_k b_k(\omega_1) \\
&\ge \sum_{k \in P} t_k (r + \delta) + \sum_{\substack{k \in N \\ b_k(\omega_1) > 0}} t_k b_k(\omega_1) \\
&\ge \sum_{k \in P} t_k (r + \delta) + \sum_{\substack{k \in N \\ b_k(\omega_1) > 0}} t_k r \\
&= \sum_{k \in P} |t_k| (r + \delta) + \sum_{\substack{k \in N \\ b_k(\omega_1) > 0}} |t_k| (-r) \\
&\ge \sum_{k \in P} |t_k| (r + \delta) + \sum_{k \in N} |t_k| (-r).
\end{aligned}$$

Summarizing, we have

$$\sum_{k=1}^n t_k b_k(\omega_1) \ge \sum_{k \in P} |t_k| (r + \delta) + \sum_{k \in N} |t_k| (-r).$$

In a similar fashion we get

$$-\sum_{k=1}^n t_k b_k(\omega_2) \ge \sum_{k \in N} |t_k| (r + \delta) + \sum_{k \in P} |t_k| (-r).$$

Adding together the right sides, we get something $\geq \delta$. It follows that one of the numbers on the left is at least $\delta/2$, and this is good enough to finish the proof. □

If we abstract from the above proof the key features, we are led to the following concept: let Ω be a set and (E_n, O_n) be a sequence of disjoint pairs of subsets of Ω. If for any finite disjoint subsets N, P of the natural numbers we have that

$$\bigcap_{n \in N} E_n \cap \bigcap_{n \in P} O_n \neq \varnothing,$$

then the sequence of pairs (E_n, O_n) is called *independent*. In turn we get the following proposition.

Proposition 3 (Rosenthal). *Let Ω be a set, (E_n, O_n) be an independent sequence of disjoint pairs of subsets of Ω, r be a real number, (b_n) be a bounded sequence in $l_\infty(\Omega)$ and $\delta > 0$. Suppose that $b_n(\omega) \geq r + \delta$ for all $\omega \in E_n$ and $b_n(\omega) \leq r$ for all $\omega \in O_n$.*

Then (b_n) is equivalent to the unit vector basis of l_1.

We conclude this section with the complex version of the above result; it is due to Leonard Dor.

Proposition 4. *Let (E_n, O_n) be an independent sequence of nonempty disjoint pairs of subsets of Ω, let D_1 and D_2 be disjoint closed disks in $\mathbb{C}$ with centers c_1 and c_2, respectively, and let (b_n) be a uniformly bounded sequence of complex-valued functions defined on Ω. Suppose that D_1, D_2 have the same diameter $\leq \frac{1}{2}(\delta =$ distance from D_1 to $D_2)$. Assume that*

$$b_n(\omega) \in \begin{cases} D_1 & \text{if } \omega \in E_n, \\ D_2 & \text{if } \omega \in O_n. \end{cases}$$

Then (b_n) is equivalent to the unit vector basis of l_1.

Proof. We will show that for any sequence (γ_n) of complex numbers and any finite set J of positive integers that

$$\frac{\delta}{8} \sum_{n \in J} |\gamma_n| \leq \left\| \sum_{n \in J} \gamma_n b_n \right\|_\infty .$$

First, we observe that we may assume of c_1 and c_2 that their difference $c_2 - c_1$ is real and positive. In fact, otherwise, just rotate until the b_n, D_1, and D_2 are properly aligned to satisfy this additional assumption; multiplication by $K = |c_2 - c_1|/(c_2 - c_1)$ will achieve this effect. If (Kb_n) is equivalent to the unit vector basis of l_1, so too is (b_n).

Now suppose $\gamma_n = \alpha_n + \beta_n i$ and assume (without any great loss in generality) that $\sum_{j\in J}|\alpha_j| \geq \sum_{j\in J}|\beta_j|$. Let

$$P = \{j \in J : \alpha_j \geq 0\}, \qquad N = \{j \in J : \alpha_j < 0\}.$$

By the independence assumption, there are points $\omega_1, \omega_2 \in \Omega$ such that

$$\omega_1 \in \bigcap_{n\in P} E_n \cap \bigcap_{n\in N} O_n, \qquad \omega_2 \in \bigcap_{n\in N} E_n \cap \bigcap_{n\in P} O_n.$$

This implies that if $\operatorname{Re}\gamma_j = \alpha_j \geq 0$, then $b_j(\omega_1) \in D_1$ and $b_j(\omega_2) \in D_2$; while if $\operatorname{Re}\gamma_j = \alpha_j < 0$, then $b_j(\omega_1) \in D_2$ and $b_j(\omega_2) \in D_1$. Note that for any $z_1 \in D_1$ and $z_2 \in D_2$ we have $\operatorname{Re}(z_2 - z_1) \geq \delta$ and $\operatorname{Im}(z_2 - z_1) \leq \operatorname{diam} D_1 = \operatorname{diam} D_2 \leq \delta$; a suitable picture will aid in explanation.

This holds in particular for $z_1 = b_j(\omega_1)$ when $j \in P$ or $z_1 = b_j(\omega_2)$ when $j \in N$ or $z_2 = b_j(\omega_1)$ when $j \in N$ or $z_2 = b_j(\omega_2)$ when $j \in P$. Whatever the case may be, we have

$$\begin{aligned}\left\|\sum_{j\in J}\gamma_j b_j\right\|_\infty &= \sup_{\omega\in\Omega}\left|\sum_{j\in J}\gamma_j b_j(\omega)\right| \\ &\geq \tfrac{1}{2}\operatorname{Re}\left(\sum_{j\in J}\gamma_j b_j(\omega_2) - \sum_{j\in J}\gamma_j b_j(\omega_1)\right)\end{aligned}$$

[since $\operatorname{Re}(uv) + |\operatorname{Im}(u)\operatorname{Im}(v)| \geq \operatorname{Re}(u)\operatorname{Re}(v)$]

$$\begin{aligned}&\geq \tfrac{1}{2}\sum_{j\in J}\alpha_j \operatorname{Re}\big(b_j(\omega_2) - b_j(\omega_1)\big) \\ &\quad - \tfrac{1}{2}\sum_{j\in J}\left|\beta_j \operatorname{Im}\big(b_j(\omega_2) - b_j(\omega_1)\big)\right|.\end{aligned}$$

[Now note that for any $j \in P$, $\alpha_j \geq 0$; so $b_j(\omega_1) \in D_1$ and $b_j(\omega_2) \in D_2$, forcing $\operatorname{Re}(b_j(\omega_2) - b_j(\omega_1))$ to be $\geq \delta$. On the other hand, if $j \in N$, $\alpha_j < 0$; so $b_j(\omega_1) \in D_2$ and $b_j(\omega_2) \in D_1$, forcing $\operatorname{Re}(b_j(\omega_2) - b_j(\omega_1)) \leq -\delta$. This makes the difference above]

$$\begin{aligned}&\geq \tfrac{1}{2}\sum_{j\in J}|\alpha_j|\delta - \tfrac{1}{2}\sum_{j\in J}|\beta_j|\frac{\delta}{2} \\ &\geq \tfrac{1}{2}\sum_{j\in J}|\alpha_j|\delta - \tfrac{1}{2}\sum_{j\in J}|\alpha_j|\frac{\delta}{2} \\ &= \frac{\delta}{4}\sum_{j\in J}|\alpha_j| \geq \frac{\delta}{8}\sum_{j\in J}|\gamma_j|.\end{aligned}$$

Done. □

The Proof of the Rosenthal-Dor l_1 Theorem

Now that we know how to recognize an l_1 if we see one, we go about the business of finding one in the most general possible circumstances; i.e., we will prove the Rosenthal-Dor theorem cited in the opening paragraph. Our proof is adapted from J. Farahat's exposition of Rosenthal's theorem as found in the notes of the Seminaire Maurey-Schwartz on "Espaces L^p, Applications Radonifiantes et Geometrie des Espaces de Banach," 1973–1974. It constitutes a beautiful variation on the original proof of Rosenthal's and an important variation because of its use of combinatorial ideas which have so recently pervaded many of the best results in Banach space theory.

Suppose (x_n) is a bounded sequence in the Banach space X and suppose (x_n) has no weakly Cauchy subsequence. Imbedding X into $l_\infty(B_{X^*})$, (x_n) has no pointwise convergent subsequence. Using Propositions 3 and 4 as our guides, we will select a subsequence of (x_n) that is the unit vector basis of l_1.

Step 1. Let $\mathscr{D}$ be the (countable) collection of all pairs (D^1, D^2) of open disks in $\mathbf{C}$ each of whose centers has rational coordinates, each disk having a rational radius and satisfying: $\operatorname{diam} D^1 = \operatorname{diam} D^2 \leq \frac{1}{2}$ distance (D^1, D^2). List the members of $\mathscr{D}$ as $((D_k^1, D_k^2))$. We make a claim: There is a $K \in \mathbf{N}$ and an infinite subset P of the natural numbers such that for any infinite subset M of P there is an x_M^* in B_{X^*} for which $(x_k x_M^*)_{k \in M}$ has points of accumulation in both D_K^1 and D_K^2.

Otherwise, for each $k \geq 1$ and each infinite subset P of $\mathbf{N}$ there would be an infinite subset M of P such that for any x^* in B_{X^*}, the sequence $(x_k x^*)_{k \in M}$ would not have accumulation points in each of D_k^1 and D_k^2.

So there is an infinite set M_1 such that for each x^* in B_{X^*}, the sequence $(x_m x^*)_{m \in M_1}$ does not have points of accumulation in D_1^1 and D_1^2. Again, there is an infinite subset M_2 of M_1 such that for each x^* in B_{X^*}, the sequence $(x_m x^*)_{m \in M_2}$ does not have points of accumulation in D_2^1 and D_2^2. Continuing in this fashion, we get a decreasing sequence (M_n) of infinite subsets of $\mathbf{N}$ for which given n if x^* is in B_{X^*}, then the sequence $(x_m x^*)_{m \in M_n}$ does not have points of accumulation in D_n^1 and D_n^2.

Let P be an infinite subset of $\mathbf{N}$ whose nth member p_n belongs to M_n; clearly we may assume the p_n form a strictly increasing sequence. Recall that no subsequence of (x_n) converges pointwise on B_{X^*}; consequently, there is an $x_0^* \in B_{X^*}$ for which $(x_{p_n} x_0^*)$ is not convergent. However, $(x_{p_n} x_0^*)$ is a bounded sequence, and so there must be (at least) two distinct numbers d^1 and d^2 that are points of accumulation for $(x_{p_n} x_0^*)$. Now d^1 and d^2 lie in some D_K^1 and D_K^2, respectively, where $(D_K^1, D_K^2) \in \mathscr{D}$. A moment's reflection reveals the fix we're in: for any $j \geq 1$ the sequence $(x_{p_n} x_0^*)_{n \geq j}$ has both d^1 and d^2 as points of accumulation yet is a subsequence of $(x_m x_0^*)_{m \in M_j}$; this forces the latter sequence to also have d^1 and d^2 as points of accumulation violating M_j's very definition. Claim established.

In the interests of "sanity in indexing" we may as well assume that $(x_n)_{n\in P}$ is in fact (x_n) and call D_K^1 and D_K^2 simply D_1 and D_2.

To recapitulate, (x_n) *is a bounded sequence in X without a subsequence that is pointwise convergent on* B_{X^*}, D_1 *and* D_2 *are disjoint disks of the same diameter,* D_1 *and* D_2 *are separated by at least twice that diameter, and given any infinite subset M of N there is an* X_M^* *in* B_{X^*} *such that* $(x_m x_M^*)_{m\in M}$ *has points of accumulation in both* D_1 *and* D_2.

In preparation for our next step we let E_n, O_n be the sets defined by

$$O_n = \{x^* \in B_{X^*} : x_n x^* \in D_1\}, \qquad E_n = \{x^* \in B_{X^*} : x_n x^* \in D_2\}.$$

Key to our discussion is the following easily observed consequence of our earlier spade work: *regardless of the subsequence* (n_k) *of positive integers one chooses, neither* $\lim_k c_{E_{n_k}}(x^*) = 0$ *nor* $\lim_k c_{O_{n_k}}(x^*) = 0$ *holds for each* x^* *in* B_{X^*}.

For notational purposes we will denote by $-E_j$ the set O_j. For each positive integer k let $\mathscr{P}_k$ denote the collection of all infinite subsets $\{n_l\}$ of $\mathbb{N}$ for which

$$\bigcap_{l=1}^{k} (-1)^l E_{n_l} \neq \varnothing.$$

We identify each subset of $\mathbb{N}$ with a point in $\{0,1\}^{\mathbb{N}}$ and claim that $\bigcap_k \mathscr{P}_k$ is a closed subset of $\mathscr{P}_\infty(\mathbb{N})$, the collection of all infinite subsets of $\mathbb{N}$, in the relative topology of $\{0,1\}^{\mathbb{N}}$. Indeed, if we fix k, then $\mathscr{P}_k = \{(n_l) \in \mathscr{P}_\infty(\mathbb{N}) : \bigcap_{l=1}^k (-1)^l E_{n_l} \neq \varnothing\}$ is itself relatively closed in $\mathscr{P}_\infty(\mathbb{N})$. To see this, let (n_l^0) be any member of $\mathscr{P}_\infty(\mathbb{N})$ in P_k's closure. Consider the basic neighborhood B of (n_l^0) given by

$$B = \bigcap_{j=1}^{n_k^0} \left\{ (\varepsilon_p) \in \{0,1\}^{\mathbb{N}} : \varepsilon_j = c_{\{n_1^0, n_2^0, \ldots, n_k^0\}}(j) \right\}.$$

B intersects $\mathscr{P}_k$ so there is a (n_l^1) in $\mathscr{P}_k$ such that (n_l^1) agrees with (n_l^0) in its first k entries, i.e., $n_1^0 = n_1^1, n_2^0 = n_2^1, \ldots, n_k^0 = n_k^1$. It follows that

$$\bigcap_{l=1}^{k} (-1)^l E_{n_l^0} = \bigcap_{l=1}^{k} (-1)^l E_{n_l^1} \neq \varnothing.$$

We now apply the following combinatorial result of C. St. J. A. Nash-Williams: *If* $\mathscr{F}$ *is a relatively closed subset of* $\mathscr{P}_\infty(\mathbb{N})$, *then given an infinite subset K of* $\mathbb{N}$ *there is an infinite subset M of K such that either* $\mathscr{P}_\infty(M) \subseteq \mathscr{F}$ *or* $\mathscr{P}_\infty(M) \subseteq \mathscr{F}^c$.

This applies in particular to $\mathscr{F} = \bigcap_k \mathscr{P}_k$.

We get then the existence of an increasing sequence (m_p) of positive integers such that either $\mathscr{P}_\infty(\{m_p\}) \subseteq \bigcap_k \mathscr{P}_k$ or $\mathscr{P}_\infty(\{m_p\}) \subseteq (\bigcap_k \mathscr{P}_k)^c$. But we have seen that given any such $M = \{m_p\}$ there is an x_M^* in B_{X^*} such that $(x_m x_M^*)_{m\in M}$ has points of accumulation in both D_1 and D_2. It follows that there is an infinite subsequence (m_{p_q}) of (m_p) such that if q is odd, $x_{m_{p_q}} x_M^*$ is in D_1, whereas if q is even, then $x_{m_{p_q}} x_M^*$ is in D_2; alternatively, for q odd, x_M^* is in $O_{m_{p_q}}$, whereas for q even, x_M^* is in $E_{m_{p_q}}$; that is, for any q,

$x_M^* \in (-1)^q E_{m_{p_q}}$. This gives $(m_{p_q}) \in \cap_k \mathscr{P}_k$, thereby ruling out the possibility that $\mathscr{P}_\infty(\{m_p'\})$ is contained in $(\cap_k \mathscr{P}_k)^c$—forcing $\mathscr{P}_\infty(\{m_p\})$ to lie in $\cap_k \mathscr{P}_k$.

We now have that (m_p) is a strictly increasing sequence of positive integers such that given any subsequence (m_p') of (m_p)

$$\bigcap_{p=1}^{k} (-1)^p E_{m_p'} \neq \varnothing .$$

for all $k \in \mathbb{N}$. Let's look at this statement for a moment. By agreement $-E_n$ and O_n are the same sets; so the sequence (m_p) has the somewhat agreeable property that if we look at any subsequence (m_p') of (m_p) and intersect $O_{m_1'}$ with $E_{m_2'}$ with $O_{m_3'}$ with $E_{m_4'}$ etc., finitely many times, then the resultant set is nonempty. Now this is almost the degree of independence for (O_{m_p}, E_{m_p}) that is required; if we could but eliminate the need to switch from O to E and back again, we would have an independent sequence of pairs of disjoint subsets of B_{X^*} and a corresponding (bounded) subsequence of (x_n) such that the action of the x on the O and E fulfills the criteria set forth in Proposition 4. To achieve this added feature, we look at the subsequence (m_{2p}). Given any subsequence (m_{2p}') of (m_{2p}), if we look at the intersection of finitely many E and O indexed by (m_{2p}'), should successive terms both be E, say $E_{m_{2p}'}$ and $E_{m_{2(p+1)}'}$, then their intersection contains $E_{m_{2p}'} \cap O_{m_k} \cap E_{m_{2(p+1)}'}$, where k is any integer such that $m_{2p}' < m_k < m_{2(p+1)}'$. Similarly, if two $O_{m_{2p}'}$ occur back to back, we can always find inside their intersection an alternating intersection of the form $O_{m_j} \cap E_{m_k} \cap O_{m_l}$, where $j < k < l$. Now falling back on the basic distinguishing property of (m_p), we see that the sequence $(E_{m_{2p}}, O_{m_{2p}})$ is an independent sequence of disjoint pairs of subsets of B_{X^*}. Furthermore, the sequence $(x_{m_{2p}})$ in X is a bounded sequence such that for x^* in $O_{m_{2p}}$, $x_{m_{2p}}(x^*) \in D_1$, whereas for x^* in $E_{m_{2p}}$, $x_{m_{2p}}(x^*) \in D_2$. It follows that $(x_{m_{2p}})$ is a Rademacher-like system in $l_\infty(B_{X^*})$, hence equivalent to the unit vector basis of l_1.

Exercises

1. *Cardinality consequences of l_1's presence.*

(i) l_∞ contains $l_1(2^{\mathbb{N}})$ isometrically.

(ii) If X contains an isomorphic copy of l_1, then X^* contains $l_1(2^{\mathbb{N}})$.

(iii) If X contains an isomorphic copy of l_1, then X^{**}'s cardinality exceeds that of the continuum.

2. *$L_1[0,1]$ in duals.*

(i) A Banach space X is isomorphic to $L_1[0,1]$ if and only if X is the closed linear span of a system $(h_{n,k})_{\substack{n \geq 1 \\ 1 \leq k \leq 2^n}}$ satisfying

(1) $(h_{n,k})_{\substack{n \geq 1 \\ 1 \leq k \leq 2^n}}$ is a dyadic tree; that is, for each n and k

$$h_{n,k} = \tfrac{1}{2} h_{n+1,2k-1} + \tfrac{1}{2} h_{n+1,2k}.$$

(2) There is a $\rho > 0$ such that for each n and each 2^n-tuple $(a_1, \ldots, a_{2^n})$ of scalars we have

$$\rho \sum_{k=1}^{2^n} |a_k| \le \left\| \sum_{k=1}^{2^n} a_k h_{n,k} \right\| \le \sum_{k=1}^{2^n} |a_k|.$$

(ii) Suppose Y is a closed linear subspace of X and $L_1[0,1]$ is isomorphic to a subspace of Y^*. Then $L_1[0,1]$ is isomorphic to a subspace of X^*, too.

(iii) l_∞ contains an isomorphic copy of $L_1[0,1]$.

(iv) If X contains an isomorphic copy of l_1, then X^* contains an isomorphic copy of $L_1[0,1]$.

3. *The Schur property.* A Banach space X has the *Schur property* if weakly convergent sequences in X are norm convergent.

Any infinite dimensional Banach space with the Schur property contains an subspace isomorphic to l_1.

4. *The Schur property for dual spaces.*

(i) The dual X^* of a Banach space X has the Schur property if and only if X has the Dunford-Pettis property and does not contain a copy of l_1.

(ii) If both X and X^* have the Schur property, then X is finite dimensional.

5. *Lohman's lifting of weakly Cauchy sequences.*

(i) Let Y be a closed linear subspace of the Banach space X and suppose that Y contains no isomorphic copy of l_1. Then each weakly Cauchy sequence in X/Y is the image under the natural quotient map of a sequence in X having a weakly Cauchy subsequence.

(ii) If X has the Dunford-Pettis property and Y is a closed linear subspace of X such that X/Y fails the Dunford-Pettis property, then Y contains a copy of l_1.

6. *Spaces with the Banach-Saks property are reflexive.* A Banach space X is said to have the *Banach-Saks property* if given a bounded sequence (x_n) in X there is a subsequence (y_n) of (x_n) such that the sequence $(\sigma_n) = (n^{-1}\sum_{k=1}^{n} x_k)$ is norm convergent.

(i) The Banach-Saks property is an isomorphic invariant.

(ii) If a Banach space X has the Banach-Saks property, then so do all of X's closed linear subspaces.

(iii) The Banach space l_1 fails to have the Banach-Saks property.

(iv) A weak Cauchy sequence (x_n) for which

$$\operatorname{norm}\lim_n n^{-1} \sum_{k=1}^{n} x_k$$

exists is weakly convergent (with an obvious weak limit).

(v) Banach spaces with the Banach-Saks property are reflexive.

7. *Dunford-Pettis subspaces of duals.*

(i) If Y is a space with the Dunford-Pettis property that does not have the Schur property and X^* contains a copy of Y, then X contains a copy of l_1.

(ii) If X^* contains an isomorphic copy of $L_1[0,1]$, then X contains a copy of l_1.

(iii) $L_1[0,1]$ and c_0 are not isomorphic to subspaces of any separable dual.

Notes and Remarks

Arising from the devastation caused by a spate of fundamental counterexamples, Rosenthal's l_1 theorem provided a rallying point for members of the Banach space faith. In its pristine form Rosenthal's l_1 theorem is basic analysis; with hardly a pause for breath, the material of the second section, The Proof of the Rosenthal-Dor l_1 Theorem, proves the following variation.

Rosenthal's Dichotomy. *Let Ω be a set and suppose (f_n) is a uniformly bounded sequence of scalar-valued functions defined on Ω. Then precisely one of the following is the case.*

1. *Every subsequence of (f_n) has in turn a pointwise convergent subsequence.*
2. *There is a subsequence (g_n) of (f_n), a tree (Ω_n) of subsets of Ω, and disjoint disks D_1, D_2 of scalars such that $g_k(\omega) \in D_1$ for $\omega \in \Omega_j$, j odd, and $g_k(\omega) \in D_2$ for $\omega \in \Omega_j$, j even.*

Plainly, Rosenthal's l_1 theorem gives the last word (in some regards) about l_1's presence in a Banach space. Others, however, have had something to say, too. Several beautiful contributions to the detection of copies of l_1 were made in the late sixties by A. Pelczynski; our treatment of the first section, How to Imbed l_1 in a Banach Space, owes an obvious debt to this work of Pelczynski (1968). We cite but one particularly noteworthy result of his.

Theorem (Pelczynski). *Let X be a separable Banach space. Then TFAE:*

1. *X contains a copy of l_1;*
2. *$C[0,1]$ is a quotient of X;*
3. *X^* contains a copy of* $\mathrm{rca}(\mathscr{B}_{[0,1]})$;
4. *X^* contains a copy of $L_1[0,1]$.*
5. *X^* contains a copy of $l_1([0,1])$.*

A very few words about the proof of Pelczynski's theorem might be in order. Actually it is easier to work with the Cantor set Δ than with $[0,1]$; no matter, $C(\Delta)$ and $C[0,1]$ are isomorphic, as are $L_1(\Delta)$ and $L_1[0,1]$, where Δ

is viewed as the compact abelian group $\{-1, +1\}^{\mathbb{N}}$ accompanied by its Haar measure. Pelczynski's approach to the implication "1 implies 2" is just this: Like all separable Banach spaces, $C(\Delta)$ is a quotient of l_1 via some bounded linear operator q, say. Viewing $C(\Delta)$ as a subspace of $l_\infty(\Delta)$, we can extend the operator q to a bounded linear operator Q from X into $l_\infty(\Delta)$. Now, $\overline{QX}$ is separable and thus is isometric to a subspace of $C(\Delta)$; without losing any sleep we may as well view Q as an operator from X into this bigger copy of $C(\Delta)$, keeping in mind the fact that our original copy of $C(\Delta)$ is contained in the range of Q! Were the original copy of $C(\Delta)$ complemented in this late entry $C(\Delta)$, all would be well—we would merely follow Q by the bounded linear projection onto the original $C(\Delta)$ and be done with the implication. Though this need not be the case, Pelczynski would not be deterred. Rather, he found a way around the difficulty. His path was cleared by the following appealing, but hard-earned, shortcut.

Theorem (Pelczynski). *Let K be any compact metric space and suppose X is a closed linear subspace of $C(K)$ which is isomorphic to $C(K)$. Then X contains a closed linear subspace Y isometric to $C(K)$ which is complemented in $C(K)$ by a norm-one projection.*

This in hand, one need only follow Q by some projection onto a more suitably located copy of $C(\Delta)$ inside the original copy of $C(\Delta)$ to obtain $C(\Delta)$ as a quotient of X.

The trees of the first section were first planted by Pelczynski with the express purpose of finding a l_1 back in X from hypotheses similar to conditions 3, 4, and 5.

J. Hagler (1973) was able to show that 1, 3, and 4 are equivalent without separability assumptions. Incidentally, the proof outlined in Exercise 2 that 1 implies 4 was shown to us by J. Bourgain.

Of course, the weight of Pelczynski's results might move one to be suspicious of the possibility that the absence of a copy of l_1 in a separable Banach space ensures the separability of its dual; after all, l_1's presence makes the dual very nonseparable. What evidence there was to support this possibility was very special indeed. R. C. James has shown that if a Banach space has an unconditional basis with no copy of l_1 inside it, then its dual is separable. C. Bessaga and A. Pelczynski (1958) extended James's result to subspaces of spaces with an unconditional basis. This was soon extended by H. Lotz to separable Banach lattices and H. P. Rosenthal showed that any quotient of $C[0,1]$ with a nonseparable dual contains a copy of $C[0,1]$. A considerable body of information had accumulated that might be viewed as supportive of the possibility that a separable Banach space with no copy of l_1 has a separable dual.

In 1973, R. C. James constructed a Banach space JT (called *James tree space*) that is separable, contains no l_1—indeed is "l_2 rich" and has nonseparable dual. A complete examination of James's construction was performed by J. Lindenstrauss and C. Stegall (1975), who also analyzed several other counterexamples discovered by Lindenstrauss, independently of James.

It is important to realize that the counterexamples discussed by James, Lindenstrauss, and Stegall *seemed* to be part of an emerging trend in Banach space theory. In the space of two years, P. Enflo found a separable Banach space without the approximation property or a basis, B. S. Tsirelson gave a scheme for producing spaces without a copy of c_0 or any l_p, new and strange complemented copies of $L_p[0,1]$ were being discovered and R. C. James had built a uniformly nonoctahedral space that is not reflexive. Pathology seemed the order of the day. Actually, as any mature mathematician realizes, pathology only highlights the natural limits of a strong and healthy subject. Nonetheless, the onslaught of counterexamples experienced in the early seventies seems to have left an impression that little could be salvaged in the general theory.

Rosenthal's l_1 theorem served notice to the doomsday soothsayers of the errors of their way.

Soon after Rosenthal's l_1 theorem hit the newstands, true understanding of l_1's absence developed. Through the combined efforts of Rosenthal, E. Odell, and R. Haydon the following characterizations were formulated and established.

Theorem. *Let X be a* separable *Banach space. Then TFAE*:

1. *X contains no copy of l_1.*
2. *Each element of $B_{X^{**}}$ is the weak* limit of a sequence from B_X.*
3. *X and X^{**} have the same cardinality.*
4. *$B_{X^{**}}$ is weak* sequentially compact.*

(Incidentally, we prove this theorem in our discussion of Banach spaces with weak* sequentially compact dual balls.)

Theorem. *Let X be* any *Banach space. Then TFAE*:

1. *X contains no copy of l_1.*
2. *Each weak* compact convex subset of X^* is the norm-closed convex hull of its extreme points.*
3. *Each $x^{**} \in X^{**}$ is measurable with respect to each regular Borel probability measure on (B_{X^*}, weak^*).*

Each of these results at some stage calls on either Proposition 3 (due to H. P. Rosenthal) or Proposition 4 (due to L. Dor). By the way, our proof of Rosenthal's l_1 theorem follows J. Farahat in his use of Ramsey theory. Rosenthal's original proof did not rely *explicitly* on any Ramsey theorems; apparently Rosenthal rederived the infinite version of Ramsey's theorem necessary in his "good-bad" description without recognizing that he had done so.

As one might suspect, a theorem that packs the punch of Rosenthal's l_1 theorem soon leads somewhere. It was not long after Rosenthal and his followers started their fruitful search for variations on the l_1 theorem that the close connections with pointwise compactness in the class of Baire-1 functions were uncovered.

Rosenthal himself initiated a penetrating investigation into pointwise compact subsets of Baire class 1. Though stymied by several problems, his work pointed the way for a real breakthrough by J. Bourgain, D. Fremlin, and M. Talagrand (1978). We quote several of their results and urge the student to carefully study their fundamental paper, which is just starting to find serious applications in integration theory and the study of operator ideals.

Theorem. *Let Ω be a Polish space (i.e., homeomorphic to a complete separable metric space). Then the space $B_1(\Omega)$ of real functions of the first Baire class is* angelic *in the topology of pointwise convergence.*

Here we must recall that a topological space T is *angelic* if relatively countably compact subsets of T are relatively compact and the closure of a relatively compact set in T is precisely the set of limits of its sequences.

Theorem. *Let Ω be a Polish space and A a countable relatively countably compact subset of the space of real Borel functions defined on Ω endowed with the topology of pointwise convergence. Then the closed convex hull of A is compact and angelic.*

Rosenthal's l_1 theorem has had a synthesizing effect on a number of problems previously attacked by pretty much ad hoc techniques.

For instance, our presentation of the Josefson-Nissenzweig theorem was made possible largely through the graces of Rosenthal's l_1 theorem. It has allowed for considerable progress in the study of the Pettis integral, starting with the discovery by K. Musial and L. Janicka that duals of spaces not containing l_1 have a kind of weak Radon-Nikodym property and continuing on through the work of J. J. Uhl and his students. Applications to operator theory, always close to the representation theory by integrals, have been uncovered by H. Fakhoury and L. Weis; the fact that operators on a $C(\Omega)$ space fixing a copy of l_1 must also fix a copy of $C[0,1]$, due to H. P. Rosenthal, is key to J. Diestel and C. J. Seifert's study of Banach-Saks phenomena for operators on $C(\Omega)$ spaces (1979). Rosenthal's l_1 theorem is an essential ingredient to B. Beauzamy's elegant presentation of "spreading models" and their attendant applications (1979). No doubt it will continue to play a greater and greater role in the synthesis of modern Banach space developments.

Finally, we must mention a recent and penetrating advance of M. Talagrand (1984), who has proved a random version of Rosenthal's l_1 theorem that incorporates the Kadec-Pelczynski theorem with Rademacher-like pathology for vector-valued functions.

Bibliography

Beauzamy, B. 1979. Banach-Saks properties and spreading models. *Math. Scand.*, **44**, 357–384.

Bessaga, C. and Pelczynski, A. 1958. A generalization of results of R. C. James concerning absolute bases in Banach spaces. *Studia Math.*, **17**, 165–174.

Bourgain, J., Fremlin, D., and Talagrand, M. 1978. Pointwise compact sets of Baire-measurable functions. *Amer. J. Math.*, **100**, 845–886.

Diestel, J. and Seifert, C. J. 1979. The Banach-Saks ideal, I. Operators acting on $C(\Omega)$. Commentationes Math. Tomus Specialis in Honorem Ladislai Orlicz, 109–118, 343–344.

Dor, L. 1975. On sequences spanning a complex l_1 space. *Proc. Amer. Math. Soc.*, **47**, 515–516.

Fakhoury, H. 1977. Sur les espaces de Banach ne contenant pas $l_1(N)$. *Math. Scand.*, **41**, 277–289.

Farahat, J. 1974. Espaces de Banach contenant l_1, d'apres H. P. Rosenthal. *Seminaire Maurey-Schwartz* (1973–1974), Ecole Polytechnique.

Hagler, J. 1973. Some more Banach spaces which contain l^1. *Studia Math.*, **46**, 35–42.

Haydon, R. G. 1976. Some more characterizations of Banach spaces containing l^1. *Math. Proc. Cambridge Phil. Soc.*, **80**, 269–276.

Heinrich, S. 1980. Closed operator ideals and interpolation. *J. Funct. Anal.*, **35**, 397–411.

James, R. C. 1974. A separable somewhat reflexive Banach space with nonseparable dual. *Bull. Amer. Math. Soc.*, **80**, 738–743.

Janicka, L. 1979. Some measure-theoretic characterizations of Banach spaces not containing l_1. *Bull. Acad. Polon. Sci.*, **27**, 561–565.

Lindenstrauss, J. and Stegall, C. 1975. Examples of separable spaces which do not contain l_1 and whose duals are non-separable. *Studia Math.*, **54**, 81–105.

Lohman, R. H. 1976. A note on Banach spaces containing l_1. *Can. Math. Bull.*, **19**, 365–367.

Musial, K. 1978. The weak Radon-Nikodym property. *Studia Math.*, **64**, 151–174.

Odell, E. and Rosenthal, H. P. 1975. A double-dual characterization of separable Banach spaces containing l^1. *Israel J. Math.*, **20**, 375–384.

Pelczynski, A. 1968. On Banach spaces containing $L_1(\mu)$. *Studia Math.*, **30**, 231–246.

Pelczynski, A. 1968. On $C(S)$ subspaces of separable Banach spaces. *Studia Math.*, **31**, 513–522.

Riddle, L. H. and Uhl, J. J., Jr. 1982. Martingales and the fine line between Asplund spaces and spaces not containing a copy of l_1. In *Martingale Theory in Harmonic Analysis and Banach spaces*. Lecture Notes in Mathematics, Vol. 939. New York: Springer-Verlag, pp. 145–156.

Rosenthal, H. P. 1970. On injective Banach spaces and the spaces $L^\infty(\mu)$ for finite measures μ. *Acta Math.*, **124**, 205–248.

Rosenthal, H. P. 1972. On factors of $C[0,1]$ with nonseparable dual. *Israel J. Math.*, **13**, 361–378; Correction: *ibid.*, **21** (1975), 93–94.

Rosenthal, H. P. 1974. A characterization of Banach spaces containing l_1. *Proc. Nat. Acad. Sci.* (*USA*), **71**, 2411–2413.

Rosenthal, H. P. 1977. Pointwise compact subsets of the first Baire class. *Amer. J. Math.*, **99**, 362–378.

Rosenthal, H. P. 1978. Some recent discoveries in the isomorphic theory of Banach spaces. *Bull. Amer. Math. Soc.*, **84**, 803–831.

Talagrand, M. 1984 Weak Cauchy sequences in $L^1(E)$. *Amer. J. Math, to appear.*

Weis, L. 1976. On the surjective (injective) envelope of strictly (co-) singular operators. *Studia Math.*, **54**, 285–290.

CHAPTER XII

The Josefson-Nissenzweig Theorem

From Alaoglu's theorem and the F. Riesz theorem, we can conclude that for infinite-dimensional Banach spaces X the weak* topology and the norm topology in X^* differ. Can they have the same convergent sequences? The answer is a resounding "no!" and it is the object of the present discussion. More precisely we will prove the following theorem independently discovered by B. Josefson and A. Nissenzweig.

Theorem. *If X is an infinite-dimensional Banach space, then there exists a weak* null sequence of norm-one vectors in X^*.*

The original proofs of both Josefson and Nissenzweig were rather unwieldy; so we follow the spirit of another proof due to J. Hagler and W. B. Johnson.

A key ingredient in our proof will be the following lemma concerned with *real* Banach spaces and l_1's appearance therein.

Lemma. *Suppose X^* contains a copy of l_1 but that no weak* null sequence in X^* is equivalent to the unit vector basis of l_1*

Then X contains a copy of l_1.

PROOF. Suppose (y_n^*) is a sequence in B_{X^*} equivalent to the unit vector basis of l_1. Define

$$\delta(y_n^*) = \sup_{\|x\|=1} \overline{\lim_n} |y_n^* x|;$$

$\delta(y_n^*) > \alpha$ just means that for some $x \in S_X$, $|y_n^* x|$ exceeds α infinitely often. Since (y_n^*) is equivalent to the unit vector basis of l_1 our hypotheses ensure that (y_n^*) is *not* weak* null; therefore, $\delta(y_n^*) > 0$.

Suppose (y_n^*) is a sequence in B_{X^*} equivalent to the unit vector basis of l_1, and suppose (z_n^*) is built from y_n^* as follows:

$$z_n^* = \sum_{i \in A_n} a_i y_i^*,$$

where (A_n) is a sequence of pairwise disjoint finite subsets of $\mathbb{N}$ and $\Sigma_{i \in A_n}|a_i| = 1$. Then we call (z_n^*) a *normalized* l_1-*block of* (y_n^*). Notice that such a sequence (z_n^*) is also equivalent to the unit vector basis of l_1 and satisfies

$$\delta(z_n^*) \leq \delta(y_n^*).$$

Finally, for a sequence (y_n^*) in B_{X^*} equivalent to the unit vector basis of l_1, we define

$$\varepsilon(y_n^*) = \inf\{\delta(z_n^*) : (z_n^*) \text{ is a normalized } l_1\text{-block of } (y_n^*)\}.$$

Plainly for any normalized l_1-block (z_n^*) of (y_n^*) we have

$$\varepsilon(y_n^*) \leq \varepsilon(z_n^*).$$

Claim. If (v_n^*) is a sequence in B_{X^*} equivalent to the unit vector basis of l_1, then we can find a normalized l_1-block (y_n^*) of (v_n^*) for which

$$\delta(y_n^*) = \delta(z_n^*)$$

holds for all normalized l_1-blocks (z_n^*) of (y_n^*).

In fact, let $(y_{n,1}^*)$ be a normalized l_1-block of (v_n^*) such that

$$\delta(y_{n,1}^*) \leq \tfrac{3}{2}\varepsilon(v_n^*).$$

Let $(y_{n,2}^*)$ be a normalized l_1-block of $(y_{n,1}^*)$ such that

$$\delta(y_{n,2}^*) \leq \tfrac{5}{4}\varepsilon(y_{n,1}^*),$$

et cetera. Let $y_n^* = y_{n,n}^*$. Then (y_n^*) is a normalized l_1-block of (v_n^*) for which it is clear that

$$\delta(y_n^*) \leq \delta(y_{n,k}^*) \quad \text{and} \quad \varepsilon(y_{n,k}^*) \leq \varepsilon(y_n^*)$$

hold for every k. Our method of selecting dictates that

$$\delta(y_n^*) \leq \varliminf_k \delta(y_{n,k}^*) \leq \varliminf_k \varepsilon(y_{n,k}^*) \leq \varepsilon(y_n^*) \leq \delta(y_n^*).$$

Claim established.

All of our building will be atop the sequence (y_n^*) resulting from our claim. Set $\delta = \delta(y_n^*)$.

Let $\varepsilon > 0$ be given.

There is an $x_1 \in S_X$ and an infinite set N_1 in $\mathbb{N}$ such that for any $n \in N_1$

$$y_n^* x_1 < -\delta + \varepsilon.$$

Suppose $0 < \varepsilon' < \varepsilon/3$. Partition N_1 into two disjoint infinite subsets enumerated by the increasing sequences (m_k) and (n_k) of positive integers. The sequence $(\frac{1}{2}(y_{n_k}^* - y_{m_k}^*))$ is a normalized l_1-block of (y_n^*); so there is an $x_2 \in S_X$ and an infinite set of k for which

$$\tfrac{1}{2}(y_{n_k}^* - y_{m_k}^*)x_2 > \delta - \varepsilon'.$$

Of course, $(y^*_{n_k})$ and $(y^*_{m_k})$ are also normalized l_1-blocks of (y^*_n) which for all but finitely many k must satisfy

$$|y^*_{n_k}x_2|, \left|y^*_{m_k}x_2\right| < \delta + \varepsilon'.$$

It follows that for those k commonly enjoying both the above estimates (and there are infinitely many of them) we have

$$y^*_{n_k}x_2 > \delta - 3\varepsilon'$$

and

$$y^*_{m_k}x_2 < -\delta + 3\varepsilon'.$$

We show the first of these; the second has a similar derivation. Suppose k satisfies

$$\tfrac{1}{2}\left(y^*_{n_k} - y^*_{m_k}\right)x_2 > \delta - \varepsilon', \qquad |y^*_{n_k}x_2| < \delta + \varepsilon', \qquad |y^*_{m_k}x_2| < \delta + \varepsilon',$$

but

$$y^*_{n_k}x_2 < \delta - 3\varepsilon'.$$

Then we would have

$$\begin{aligned}\delta - \varepsilon' &< \tfrac{1}{2}\left(y^*_{n_k} - y^*_{m_k}\right)x_2\\ &= \tfrac{1}{2}\left(y^*_{n_k}x_2 - y^*_{m_k}x_2\right)\\ &< \tfrac{1}{2}(\delta - 3\varepsilon' + \delta + \varepsilon')\\ &= \delta - \varepsilon',\end{aligned}$$

a contradiction.

Keeping in mind the choice of $\varepsilon' < \varepsilon/3$, we see that the sets

$$\begin{aligned}N_2 &= \left\{n_k : y^*_{n_k}x_2 > \delta - \varepsilon\right\},\\ N_3 &= \left\{m_k : y^*_{m_k} < -\delta + \varepsilon\right\}\end{aligned}$$

are infinite disjoint subsets of N_1.

Let $0 < \varepsilon' < \varepsilon/7$.

We can decompose N_2 into two disjoint infinite subsets which we enumerate as increasing sequences $(n_k(1)), (n_k(2))$ of positive integers and similarly decompose N_3 into sequences $(m_k(1)), (m_k(2))$. Then the sequence $(\frac{1}{4}(y^*_{n_k(1)} - y^*_{n_k(2)} + y^*_{m_k(1)} - y^*_{m_k(2)}))$ is a normalized l_1-block of (y^*_n). So there is an $x_3 \in S_X$ such that for infinitely many k

$$\tfrac{1}{4}\left(y^*_{n_k(1)} - y^*_{n_k(2)} + y^*_{m_k(1)} - y^*_{m_k(2)}\right)(x_3) > \delta - \varepsilon'.$$

Of course, each of the sequences $(y^*_{n_k(1)})$, $(y^*_{n_k(2)})$, $(y^*_{m_k(1)})$, and $(y^*_{m_k(2)})$ are normalized l_1-blocks of (y^*_n), and so for all but a finite number of k we must have

$$\left|y^*_{n_k(1)}x_3\right|, \left|y^*_{n_k(2)}x_3\right|, \left|y^*_{m_k(1)}x_3\right|, \left|y^*_{m_k(2)}x_3\right| < \delta + \varepsilon'.$$

It follows that for those k satisfying all the above relationships simultaneously (and, again, there are infinitely many such k) we have

$$y^*_{n_k(1)}x_3,\ y^*_{m_k(1)}x_3 > \delta - 7\varepsilon',$$

and

$$y^*_{n_k(2)}x_3,\ y^*_{m_k(2)}x_3 < -\delta + 7\varepsilon'.$$

As before, we establish the first of these relationships; the rest follow a similar path. Suppose k were such that

$$\tfrac{1}{4}\left(y^*_{n_k(1)} - y^*_{n_k(2)} + y^*_{m_k(1)} - y^*_{m_k(2)}\right)x_3 > \delta - \varepsilon',$$

$$\left|y^*_{n_k(1)}x_3\right|, \left|y^*_{n_k(2)}x_3\right|, \left|y^*_{m_k(1)}x_3\right|, \left|y^*_{m_k(2)}x_3\right| < \delta + \varepsilon'$$

yet

$$y^*_{n_k(1)}x_3 \leq \delta - 7\varepsilon'.$$

Then

$$\begin{aligned}\delta - \varepsilon' &< \tfrac{1}{4}\left(y^*_{n_k(1)}x_3 - y^*_{n_k(2)}x_3 + y^*_{m_k(1)}x_3 - y^*_{m_k(2)}x_3\right)\\ &\leq \tfrac{1}{4}(\delta - 7\varepsilon' + \delta + \varepsilon' + \delta + \varepsilon' + \delta + \varepsilon')\\ &= \delta - \varepsilon',\end{aligned}$$

a contradiction.

Keeping in mind the choice of $\varepsilon' < \varepsilon/7$, we have that the sets

$$N_4 = \left\{n_k(1) : y^*_{n_k(1)}x_3 > \delta - \varepsilon\right\},$$

$$N_5 = \left\{n_k(2) : y^*_{n_k(2)}x_3 < -\delta + \varepsilon\right\}$$

are disjoint infinite subsets of N_2 and the sets

$$N_6 = \left\{m_k(1) : y^*_{m_k(1)}x_3 > \delta - \varepsilon\right\},$$

$$N_7 = \left\{m_k(2) : y^*_{m_k(2)}x_3 < -\delta + \varepsilon\right\}$$

are disjoint infinite subsets of N_3.

The continuing procedure is clear. Where does it lead us to? Well, letting $\Omega_n = \{y^*_k : k \in N_n\}$, we get a tree of subsets of B_{X^*}. Furthermore, (x_n) has been so selected from S_X that if $2^{n-1} \leq k < 2^n$, then $(-1)^k x_n(y^*) \geq \delta - \varepsilon$ for all $y^* \in \Omega_k$. We have created a Rademacher-like sequence (x_n); in other words, we have a l_1 unit vector basis in X and X contains a copy of l_1. □

Let's prove the Josefson-Nissenzweig theorem.

First we suppose X is a real Banach space. Suppose that in X^*, weak* null sequences are norm null. Then either X^* contains an isomorph of l_1 or it does not. If not, then each bounded sequence in X^* has a weakly Cauchy subsequence which, with our supposition in place, is a fortiori norm convergent; X^* (and hence X) is finite dimensional.

Okay, our result is so if X^* does not contain l_1; what if l_1 is isomorphic to a subspace of X^*? Clearly in this case no weak* null sequence can be equivalent to the unit vector basis of l_1, since weak* null sequences are norm null. Therefore, X contains an isomorphic copy of l_1—thanks to our lemma. Let's produce a weak* null sequence in X^* that is not norm null!

To produce such sequences we will use the following correspondence: weak* null sequences (y_n^*) on any Banach space Y are in one-to-one correspondence with the bounded linear operators from Y to c_0; this is an easy exercise, and we omit the few details needed to prove it. We make the following claim: Regardless of where l_1 finds itself inside a Banach space X the natural inclusion map i of l_1 into c_0 extends in a bounded linear fashion to an operator T from X to c_0. Of course, $x_n^*x = (Tx)_n$ defines the sequence (x_n^*) sought after.

Let's see why $i: l_1 \to c_0$ extends to any superspace. First, look at the operator $R: l_1 \to L_\infty[0,1]$ defined by $Re_n = r_n$, where e_n is the nth unit vector and r_n is the nth Rademacher function; R is an isomorphism of l_1 into $L_\infty[0,1]$. Notice that the operator $L: L_\infty[0,1] \to c_0$ defined by $Lf = (\int_0^1 f(t) r_n(t)\,dt)$ is a bounded linear operator, well-defined because of the orthonormality of (r_n). Moreover,

$$i = LR.$$

Recall now the standard proof of the Hahn-Banach theorem: to extend a linear continuous functional, you use the order completeness of the reals and the fact that the closed unit ball of the reals has a biggest element. These ingredients are also supplied by the lattice $L_\infty[0,1]$. The Hahn-Banach conclusion applies to $L_\infty[0,1]$-valued operators. In particular, R extends to a bounded operator N from X to $L_\infty[0,1]$. By construction $LN = i|_{l_1}$.

For complex Banach spaces we proceed as follows: if X is a complex Banach space, then X is a real Banach space as well; let (x_n^*) be a weak* null sequence of real linear functionals of norm 1 and define z_n^* by $z_n^*(x) = x_n^*(x) - ix_n^*(ix)$. (z_n^*) is weak* but not norm null.

Exercises

1. *The existence of noncompact operators into c_0.* For any infinite-dimensional Banach space X there exists a noncompact bounded linear operator $T: X \to c_0$.

2. *Weak sequential density of spheres.* Let X be an infinite-dimensional Banach space.

   (i) S_{X^*} is weak* sequentially dense in B_{X^*}.

   (ii) S_X is weakly sequentially dense in B_X if and only if X does not have the Schur property.

3. *Fixing l_1.*

(i) Let $T: X \to Y$ be a bounded linear operator whose adjoint $T^*: Y^* \to X^*$ fixes a copy of l_1. Suppose, however, that whenever (y_n^*) is the unit vector basis of a copy of l_1 in Y^* that is fixed by T^* the sequence $(T^*y_n^*)$ is *not* weak* null. Then T fixes a copy of l_1. (*Hint*: Look very carefully at the proof of the Josefson-Nissenzweig theorem.)

(ii) Suppose weak* null sequences in X^* are weakly null. Show that any non-weakly compact operator $T: X \to Y$ fixes a copy of l_1.

(iii) A bounded linear operator $T: X \to Y$ is called *strictly cosingular* if given any Banach space Z, if there are quotient operators $\varphi_X: X \to Z$, $\varphi_Y: Y \to Z$ for which

$$\varphi_Y T = \varphi_X,$$

then $\dim Z < \infty$. If X is a Banach space with the Dunford-Pettis property in whose dual weak* null sequences are weakly null and if $T: X \to Y$ fixes no copy of l_1, then T is strictly cosingular.

4. *Relative weak compactness of limited sets.*

(i) If (x_n) is a sequence from the limited subset K of a Banach space X, then (x_n) has a weak Cauchy subsequence.

(ii) Limited subsets of weakly sequentially complete spaces are relatively weakly compact.

(iii) Limited subsets of spaces containing no copy of l_1 are relatively weakly compact.

5. *The Szlenk index.* Let X be an infinite-dimensional Banach space.

(i) If X^* is separable, then there exist a weakly null sequence in X and a weak* null sequence (x_n^*) in X^* such that $\inf_n |x_n^* x_n| > 0$.

(ii) Let K and K^* be nonempty subsets of X and X^*, respectively, with K bounded and K^* weak* compact. For $\varepsilon > 0$ define the set $P(\varepsilon, K; K^*)$ to be the totality of all x^* in K^* for which there are sequences (x_n) in K and (x_n^*) in K^* such that $\text{weak} \lim_n x_n = 0$, $x^* = \text{weak}^* \lim_n x_n^*$, and $|x_n^* x_n| \geq \varepsilon$ for all n.

$P(\varepsilon, K; K^*)$ is a weak* closed, weak* nowhere dense subset of K^*.

(iii) For each ordinal number α we define the sets $S_\alpha(\varepsilon)$ as follows:
(a) $S_0(\varepsilon) = P(\varepsilon, B_X; B_{X^*})$.
(b) $S_{\alpha+1}(\varepsilon) = P(\varepsilon, B_X; S_\alpha(\varepsilon))$.
(c) If α is a limit ordinal, then $S_\alpha(\varepsilon) = \cap_{\gamma < \alpha} S_\gamma(\varepsilon)$.Let

$$\sigma(\varepsilon) = \sup\{\alpha : S_\alpha(\varepsilon) \neq \varnothing\}$$

and define the Szlenk index of X to be the ordinal $\sigma(X)$ given by

$$\sigma(X) = \sup_{\varepsilon > 0} \sigma(\varepsilon).$$

If X^* is separable, then $\sigma(X) < \omega_1$, the first uncountable ordinal.

Notes and Remarks

The object of our attentions in this chapter was conquered independently by B. Josefson and A. Nissenzweig. Our proof follows the instructions set forth by J. Hagler and W. B. Johnson (1977) with a few variations in execution aimed at lightening the necessary background. Incidentally, although we do not present them, the original solutions are all the more impressive because of their bare knuckles frontal attacks.

Josefson's interests were sparked by problems arising in infinite-dimensional holomorphy. An excellent description of those problems that aroused Josefson, as well as their present status, can be found in the monograph of S. Dineen (1981). It was Dineen who wanted to know if the closed unit ball of a Banach space is ever limited; the Josefson-Nissenzweig theorem provides a negative answer. A related question of Dineen remains open: *In which Banach spaces X are limited subsets relatively compact*?

The observation that limited sets are conditionally weakly compact is due to J. Bourgain and J. Diestel, who also noted that in spaces with no copy of l_1 limited sets are relatively weakly compact.

The Szlenk index was invented by W. Szlenk (1968) in his solution of a problem from the Scottish book. We have taken our exercise from Szlenk's paper and a related note of P. Wojtaszczyk (1970). The upshot of their efforts is the nonexistence of a universal object in the classes of separable reflexive Banach spaces (Szlenk) or separable dual spaces (Wojtaszczyk).

Bibliography

Bourgain, J. and Diestel, J. 1984. Limited operators and strict cosingularity. *Math. Nachrichten, to appear.*

Dineen, S. 1981. *Complex Analysis in Locally Convex Spaces*. North-Holland Mathematics Studies, Vol. 57. New York: North-Holland.

Hagler, J. and Johnson, W. B. 1977. On Banach spaces whose dual balls are not weak* sequentially compact. *Israel J. Math.*, **28**, 325–330.

Josefson, B. 1975. Weak sequential convergence in the dual of a Banach space does not imply norm convergence. *Ark. mat.*, **13**, 79–89.

Nissenzweig, A. 1975. ω* sequential convergence. *Israel J. Math.*, **22**, 266–272.

Szlenk, W. 1968. The non-existence of a separable reflexive Banach space universal for all separable reflexive Banach spaces. *Studia Math.*, **30**, 53–61.

Wojtaszczyk, P. 1970. On separable Banach spaces containing all separable reflexive Banach spaces. *Studia Math.*, **37**, 197–202.

CHAPTER XIII

Banach Spaces with Weak* Sequentially Compact Dual Balls

Alaoglu's theorem ensures that every bounded sequence (x_n^*) in X^* has a weak* convergent *subnet*. When can one actually extract a weak* convergent *subsequence*? As yet, no one knows. In this chapter a few of the most attractive conditions assuring the existence of such subsequences are discussed.

To be sure, it is not always possible. Consider $X = l_1(\mathbf{R})$. B_{X^*} is not weak* sequentially compact. To see this, let $\mathcal{P}_\infty(\mathbf{N})$ denote the collection of infinite subsets of the natural numbers; the cardinality of $\mathcal{P}_\infty(\mathbf{N})$ is that of $\mathbf{R}$. Let $\varphi: \mathbf{R} \to \mathcal{P}_\infty(\mathbf{N})$ be a one-to-one onto mapping. Consider the sequence (x_n^*) in $B_{l_\infty(\mathbf{R})} = B_{l_1(\mathbf{R})^*}$ given by

$$x_n^*(t) = \begin{cases} 1 & \text{if } n \in \varphi(t), \\ 0 & \text{if } n \notin \varphi(t). \end{cases}$$

Evaluating a subsequence $(x_{n_k}^*)$ of (x_n^*) at the real number r whose image under φ is $\{n_{2k}\}$, we see that

$$x_{n_k}^*(r) = \begin{cases} 1 & \text{if } k \text{ is even}, \\ 0 & \text{if } k \text{ is odd}. \end{cases}$$

Therefore, $(x_{n_k}^*)$ is not pointwise convergent. Alternatively, (x_n^*) admits of no subsequence that converges at each of the (continuum of) unit vectors in $l_1(\mathbf{R})$.

It follows from this (and the stability results presented below) that any space containing an isomorph of $l_1(\mathbf{R})$ cannot have a weak* sequentially compact dual ball. So, in particular, if X contains a copy of l_∞, then B_{X^*} is not weak* sequentially compact. What about some positive results?

If X is a separable Banach space then B_{X^*} is weak metrizable and so, being also weak* compact, is weak* sequentially compact.* This sometimes happens even in nonseparable situations. For instance, if bounded sequences in X^* were assured of weakly Cauchy subsequences, the would be sure of weak* convergent subsequences as well; in other words, *B_{X^*} is weak* sequentially compact whenever X^* does not contain a copy of l_1.* In particular, *B_{X^*} is weak* sequentially compact whenever X is reflexive.* Of course, this last fact can also be seen to follow (more easily, in fact) from Eberlein's theorem.

We begin our more serious discussion by noting a few of the basic stability properties enjoyed by the class under investigation.

Lemma 1. *The class of Banach spaces having weak* sequentially compact dual ball is closed under the following operations:*

1. *Taking dense continuous linear images*
2. *Quotients*
3. *Subspaces*

PROOF. If $T: X \to Y$ is a bounded linear operator with dense range, then $T^*: Y^* \to X^*$ is a bounded linear operator that's one to one. It follows that T^* is a weak* homeomorphism between B_{Y^*} and $T^*B_{Y^*}$. This proves 1 from which 2 follows. Part 3 is a consequence of the Hahn-Banach theorem. □

Lemma 2 (A. Grothendieck). *Let K be a weakly closed subset of the Banach space X. Suppose that for each $\varepsilon > 0$ there is a weakly compact set K_ε in X such that*

$$K \subseteq K_\varepsilon + \varepsilon B_X.$$

Then K is weakly compact.

PROOF. Let $\overline{K}^{\text{weak}^*}$ denote K's weak* closure up in X^{**}. If $\overline{K}^{\text{weak}^*}$ should find itself back in X, then we are done. In fact, K, sitting as it does in $K_1 + B_X$ for some weakly compact set K_1 corresponding to $\varepsilon = 1$, must be bounded; so $\overline{K}^{\text{weak}^*}$ is weak* compact. Consequently, if $\overline{K}^{\text{weak}^*}$ lies in X it is weakly compact and is nothing but K's weak closure, i.e., K.

Now each hypothesized K_ε is weakly compact and so for each $\varepsilon > 0$ $\overline{K_\varepsilon}^{\text{weak}^*} = K_\varepsilon$. This, plus the continuity of addition, gives

$$\begin{aligned}\overline{K}^{\text{weak}^*} &\subseteq \text{weak* closure}\,(K_\varepsilon + \varepsilon B_X)\\ &\subseteq \overline{K_\varepsilon}^{\text{weak}^*} + \overline{\varepsilon B_X}^{\text{weak}^*}\\ &\subseteq K_\varepsilon + \varepsilon B_{X^{**}}.\end{aligned}$$

Consequently,

$$\begin{aligned}\overline{K}^{\text{weak}^*} &\subseteq \cap_{\varepsilon > 0}(K_\varepsilon + \varepsilon B_{X^{**}})\\ &\subseteq X.\end{aligned}$$

□

Lemma 3 (W. J. Davis, T. Figiel, W. B. Johnson, A. Pelczynski). *Let K be a weakly compact absolutely convex subset of the Banach space X. Then there is a weakly compact absolutely convex set $C \subseteq X$ that contains K such that X_C is reflexive, where X_C is the linear span of C with closed unit ball C.*

Remark. It is well known and easily verified that X_C is in fact a Banach space.

PROOF. For each n let $B_n = 2^n K + 2^{-n} B_X$. Since K is weakly compact and absolutely convex, B_n is weakly closed and absolutely convex. Further, B_n has a nonempty interior since it contains $2^{-n} B_X$. In other words, renorming X to have B_n as a closed unit ball leads to an equivalent norm $\| \|_n$ on X.

Let $C = \{x \in X : \Sigma_n \|x\|_n^2 \le 1\}$. Then $C = \cap_n \{x \in X : \Sigma_{k=1}^n \|x\|_k^2 \le 1\}$; so C is closed and absolutely convex, hence, weakly closed. Moreover, if we look in X_C, then the Minkowski functional $\| \|_C$ of an x is given by $\|x\|_C = (\Sigma_n \|x\|_n^2)^{1/2}$. Some noteworthy points to be made in favor of C:

1. $K \subseteq C$. Indeed, if $x \in K$, then $2^n x \in 2^n K$ which is contained in B_n. Therefore, $\|2^n x\|_n \le 1$ or $\|x\|_n \le 2^{-n}$. It follows that $\|x\|_C \le (\Sigma_n 2^{-2n})^{1/2} \le 1$.
2. C is weakly compact in X; this follows from Grothendieck's Lemma if you just notice that C is weakly closed and $C \subseteq 2^n K + 2^{-n} B_X$ for each n.
3. On C the weak topology of X and that of X_C agree. Here we observe that the map from X_C to $(\Sigma_n X_n)_{l_2}$ that takes $x \in X_C$ to $(x, x, \ldots, x, \ldots)$ in $(\Sigma_n X_n)_{l_2}$ is an isometric imbedding. It is easy to believe (and not much harder to prove) that $(\Sigma_n X_n)_{l_2}^*$ is just $(\Sigma_n X_n^*)_{l_2}$ and that in $(\Sigma_n X_n^*)_{l_2}$ the subspace $(\Sigma_n X_n^*)_\varphi$ of finitely nonzero sequences is dense. Therefore, on the bounded set C in $(\Sigma_n X_n)_{l_2}$, the weak topology generated by $(\Sigma_n X_n^*)_{l_2}$ and the topology of pointwise convergence on members of $(\Sigma_n X_n^*)_\varphi$ are the same. But *on* C the topology of pointwise convergence on members of $(\Sigma_n X_n^*)_\varphi$ is just the weak topology of X! This follows from the fact that $x \in C$ corresponds to an $(x, x, \ldots, x, \ldots)$ in $(\Sigma_n X_n)_{l_2}$ so the action of $(x_1^*, \ldots, x_n^*, 0, 0, \ldots)$—typical member of $(\Sigma_n X_n^*)_\varphi$—on x is given by $(\Sigma_{k=1}^n x_k^*)(x)$.

In all, 1 through 3 add up to a proof of Lemma 3. □

A Banach space X is said to be *weakly compactly generated* if X contains a *weakly compact* absolutely convex set whose linear span is dense in X.

Theorem 4 (D. Amir, J. Lindenstrauss). *Any subspace of a weakly compactly generated Banach space has a weak* sequentially compact dual ball.*

PROOF. Suppose first that X is a weakly compactly generated Banach space and assume that K is a weakly compact absolutely convex set in X, whose linear span is dense in X. Let C be the weakly compact absolutely convex set produced in Lemma 3. The linear operator $S : X_C \to X$ defined by $Sx = x$ is easily seen to be bounded, with dense range. Since reflexive spaces have weak* sequentially compact dual balls, this shows that weakly compactly generated spaces do too by using Lemma 1 (part 1). To finish fully the proof of the whole assertion in the theorem, apply Lemma 1 (part 3). □

Before presenting our next result, we establish some notation. If $A \subseteq X^*$, then the weak* closed convex hull of A will be denoted by $\overline{\text{co}}^* A$ and the set of weak* points of accumulation of A will be denoted by $A^{\sim}$. One further notational device will allow us to dispense with a great deal of sub- and superscripting. This device was used already in proving Rosenthal's l_1 theorem. Instead of denoting a subsequence of a sequence (x_n^*) by directly indexing the x_n^* belonging to the subsequence, we restrict the subscripts to infinite subsets of $\mathbb{N}$ keeping in mind that such subsets always carry with them a natural ordering between their elements. For instance, if (k_n) is a subsequence of the natural numbers, then instead of listing the corresponding subsequence of (x_n^*) by $(x_{k_n}^*)$, we list it as $(x_n^*)_{n \in M}$, where $M = \{k_n\}$, or simply as $(x_n^*)_M$.

Okay, the notation has been cared for, let's see how a given bounded sequence in X^* can *fail* to have a weak* convergent subsequence. Suppose (x_n^*) is such a poor specimen. Then it must be that given any subsequence $(x_n^*)_{n \in M}$ of (x_n^*), there are at least two distinct weak* points of accumulation—say, y^* and z^*. Let $W(y^*)$ and $W(z^*)$ be disjoint weak* closed convex neighborhoods of y^* and z^*, respectively. For infinitely many n in M we have x_n^* in $W(y^*)$, and for infinitely many n in M we have x_n^* in $W(z^*)$. List the n of the first case as M_0 and the n of the second case as M_1. Then $\overline{\text{co}}^*\{x_n^*\}_{M_0}$ and $\overline{\text{co}}^*\{x_n^*\}_{n \in M_1}$ are disjoint weak* compact convex sets, the first being contained in $W(y^*)$, the second in $W(z^*)$. Consequently, $\overline{\text{co}}^*\{x_n^*\}_{M_0}$ and $\overline{\text{co}}^*\{x_n^*\}_{M_1}$ can be separated by at least ε by some *weak* continuous* norm-one functional and some $\varepsilon > 0$. We have proved the following lemma.

Lemma 5. *Let (x_n^*) be a sequence in B_{X^*} with no weak* convergent subsequence. Then for any subsequence M of $\mathbb{N}$*

$$0 < \delta(M) = \sup_{y^*, z^*} \inf |y^*(x) - z^*(x)|,$$

where the supremum is taken over all the subsequences M_0 and M_1 of M and all $y^* \in \overline{\text{co}}^*[(x_n^*)_{\tilde{M}_0}]$ and $z^* \in \overline{\text{co}}^*[(x_n^*)_{\tilde{M}1}]$, and all $x \in B_X$.

The modulus $\delta(\cdot)$ gives an estimate as to just how weak* divergent a given subsequence of (x_n^*) is.

Lemma 5 will be our key "splitting" tool. What we do is start with a bounded sequence (x_n^*) admitting no weak*-convergent subsequence. We split (x_n^*) by extracting two subsequences according to the dictates of Lemma 5. Neither of these subsequences are weak* convergent, nor do they admit of any weak* convergent subsequence. So we can split them as well. We continue this process. Keep in mind that at each stage we get pairs of subsequences and an element of B_X on which the value of the differences of the pairs of subsequences stay far away from zero. The hope is that with a bit of judicious pruning we can find a branch of the splitting along which the different pairs act on the corresponding members of B_X to have differences that stay uniformly away from zero. The attentive reader will observe that the diagonalization procedure employed here is the same in spirit as that used in proving the Josefson-Nissenzweig theorem. A derived benefit will be the following theorem.

Theorem 6 (J. Hagler, W. B. Johnson). *If X^* contains a bounded sequence without a weak* convergent subsequence, then X contains a separable subspace with nonseparable dual.*

PROOF. We start with more notation! Before entering the notation be assured that the notation already introduced and that about to be introduced will indeed soften the proof considerably. In fact, a worthwhile lesson in the value of clever notation may be gained if the reader will attempt to redo this proof using the standard subscripting and superscripting associated with multiple passage to subsequences.

Suppose $\mathscr{F}$ denotes the set of finite sequences of 0's and 1's. If $\varphi, \chi \in \mathscr{F}$, we say $\varphi \geq \chi$ if φ is as long or longer than χ and the first $|\chi|$ (= number of entries in χ) members of φ are χ. Given $\varphi \in \mathscr{F}$ the member of $\mathscr{F}$ whose first $|\varphi|$ terms are just φ and whose next term is i (= 0 or 1) is denoted by φ, i. $\varnothing \in \mathscr{F}$ denotes the empty sequence. We denote by Δ the set of all infinite sequences of 0's and 1's. If $\xi \in \Delta$, then we can associate with ξ the sequence (φ_n) from $\mathscr{F}$ where φ_n is formed by taking the first n terms of ξ.

Let $(x_n^*)_N$ be a sequence from B_{X^*} without a weak* convergent subsequence. By Lemma 5 there are subsequences N_0 and N_1 of $\mathbb{N}$ and an element $x_\varnothing$ in B_X such that

$$(y^* - z^*)(x_\varnothing) \geq \frac{\delta(N)}{2}$$

for any $y^* \in \overline{\text{co}}^*[(x_n^*)_{\tilde{N}_0}]$ and $z^* \in \overline{\text{co}}^*[(x_n^*)_{\tilde{N}_1}]$.

Again by Lemma 5 there are subsequences $N_{0,0}$ and $N_{0,1}$ of N_0 and subsequences $N_{1,0}$ and $N_{1,1}$ of N_1 as well as elements x_0 and x_1 in B_X such that

$$(y^* - z^*)(x_i) \geq \frac{\delta(N_i)}{2}$$

for any $y^* \in \overline{\text{co}}^*[(x_n^*)_{\tilde{N}_{i,0}}]$ and $z^* \in \overline{\text{co}}^*[(x_n^*)_{\tilde{N}_{i,1}}]$.

Continuing in this fashion, we see that given any $\varphi \in \mathscr{F}$ we get a subsequence N_φ of $\mathbb{N}$ from which we can extract subsequences $N_{\varphi,0}$ and $N_{\varphi,1}$ for which there is an $x_\varphi \in B_X$ such that

$$(y^* - z^*)(x_\varphi) \geq \frac{\delta(N_\varphi)}{2}$$

for any $y^* \in \overline{\text{co}}^*[(x_n^*)_{\tilde{N}_{\varphi,0}}]$ and $z^* \in \overline{\text{co}}^*[(x_n^*)_{\tilde{N}_{\varphi,1}}]$.

Claim. There is a $\varphi_0 \in \mathscr{F}$ and a $\delta > 0$ so that for all $\chi \geq \varphi_0$ we have $\delta(N_\chi) \geq \delta$.

Once this claim is established the proof of Theorem 6 follows easily. Indeed, the claim in hand, by reindexing, we can assume $\varphi_0 = \varnothing$ so that for all $\varphi \in \mathscr{F}$ if $y^* \in \overline{\text{co}}^*[(x_n^*)^{\sim}_{N_{\varphi,0}}]$ and $z^* \in \overline{\text{co}}^*[(x_n^*)^{\sim}_{N_{\varphi,1}}]$, then

$$(y^* - z^*)(x_\varphi) \geq \delta$$

for some $x_\varphi \in B_X$. Since $\mathscr{F}$ is countable, the closed linear span X_0 of the x_φ (φ running through $\mathscr{F}$) is separable. On the other hand, if we take a $\xi \in \Delta$ and let (φ_n) be the corresponding sequence of members of $\mathscr{F}$, then we can find an $x_\xi^* \in \bigcap_{j=0}^{\infty}(x_n^*)^{\sim}_{N\varphi_j}$; this follows from the fact that the sequence $((x_n^*)^{\sim}_{N_{\varphi_j}})$ is a decreasing sequence of nonempty weak* compact sets. Now if ξ and η are distinct members of Δ with corresponding sequences (φ_n) and (χ_n) of members of $\mathscr{F}$, then eventually $\varphi_j \neq \chi_j$; let $j_0 = \max\{j : \varphi_j = \chi_j\}$ and let $\varphi = \varphi_{j_0} = \chi_{j_0}$. then $x_\xi^* \in (x_n^*)^{\sim}_{N_{\varphi,0}}$ and $x_\eta^* \in (x_n^*)^{\sim}_{N_{\varphi,1}}$ (or vice versa). Thus,

$$\|x_\xi^* - x_\eta^*\|_{X_0} \geq (x_\xi^* - x_\eta^*)(x_\varphi) \geq \delta,$$

and X_0^* is nonseparable since Δ is uncountable.

So we are left with establishing our claim. Were the claim groundless, there'd be a sequence (φ_n) of members of $\mathscr{F}$, $\varphi_1 \leq \varphi_2 \leq \cdots$ such that for each k

$$\delta(N_{\varphi_k}) < \frac{1}{k}.$$

If we now choose M so that its nth term is the nth term of N_{φ_n}, then the nth tail end of M would be a subsequence of N_{φ_n} from which it follows that $\delta(M) \leq \delta(N_{\varphi_n}) < 1/n$ for all n. But this says that $\delta(M) = 0$ contradicting Lemma 5 and establishing our claim. □

A number of other conditions related to those presented in Theorem 4 and Theorem 6 are discussed in the Notes and Remarks section at the end of the chapter. All these are primarily concerned with B_{X^*}'s weak* sequential compactness. What of $B_{X^{**}}$? Of course, it is too much to expect much in such a case, but surprisingly there is a very sharp result even here.

We saw in Goldstine's theorem that every Banach space is weak* dense in its bidual; in fact, for each X, B_X is weak* dense in $B_{X^{**}}$. Of course, this says that given $x^{**} \in B_{X^{**}}$ there is a *net* (x_α) in B_X converging to x^{**} in the weak* topology. When can one find a *sequence* (x_n) in B_X that converges to x^{**} in the weak* topology? The rest of this chapter is devoted to characterizing those *separable* Banach spaces X in which B_X is weak* sequentially dense in $B_{X^{**}}$. We see that the absence of a copy of l_1 is precisely the catalyst for approximating members of $B_{X^{**}}$ by sequences in B_X. Along the way we also characterize those separable X having weak* sequentially compact *second* dual balls.

A key role in our presentation is played by Baire's characterization of functions of the so-called first Baire class, presented in the first section of Chapter VII. Let us recall that theorem in the form we find useful for the present setup.

Baire's Characterization Theorem. *Suppose* Ω *is a compact Hausdorff space.*

I. *If* (f_n) *is a sequence in* $C(\Omega)$ *for which*

$$f(\omega) = \lim_n f_n(\omega)$$

exists for each $\omega \in \Omega$, *then for every nonempty closed subset* F *of* Ω, $f|_F$ *has a point of continuity relative to* F.

II. *Supposing* Ω *to be metrizable, any scalar-valued function* f *on* Ω *having the property that* $f|_F$ *has a point of continuity relative to* F *for each nonempty closed subset* F *of* Ω *is the pointwise limit of a sequence* (f_n) *of functions in* $C(\Omega)$.

Let's get some terminology straight. An element of X^{**} is called a *Baire-*1 *functional* provided it is the weak* limit of a sequence of elements of X. Denote by $\mathscr{B}_1(X)$ the set of all Baire-1 functionals in X^{**}.

To get some idea of what $\mathscr{B}_1(X)$ entails, suppose X is the space $C(\Omega)$ of continuous real-valued functions on the compact Hausdorff space Ω. We can imbed Ω into $B_{C(\Omega)^*}$ in its weak* topology by the map $\delta: \omega \to \delta_\omega$. Here $\delta_\omega(f) = f(\omega)$, for $f \in C(\Omega)$ and $\omega \in \Omega$. Notice that the map $\delta^0: C(\Omega)^{**} \to l_\infty(\Omega)$ defined by $(\delta^0 x^{**})(\omega) = x^{**}(\delta_\omega)$ assigns to each $x^{**} \in \mathscr{B}_1(C(\Omega))$ a bounded function on Ω that belongs to the first Baire class. Conversely, if g is a bounded function on Ω belonging to the first Baire class, then there is a sequence (g_n) of members of $C(\Omega)$ such that $g_n(\omega) \to g(\omega)$ for each $\omega \in \Omega$. Without any loss of generality we may assume that $\|g_n\|_\infty \le \sup\{|g(\omega)|: \omega \in \Omega\}$ for all n; a suitable truncation may be necessary here, but it won't seriously injure anything. Now the bounded convergence theorem steps forward to say "for each $\mu \in M(\Omega)$ we have $\int g_n \, d\mu \to \int g \, d\mu$." In other words, g is the weak* limit of a sequence of members of $C(\Omega)$, where g is viewed herein as a continuous linear functional on $M(\Omega) = C(\Omega)^*$.

In this way we see that $\mathscr{B}_1(C(\Omega))$ *and the class of bounded functions of the first Baire class on* Ω *are identifiable.*

For our purposes we need more. Another natural compact Hausdorff space is on the horizon, namely, $B_{C(\Omega)^*}$ in the weak* topology, and it is the one that allows a general argument to be brought to bear on the question at hand. For the rest of this section we refer to $B_{C(\Omega)^*}$ in the weak* topology as BIG Ω.

Ω is of course imbedded into BIG Ω by the map $\delta: \Omega \to \text{BIG } \Omega$ that takes $\omega \in \Omega$ to δ_ω. It is a special feature of this imbedding that *an element* $x^{**} \in C(\Omega)^{**}$ *belongs to* $\mathscr{B}_1(C(\Omega))$ *if and only if* $x^{**}|_{\text{BIG } \Omega}$ *is a bounded function of the first Baire class on* BIG Ω.

Let's see why this is so.

Let $\mu \in M(\Omega)$. Denote by *supp* μ the set of all $\omega \in \Omega$ for which $|\mu|(U) > 0$ for any open set U containing ω; supp μ is a closed subset of Ω. If S is a closed subset of Ω we let $\mathscr{P}(S)$ denote the set of all probability measures $\mu \in M(\Omega)$ whose support, supp μ, is contained in S: *$\mathscr{P}(S)$ is a closed subset of* BIG Ω.

We're going to show that if x^{**}'s restriction to BIG Ω is a bounded function of first Baire class on BIG Ω, then x^{**} is a member of $\mathscr{B}_1(C(\Omega))$; since the converse is so, we have accomplished what we set forth to do.

Our setup: We have an x^{**} in X^{**} that is not in $\mathscr{B}_1(C(\Omega))$, and we want to show that x^{**}'s restriction to BIG Ω must fail to be in the first Baire class over BIG Ω. Try the contrary. Then $x^{**} \notin \mathscr{B}_1(C(\Omega))$, yet $x^{**}|_{\text{BIG}\,\Omega}$ belongs to the first Baire class of functions defined on BIG Ω; of course, this gives us the immediate consequence that $x^{**}\,\delta$ is in the first Baire class of functions defined on Ω. From this it is easy to see that the functional $y^{**} \in M(\Omega)^*$ defined by

$$y^{**}(\mu) = \int_\Omega x^{**}\delta_\omega \, d\mu(\omega)$$

belongs to $\mathscr{B}_1(C(\Omega))$. Look at $z^{**} = x^{**} - y^{**}$. z^{**} is a function of the first Baire class on BIG Ω.

Some noteworthy observations about z^{**}:

First, for any point $\omega_0 \in \Omega$, $z^{**}(\delta_{\omega_0}) = 0$; indeed $y^{**}\delta_{\omega_0} = \int_\Omega x^{**}(\delta_\omega)\, d\delta_{\omega_0}(\omega) = x^{**}\delta_{\omega_0}$. Since every purely atomic member of $M(\Omega)$ is in the norm-closed linear span of the point charges, it follows that z^{**} *vanishes on the subspace of purely atomic members of* $M(\Omega)$.

Next, $z^{**} \neq 0$; otherwise, $x^{**} = y^{**} \in \mathscr{B}_1(C(\Omega))$. Therefore, $z^{**}(\nu) \neq 0$ for some $\nu \in M(\Omega)$. We can assume $\nu \geq 0$ since if z^{**} vanishes on all nonnegative measures, it will vanish on all differences of such; as is well known, this takes into account all measures. By normalization and possibly by multiplying z^{**} by -1, *we may* as well *assume* $z^{**}(\nu) > 0$ *for some* $\nu \in \mathscr{P}(\Omega)$.

Let $Z = \{\lambda \in M(\Omega) : \lambda \ll \nu\}$. By the Radon-Nikodym theorem Z may be identified with $L_1(\nu)$. Therefore, if we restrict z^{**} to Z, we get a member of $Z^* = L_\infty(\nu)$. Consequently, there is a bounded Borel-measurable function φ for which

$$z^{**}(\lambda) = \int \varphi \, d\lambda$$

for each $\lambda \ll \nu$. In particular,

$$z^{**}(\nu) = \int \varphi \, d\nu > 0.$$

This gives us that

$$\int \varphi^+ \, d\nu > 0,$$

where $\varphi^+ = \max\{\varphi, 0\}$. Let $c > 0$ be such that

$$\nu[\varphi(\omega) \geq c] > 0.$$

Notice that if $\lambda \in \mathscr{P}(\Omega)$ vanishes on $[\varphi(\omega) < c]$, then

$$\int \varphi \, d\lambda \int_{[\varphi(\omega) \geq c]} \varphi \, d\lambda \geq c.$$

Therefore, letting $\mu \in \mathscr{P}(\Omega)$ be defined by

$$\mu(B) = \frac{\nu[\omega \in B \text{ and } \varphi(\omega) \geq c]}{\nu[\varphi(\omega) \geq c]}$$

we know that if $\lambda \in \mathscr{P}(\operatorname{supp}\mu)$ and $\lambda \ll \mu$, then

$$z^{**}(\lambda) = \int_{[\varphi(\omega) \geq c]} \varphi \, d\lambda \geq c > 0.$$

The result: $z^{**}(\lambda) \geq c > 0$ *for each* $\lambda \in \mathscr{P}(\operatorname{supp}\mu)$ *such that* $\lambda \ll \mu$.

z^{**} is a most interesting character. It vanishes on the purely atomic members of $M(\Omega)$ and is bigger than c on those probability measures on the support of μ that are μ-continuous. Can z^{**} have any points of continuity in the set $\mathscr{P}(\operatorname{supp}\mu)$? No, it cannot. In fact, both $\{\lambda \in \mathscr{P}(\operatorname{supp}\mu) : \lambda \text{ is purely atomic}\}$ and $\{\lambda \in \mathscr{P}(\operatorname{supp}\mu) : \lambda \ll \mu\}$ are weak* dense in $\mathscr{P}(\operatorname{supp}\mu)$, and z^{**} behaves much too loosely on these sets to have any points of continuity on $\mathscr{P}(\operatorname{supp}\mu)$. Since $\mathscr{P}(\operatorname{supp}\mu)$ is a closed subset of BIG Ω, we see that z^{**} could not have been of the first Baire class on BIG Ω—we have done what we said we could.

To push this a bit further we need the next lemma.

Lemma 7. *Suppose X is a subspace of the Banach space Y. Identify X^{**} with the subspace $X^{\perp\perp}$ in Y^{**}. Let $G \in X^{**}$ be a Baire-1 member of Y^{**}. Then G is a Baire-1 member of X^{**}; in fact, if $\|G\| = 1$, there is a sequence of norm ≤ 1 members of X converging weak* to G.*

PROOF. Let $y_n \in Y$ be chosen so that $G = \text{weak}^*\lim_n y_n$. We claim that distance $(B_X, \overline{\text{co}}\{y_n, y_{n+1}, \ldots\}) = 0$ for all n. In fact, were this not so then there would be an n so that distance $(B_X, \overline{\text{co}}\{y_n, y_{n+1}, \ldots\}) > 0$. By the Hahn-Banach theorem this would imply the existence of a $y^* \in Y^*$ such that

$$0 \leq \sup y^* B_X < \inf_{n \leq k} y^* y_k.$$

But Goldstine's theorem applies to give

$$\begin{aligned} |G(y^*)| &\leq \sup |y^* B_X| \\ &< \inf_{n \leq k} |y^* y_k| \\ &\leq \lim_n y^* y_k = G y^*. \end{aligned}$$

The obvious contradiction proves that the distance from B_X to $\overline{\text{co}}\{y_n, y_{n+1}, \dots\}$ is zero for each n.

Therefore, for each n we can find $x_n \in B_X$ and a σ_n in the convex hull of $\{y_n, y_{n+1}, \dots\}$ so that $\|x_n - \sigma_n\|$ tends to 0 as $n \to \infty$. Since $y_n \to G$ weak* in Y^{**}, $\sigma_n \to G$ weak* in Y^{**}. But this implies that $x_n \to G$ weak* in Y^{**} and of course by the Hahn-Banach Theorem this implies that $x_n \to G$ weak* in X^{**}. □

One now need only imbed a Banach space X into $C(B_{X^*}, \text{weak*})$ and apply the results above to deduce the next lemma.

Basic Lemma. *If X is any Banach space and $x^{**} \in X^{**}$, then $x^{**} \in \mathscr{B}_1(X)$ if and only if x^{**}'s restriction to B_{X^*} (in its weak* topology) is a function of the first Baire class on this compact space.*

Our strategy now will be to show that if X is separable and there is an $x^{**} \in X^{**}$ that is not a $\mathscr{B}_1(X)$ functional, then X contains a copy of l_1. We know, of course, from the "if" part of the Baire characterization theorem and the Basic Lemma above that a non-$\mathscr{B}_1(X)$ functional x^{**} gives rise to a weak*-closed subset K of B_{X^*} such that x^{**} is everywhere weak* discontinuous on K. That x^{**} is actually quite radical follows from the next lemma.

Lemma 8. *If K is a compact Hausdorff space and $f: K \to R$ is a bounded function with no points of continuity, then there exist a nonempty closed subset L of K and real numbers r, δ with $\delta > 0$ so that*

(*) *For every nonempty relatively open subset U of L there are $y, z \in U$ such that $f(y) > r + \delta$ and $f(z) < r$.*

holds.

Proof. For each n let

$$C_n = \left\{ x \in K : \text{if } U \text{ is open and contains } x, \text{ there are } y, z \in U \text{ with } f(y) - f(z) > \frac{1}{n} \right\}.$$

Since f is nowhere continuous, $K = \cup_n C_n$. It is easy to see (using nets for instance) that each C_n is closed. By the Baire category theorem, one of the C_n, say C_N, has nonvoid interior U_N. Let $K_N = \overline{U}_N$ and let $\delta = 1/N$. We now have that *if V is a nonempty relatively open subset of K_N, then $U_N \cap V$ is a nonempty open subset of K_N, and so there are $y, z \in U_N \cap V$ for which $f(y) - f(z) > \delta$.*

Let (r_n) be an enumeration of all the rational numbers. For each n let F_n be the set

$$\{x \in K_N : \text{if } U \text{ is open and contains } x \text{ there are } y, z \text{ in } U \cap K_N \text{ with } f(z) < r_n < r_n + \delta < f(y)\}.$$

Again it is easy to see that each F_n is closed and by the first paragraph $K_N = U_n F_n$. Applying the Baire category theorem once more, we derive the existence of an F_M with a nonempty interior V_M. Letting $L = \bar{V}_M$ and $r = r_M$ we get (*). □

This lemma in hand, we know that if $x^{**} \in X^{**}$ is not a Baire-1 functional, then there is a weak*-closed subset of B_{X^*} on which not only is x^{**} totally discontinuous but somewhere x^{**} is oscillating very rapidly. In tandem with Goldstine's theorem this will allow us to build a Rademacher-like system back in X and so conclude that X contains l_1. The technical vehicle for such a construction is the next lemma.

Lemma 9. *Let L be a compact Hausdorff space and $f: L \to R$ be a bounded function. Suppose r, δ are real numbers with $\delta > 0$ and assume that*

(*) *For each nonempty relatively open subset U of L there are $y, z \in U$ with $f(z) < r$ and $f(y) > r + \delta$*

holds. Assume further that f is in the pointwise closure of some bounded family $\mathscr{G}$ of $C(L)$. Then there exists a sequence $(g_n) \subset \mathscr{G}$ such that the sequence of pairs of sets $([g_n(x) < r], [g_n(x) > r + \delta])$ is an independent sequence.

Proof. By (*) there are $y_1, y_2 \in L$ with $f(y_1) > r + \delta$ and $f(y_2) < r$.

Choose $g_1 \in \mathscr{G}$ so that $g_1(y_1) > r + \delta$ and $g_1(y_2) < r$. Consider the nonempty open subsets $A_1 = [g_1(x) > r + \delta]$ and $B_1 = [g_1(x) < r]$ of L. There are points $\tilde{y}_1, \tilde{y}_2$ in A_1 and $\tilde{\tilde{y}}_1, \tilde{\tilde{y}}_2$ in B_1 for which $f(\tilde{y}_1), f(\tilde{\tilde{y}}_1) > r + \delta$ and $f(\tilde{y}_2), f(\tilde{\tilde{y}}_2) < r$.

Choose $g_2 \in \mathscr{G}$ so that $g_2(\tilde{y}_1), g_2(\tilde{\tilde{y}}_1) > r + \delta$ and $g_2(\tilde{y}_2), g_2(\tilde{\tilde{y}}_2) < r$. Consider the disjoint nonempty open sets $A_2 = [g_2(x) > r + \delta]$ and $B_2 = [g_2(x) < r.]$ Notice that $A_1 \cap A_2$, $A_1 \cap B_2$, $B_1 \cap A_2$, $B_1 \cap B_2$ are all nonempty containing $\tilde{y}_1$, $\tilde{\tilde{y}}_2$, $\tilde{\tilde{y}}_1$, and $\tilde{y}_2$, respectively. Now there are points $y_1^{\#}, y_2^{\#}, y_3^{\#}, y_4^{\#}$ in A_2, where f is bigger than $r + \delta$, and points $y_1^{\#\#}, y_2^{\#\#}, y_3^{\#\#}, y_4^{\#\#}$ in B_2, where f is less than r. The procedure is clear from here or should be. □

We are now ready to prove the following.

Theorem 10 (Odell-Rosenthal). *Let X be a separable Banach space. Then the following are equivalent*:

1. *X contains no isomorph of l_1.*
2. *B_X is weak* sequentially dense in $B_{X^{**}}$.*
3. *$B_{X^{**}}$ is weak* sequentially compact.*

Proof. Suppose $x^{**} \in B_{X^{**}}$ is *not* the weak* limit of any sequence of terms from X. Since X is separable, B_{X^*} is weak* compact, weak* metrizable; therefore, our Basic Lemma tells us that $x^{**}|_{(B_{X^*}, \text{weak}^*)}$ is not of the first Baire class. Baire's characterization theorem implies that there is a nonvoid weak* compact subset M of B_{X^*} such that $x^{**}|_M$ has no points of (weak*) continuity. Lemmas 8 and 9 in tandem with Proposition 3 of Chapter XI allow us to conclude that X contains a copy of l_1; after all x^{**} is in the pointwise closure of B_X. Taken in toto the above argues that "1 implies 2."

It is plain from Rosenthal's l_1 theorem that "3 implies 1," and "2 implies 3" is an easy diagonal argument, whose details we leave to the imagination of the student. □

Exercises

1. *Factoring weakly compact operators.* Every weakly compact linear operator factors through a reflexive Banach space; i.e., if $T: X \to Y$ is a weakly compact linear operator, then there is a reflexive Banach space Z and bounded linear operators $S: X \to Z$, $R: Z \to Y$ such that $RS = T$.

2. *Conditionally weakly compact sets.* A subset K of a Banach space is *conditionally weakly compact* if each sequence in K has a weakly Cauchy subsequence.

   (i) A closed bounded set is conditionally weakly compact if and only if it contains no basic sequence equivalent to the unit vector basis of l_1.

   (ii) Let K be a closed bounded subset of the Banach space X. Suppose that for each $\varepsilon > 0$ there is a conditionally weakly compact set K_ε such that
   $$K \subseteq K_\varepsilon + \varepsilon B_X.$$
   Then K is conditionally weakly compact.

   (iii) A bounded linear operator $R: X \to Y$ fixes no copy of l_1 if and only if there is a Banach space Z containing no copy of l_1 and bounded linear operators $P: X \to Z$, $H: Z \to Y$ such that
   $$HP = R.$$

3. *Having weak* sequentially compact dual ball is not a three-space property.*

   (i) There is a well-ordered set $(I, <)$ and a collection $(M_\alpha)_{\alpha \in I}$ of infinite subsets of the set $\mathbb{N}$ of natural numbers such that (1) for $\alpha < \beta$ either $M_\alpha \cap M_\beta$ have only finitely many members in common or all but finitely many members of M_α belong to M_β and (2) if $M \in \mathcal{P}_\infty(\mathbb{N})$, then there is $\alpha \in I$ such that both $M \cap M_\alpha$ and $M \setminus M_\alpha$ are infinite.

   (ii) Let X be the closed linear subspace of l_∞ spanned by the set
   $$\{\chi_{M_\alpha} : \alpha \in I\} \cup c_0.$$
   Then (δ_n) is a sequence in B_{X^*} without a weak* convergent subsequence.

(iii) $B_{(X/c_0)^*}$ is weak* sequentially compact. (*Hint*: Any norm-one sequence in $B_{(X/c_0)^*}$ without a weak* convergent subsequence would have to have a subsequence that acts in a Rademacher-like fashion on the set $\{\chi_{M_\alpha}: \alpha \in I\}$.)

4. *Limited sets and sequential compactness.*

(i) If B_{X^*} is weak* sequentially compact, then limited subsets of X are relatively compact.

(ii) If Ω is a compact, sequentially compact Hausdorff space, then limited subsets of $C(\Omega)$ are relatively compact.

5. $\mathscr{B}_1(X)$. Let X be any Banach space. then $\mathscr{B}_1(X)$ is a *norm-closed* linear subspace of X^{**}.

6. *Pettis integrability in dual spaces.* Let (Ω, Σ, μ) be a probability measure space, X be a separable Banach space, and $F: \Sigma \to X^*$ be a countably additive measure for which $\|F(E)\| \leq \mu(E)$ for each $E \in \Sigma$.

(i) There exists an $f: \Omega \to X^*$ such that $f(\cdot)(x) \in L_\infty(\mu)$ for each $x \in X$ and

$$F(E)(x) = \int_E f(\omega)(x)\, d\mu(x)$$

for each $x \in X$ and each $E \in \Sigma$.

(ii) If X contains no copy of l_1, then f in (i) is Pettis integrable.

(iii) If X^* is separable, then f is Bochner integrable.

7. *Weak* continuity on B_{l_1}.* If $F \in l_\infty \setminus c_0$, then any point of $B_{l_1} = B_{c_0^*}$ at which F is weak* continuous (relative to B_{l_1}) lies in S_{l_1}.

Notes and Remarks

Our presentation of Lemma 3 resulted from a conversation with Walter Schachermayer, in which he made the point of noting just how overlooked Grothendieck's observation (Lemma 2) really is. Anyone who has read the Davis-Figiel-Johnson-Pelczynski factorization paper (1974) will recognize the purely cosmetic changes made in the original proof. There are a number of consequences of Lemma 3, including Exercise 1, to be found in Davis et al. (1974); the paper is so elegantly conceived and executed that it would be a shame if the earnest student did not spend a reasonable amount of time in its mastery. Exercise 2 is essentially due to Davis, Figiel, Johnson, and Pelczynski, too.

Theorem 4, due to D. Amir and J. Lindenstrauss (1968), is but a simple example of the good life enjoyed by the weakly compact subsets of a Banach space and subspaces of the spaces weakly compact sets generate. Lindenstrauss's survey paper (1972) on weakly compact sets presents a plethora of fascinating faces of weak compactness and the review of Lindenstrauss's survey gives a lead to many of the later works on the subject.

Theorem 6 is in the same paper of J. Hagler and W. B. Johnson (1977) that formed the basis of our presentation of the Josefson-Nissenzweig theorem.

There are two other particularly noteworthy conditions in which the dual ball is weak* sequentially compact: if a Banach space X has an equivalent *smooth norm* or if a Banach space X is a *weak Asplund* space, then B_{X^*} is weak* sequentially compact. Recall that a norm is smooth if each element of the unit sphere has a unique support functional; J. Hagler and F. Sullivan (1980) gave an elegant argument that *the dual ball of a smoothly normed Banach space is weak* sequentially compact* based, ultimately, on some rough ideas of E. Leach and J. H. M. Whitfield (1972). Following I. Namioka and R. R. Phelps (1975), we say a Banach space X is a weak Asplund space if every convex extended real-valued function defined on X is Gateaux differentiable on a dense G_δ subset of its domain of finiteness; C. Stegall (1971) showed that if X is a weak Asplund space, then there exists a Banach space Y, each separable subspace of which has a separable dual, and a bounded linear operator $T: Y \to X$ with the dense range—an appeal to Theorem 6 and part 1 of Lemma 1—consequently, *weak Asplund spaces have weak* sequentially compact dual balls*. D. Larman and R. R. Phelps (1979) have given an intriguing view of Stegall's theorem.

To date, *there is no characterization of those Banach spaces X having weak* sequentially compact dual balls*. Furthermore, it appears that none of the classes of Banach spaces presently under study offers any hope of a viable candidate for the characterization of spaces with sequentially compact dual balls.

One class of spaces that *ought* to have weak* sequentially compact dual balls consists of spaces without a copy of l_1. However, J. Hagler and E. Odell (1978) have provided an example of a space without a copy of l_1 whose dual ball is not weak* sequentially compact. Their construction is not unlike that of the example of Exercise 3, which is due, incidentally, to J. Bourgain.

A problem related to the characterization of spaces with weak* sequentially compact dual balls but conceivably more tractable is to determine *which Banach spaces X have weak* angelic dual balls*. This too remains open and surprisingly untested.

The characterization of which separable Banach spaces have weak* sequentially compact dual balls is due to E. Odell and H. P. Rosenthal. Actually, it is the equivalence of the noncontainment of l_1 with the weak* sequential density of X in X^{**} that elicits the greatest interest in the Odell-Rosenthal theorem; the fact that the second dual ball's weak* sequential compactness fits so nicely into the scheme of things just provides us with a ready-made excuse to include a discussion of this theorem in the text. The application of the Odell-Rosenthal theorem to the Pettis integral cited in Exercise 6(ii) is due to K. Musial; parts (i) and (iii) were known to N. Dunford, B. J. Pettis, and I. M. Gelfand.

The results of Exercies 4 are due to I. M. Gelfand (1938). The fact that $B_1(X)$ is always a Banach space (Exercise 5) is due to R. D. McWilliams (1968).

BIBLIOGRAPHY

Amir, D. and Lindenstrauss, J. 1968. The structure of weakly compact sets in Banach spaces. *Ann. Math.*, **88**, 35–46.

Davis, W. J., Figiel, T., Johnson, W. B., and Pelczynski, A. 1974. Factoring weakly compact operators. *J. Funct. Anal.*, **17**, 311–327.

Gelfand, I. M. 1938. Abstrakte Funktionen und lineare Operatoren. *Mat. Sbornick*, **46**, 235–286.

Hagler, J. and Johnson, W. B. 1977. On Banach spaces whose dual balls are not weak* sequentially compact. *Israel J. Math.*, **28**, 325–330.

Hagler, J. and Odell, E. 1978. A Banach space not containing l_1 whose dual ball is not weak* sequentially compact. *Illinois J. Math.*, **22**, 290–294.

Hagler, J. and Sullivan, F. 1980. Smoothness and weak* sequential compactness. *Proc. Amer. Math. Soc.*, **78**, 497–503.

Larman, D. G. and Phelps, R. R. 1979. Gateaux differentiability of convex functions on Banach spaces. *J. London Math. Soc.*, **20** (2), 115–127.

Leach, E. B. and Whitfield, J. H. M. 1972. Differentiable functions and rough norms on Banach spaces. *Proc. Amer. Math. Soc.*, **33**, 120–126.

Lindenstrauss, J. 1972. Weakly compact sets—their topological properties and the Banach spaces they generate. In *Proceedings of Symposium on Infinite Dimensional Topology. Ann. Math. Studies*, **69**, 235–293 [MR 49 No. 11514].

McWilliams, R. D. 1968. On the w^*-sequential closure of a Banach space in its second dual. *Duke Math. J.*, **35**, 369–373.

Namioka, I. and Phelps, R. R. 1975. Banach spaces which are Asplund spaces. *Duke Math. J.*, **42**, 735–750.

Stegall, C. 1981. The Radon-Nikodym property in conjugate Banach spaces, II. *Trans. Amer. Math. Soc.*, **264**, 507–519.

CHAPTER XIV

The Elton-Odell $(1+\varepsilon)$-Separation Theorem

In this concluding chapter we prove the following separation theorem, due to J. Elton and E. Odell.

Theorem. *If X is an infinite-dimensional normed linear space, then there is an $\varepsilon > 0$ and a sequence (x_n) of members of S_X such that $\|x_m - x_n\| \geq 1 + \varepsilon$ whenever $m \neq n$.*

Once again, Ramsey set theatrics dominate a good part of the action.

To get us on our way notice that in c_0 the sequence (x_n) given by $x_n = \Sigma_{k=1}^n e_k - e_{n+1}$ is ideal for the role described in the theorem with $\varepsilon = 1$. Of course, this says that whenever a Banach space contains "very good" copies of c_0, then a substitute sequence may be constructed that will still do the trick. Actually whenever c_0 is isomorphic to a subspace of X, X contains very good copies of c_0.

Theorem 1 (R. C. James). *If a Banach space X contains a subspace isomorphic to c_0, then for any $\delta > 0$ there is a sequence $(u_n) \subseteq B_X$ such that*

$$(1-\delta)\sup|a_i| \leq \|\Sigma a_i u_i\| \leq \sup|a_i|$$

holds for any $(a_i) \in c_0$.

Proof. Let (x_n) be the preordained unit vector basis of c_0 in X; that is, suppose $m, M > 0$ exist for which

$$m \sup|a_i| \leq \|\Sigma a_i x_i\| \leq M \sup|a_i|$$

holds for any $(a_i) \in c_0$.

For each $n \in N$, let's define

$$K_n = \sup\left\{\left\|\sum_i a_i x_i\right\| : \|(a_i)\|_{c_0} = 1, (a_i) \text{ finitely nonzero}, \right.$$

$$\left. \times a_1 = a_2 = \cdots = a_{n-1} = 0\right\}.$$

Observe that (K_n) is a monotone sequence of reals all values of which lie between m and M; it follows that $m \le K = \lim_n K_n \le M$. Let $0 < \theta < 1 < \theta'$ and suppose we pick p_1 so that $K_{p_1} < \theta' K$. Take advantage of the definition of the K_n to select $p_2 < p_3 < \cdots < p_n < \cdots$ (all greater than p_1, of course) and $a^n_{p_n}, \ldots, a^n_{p_{n+1}-1}$ scalars such that $\|(0,\ldots,0, a^n_{p_n}, \ldots, a^n_{p_{n+1}-1}, 0, \ldots)\|_{c_0} = 1$ as well as

$$\left\| y_n = \sum_{i=p_n}^{p_{n+1}-1} a_i^n x_i \right\| > \theta K.$$

Once this is done notice that for any sequence $(a_i) \in c_0$

$$\|\Sigma a_i y_i\| \le K_{p_1} \sup|a_i| \le \theta' K \sup|a_i|.$$

If we now let

$$u_i = \frac{y_i}{\theta' K},$$

the result will be that

$$\|\Sigma a_i u_i\| \le \sup|a_i|$$

holds for any $(a_i) \in c_0$. Let's check to see what else comes from one choice of u_i. Let $a_1, \ldots, a_n$ be scalars with $\sup_{1 \le i \le n}|a_i| = 1$. Pick a_k so that $|a_k| = 1$. Write

$$w = a_k y_k + \sum_i a_i' y_i,$$

where $a_i' = a_i$ for $i \ne k$ and $a_k' = 0$. Then

$$\begin{aligned} 2\theta K < 2\|y_k\| &= \|2a_k y_k\| \\ &= \left\| w + a_k y_k - \sum_i a_i' y_i \right\| \\ &\le \|w\| + \left\| a_k y_k - \sum_i a_i' y_i \right\| \\ &\le \|w\| + \theta' K \sup\{|a_1|, \ldots, |a_n|\} \\ &\le \|w\| + \theta' K. \end{aligned}$$

Therefore,

$$\|w\| > (2\theta - \theta')K.$$

A look at the scaling that changes y_i to u_i tells us that

$$\|\Sigma a_i u_i\| > \frac{2\theta - \theta'}{\theta'} \sup|a_i|;$$

it remains only to choose θ, θ' so that $(2\theta - \theta')/\theta' > 1 - \delta$. □

The upshot of Theorem 1 is that the Elton-Odell theorem holds for Banach spaces containing isomorphic copies of c_0.

Our next four lemmas provide the backdrop for the combinatorial softshoe needed to set up the proof of the Elton-Odell theorem. These lemmas aim at providing a sufficiently sharp criterion for determining c_0's presence in a Banach space that any alternative to the Elton-Odell theorem's verity will prove untenable.

Lemma 2. *Let (x_n) be a sequence in the Banach space X. For any $K > 0$, the set*

$$\mathscr{B}_K = \left\{ M = (m_i) \in \mathscr{P}_\infty(\mathbb{N}) : \sup_n \left\| \sum_{i=1}^n x_{m_i} \right\| \le K \right\}$$

is relatively closed in $\mathscr{P}_\infty(\mathbb{N})$.

Proof. We show that $\mathscr{P}_\infty(\mathbb{N}) \setminus \mathscr{B}_K$ is relatively open. Take an $M = (m_i) \in \mathscr{P}_\infty(\mathbb{N}) \setminus \mathscr{B}_K$. There must be some n such that

$$\left\| \sum_{i=1}^n x_{m_i} \right\| > K.$$

If $L = (l_i) \in \mathscr{P}_\infty(\mathbb{N})$ and L satisfies

$$l_1 = m_1, \quad l_2 = m_2, \quad \ldots, \quad l_n = m_n,$$

then

$$\left\| \sum_{i=1}^n x_{l_i} \right\| = \left\| \sum_{i=1}^n x_{m_i} \right\| > K,$$

and so L too belongs to $\mathscr{P}_\infty(\mathbb{N}) \setminus \mathscr{B}_K$. The set of L of the above prescribed form constitute a relatively open neighborhood of M in $\mathscr{P}_\infty(\mathbb{N})$. Enough said.

A technical tool is needed before we proceed much further: bimonotone basic sequences. A basic sequence (x_n) is called *bimonotone* if for each n,

$$\max_n \left\{ \left\| \sum_{i=1}^k a_i x_i \right\|, \left\| \sum_{i=k+1}^\infty a_i x_i \right\| \right\} \le \left\| \sum_{i=1}^\infty a_i x_i \right\|$$

holds for any sequence (a_n) of scalars that let $\sum_n a_n x_n$ converge. It is not difficult to see that if (x_n) is a basic sequence with closed linear span $[x_n]$, then $[x_n]$ can be equivalently renormed so as to make (x_n) bimonotone; indeed, if (a_n) is a sequence of scalars for which $\sum_n a_n x_n$ converges, define $|||\sum_n a_n x_n||| = \sup_{n \ge 1} \{ \|\sum_{i=1}^n a_i x_i\|, \|\sum_{i=n+1}^\infty a_i x_i\| \}$. You will notice that if each x_n started with $\|x_n\| = 1$, the renorming still leaves $|||x_n||| = 1$; in fact, $|||x_n||| = \|x_n\|$ regardless. □

The bimonotonicity of a basic sequence has combinatorial consequences as we see in the next lemma.

Lemma 3. *Suppose (x_n) is a bimonotone basic sequence in the Banach space X and for each $K > 0$ let*

$$\mathscr{B}_K = \left\{ M = (m_i) \in \mathscr{P}_\infty(\mathbb{N}) : \sup_n \left\| \sum_{i=1}^n x_{m_i} \right\| \leq K \right\}.$$

Suppose $M \in \mathscr{P}_\infty(\mathbb{N})$ satisfies

$$\mathscr{P}_\infty(M) \subseteq \bigcup_{K>0} \mathscr{B}_K.$$

Then there is an $M' \in \mathscr{P}_\infty(M)$ and a $K_M > 0$ such that

$$\mathscr{P}_\infty(M') \subseteq \mathscr{B}_{K_M}.$$

PROOF. By Lemma 2, the set $\mathscr{B}_1$ is relatively closed in $\mathscr{P}_\infty(\mathbb{N})$; therefore, $\mathscr{B}_1$ is completely Ramsey. It follows that there is an $M_1 \in \mathscr{P}_\infty(M)$ for which

$$\mathscr{B}_\infty(M_1) \subseteq \mathscr{B}_1 \quad or \quad \mathscr{P}_\infty(M_1) \subseteq \mathscr{B}_1^c.$$

In case $\mathscr{P}_\infty(M_1) \subseteq \mathscr{B}_1$, we are done; if, on the other hand, $\mathscr{P}_\infty(M_1)$ and $\mathscr{B}_1$ are disjoint, then we look to $\mathscr{B}_2$. By Lemma 2 we know that $\mathscr{B}_2$ is relatively closed in $\mathscr{P}_\infty(\mathbb{N})$ and so $\mathscr{B}_2$ is completely Ramsey. It follows that there is an $M_2 \in \mathscr{P}_\infty(M_1)$ for which

$$\mathscr{P}_\infty(M_2) \subseteq \mathscr{B}_2 \quad or \quad \mathscr{P}_\infty(M_2) \subseteq \mathscr{B}_2^c.$$

In case $\mathscr{P}_\infty(M_2) \subseteq \mathscr{B}_2$ we are done; if, on the other hand, $\mathscr{P}_\infty(M_2)$ and $\mathscr{B}_2$ are disjoint then we look to $\mathscr{B}_3$.

The way seems clear. Unless we come to a sudden stop, we produce a sequence (M_n) of infinite subsets of $\mathbb{N}$ satisfying

$$M_{n+1} \in \mathscr{P}_\infty(M_n) \quad and \quad \mathscr{P}_\infty(M_n) \subseteq \mathscr{B}_n^c$$

for all $n \in \mathbb{N}$. Now we can pick a pathological subsequence of (x_n). Indeed let $(k_n) \in \mathscr{P}_\infty(\mathbb{N})$ be selected so as to satisfy $(k_n)_{n \geq j} \in \mathscr{P}_\infty(M_j)$ for each j. Then for each j we have [on applying our construction to the bimonotone basic sequence $(x_{k_j}, x_{k_{j+1}}, \ldots, x_{k_{j+n-1}}, \ldots)$]

$$\begin{aligned} j &< \sup_n \left\| \sum_{i=j}^n x_{k_i} \right\| \\ &\leq \sup_n \left\| \sum_{i=1}^n x_{k_i} \right\| \quad \text{(by bimonotonicity).} \end{aligned}$$

It follows that

$$\sup_n \left\| \sum_{i=1}^n x_{k_i} \right\| = \infty$$

even though $(k_n) \in \mathscr{P}_\infty(M) \subseteq \cup_{K>0} \mathscr{B}_K$. This is a contradiction, which completes the proof. □

Though we are not quite through with our combinatorial trials, it is worth observing here that Lemmas 2 and 3 are almost pure combinatorics. The easy method used in their proof ought to be closely studied. We've seen it before. It made its first appearance in these discussions in the proof of Rosenthal's dichotomy. As in that appearance, here too combinatorics are not the whole show, but they are a featured part of the program. To prepare us for the finale, we need to start mixing combinatorics with things having a real Banach space flavor. This we start doing shortly.

The next lemma is key to the success of our program.

Lemma 4 (W. B. Johnson). *Let (x_n) be a seminormalized sequence in the Banach space X each subsequence of which admits of a subsequence (y_n) for which $\sup_n\|\sum_{i=1}^n y_i\|<\infty$. Then (x_n) has a subsequence equivalent to the unit vector basis of c_0.*

Proof. First, we set the stage for the Bessaga-Pelczynski selection principle by showing that a sequence (x_n) satisfying the hypotheses of this lemma is necessarily weakly null. If not, there would be an $x^*\in S_{X^*}$ and an $\varepsilon>0$ such that

$$|x^*x_n|>\varepsilon$$

for infinitely many n. In case X is a real Banach space this means that (on maybe doctoring x^* by a sign) we can assume that

$$x^*x_{m_i}>\varepsilon$$

for each i, where $(m_i)=M\in\mathscr{P}_\infty(\mathbb{N})$; this in turn gives the contradictory conclusion that for some $(k_i)\in\mathscr{P}_\infty(M)$

$$n\varepsilon\leq x^*\sum_{i=1}^n x_{k_i}\leq\left\|\sum_{i=1}^n x_{k_i}\right\|\leq\sup_n\left\|\sum_{i=1}^n x_{k_i}\right\|<\infty,$$

for all n. In case X is a complex Banach space we can conclude that there is an $(m_i)=M\in\mathscr{P}_\infty(\mathbb{N})$ such that either $|\mathrm{Re}\,x^*x_{m_i}|$ or $|\mathrm{Im}\,x^*x_{m_i}|$ exceed $\varepsilon/2$ for all i; whichever the case might be, similar reasoning to that of the real case leads to a contradiction of similar proportions.

Okay, we can apply the Bessaga-Pelczynski selection principle. On doing so, we can assume that (x_n) is a seminormalized weakly null basic sequence. More can be assumed: since the hypotheses pertaining to (x_n)'s subsequential behavior and the existence of a subsequence of (x_n) equivalent to c_0's unit vector basis are plainly invariant under (equivalent) changes of norm in the closed linear span of the x_n, *we can assume that (x_n) is also a bimonotone, seminormalized weakly null basic sequence*! Now Lemmas 2 and 3 are standing by ready to get in on the action. We use the notation of these lemmas. By Lemma 2, $\cup_{K=1}^\infty\mathscr{B}_K$ is an $\mathscr{F}_\sigma$-subset of $\mathscr{P}_\infty(\mathbb{N})$; hence $\cup_{K=1}^\infty\mathscr{B}_K$

is completely Ramsey. There is an $M \in \mathscr{P}_\infty(\mathbb{N})$ for which

$$\mathscr{P}_\infty(M) \subseteq \bigcup_{K=1}^{\infty} \mathscr{B}_K \quad \text{or} \quad \mathscr{P}_\infty(M) \subseteq \left(\bigcup_{K=1}^{\infty} \mathscr{B}_K \right)^c.$$

Our hypotheses preclude the latter possibility, and so we can find $M \in \mathscr{P}_\infty(\mathbb{N})$ such that

$$\mathscr{P}_\infty(M) \subseteq \bigcup_{K=1}^{\infty} \mathscr{B}_K.$$

Lemma 3 now alerts us to the fact that there's an $M' = (m_i') \in \mathscr{P}_\infty(M)$ and a $K' > 0$ such that

$$\mathscr{P}_\infty(M') \subseteq \mathscr{B}_{K'}.$$

What does this mean? Well, it means that if (y_n) is the subsequence of (x_n) indexed by the members of M', then given any subsequence (z_n) of (y_n),

$$\sup_n \left\| \sum_{k=1}^{n} z_k \right\| \leq K'.$$

It is easy to see from this that for any finite set Δ in N and any signs $\pm$,

$$\left\| \sum_{n \in \Delta} \pm y_n \right\| \leq 2K'.$$

Of course, this just says that (y_n) satisfies the Bessaga-Pelczynski criterion for equivalence to c_0's unit vector basis. □

Much in the same spirit as Lemma 4 and dependent on it is the next lemma, also due to Bill Johnson. It is precisely the criterion we are seeking.

Lemma 5. *Let (x_n) be a normalized basic sequence in X. Suppose that each subsequence of (x_n) has a subsequence (y_n) such that*

$$\sup_n \left\| \sum_{i=1}^{n} (-1)^i y_i \right\| < \infty.$$

Then the closed linear span $[x_n]$ of (x_n) contains a copy of c_0.

Proof. Plainly the statement of the lemma allows us to equivalently renorm $[x_n]$ to achieve our goal: locate a copy of c_0 inside $[x_n]$. We do so to make (x_n) a bimonotone normalized basic sequence. □

Now looking at the proofs of Lemmas 2 and 3, we see quite easily that the following can be proved

Lemma 2′. *For any $k \in \mathbb{N}$, the set*

$$\mathscr{A}_k = \left\{ M = (m_i) \in \mathscr{P}_\infty(\mathbb{N}) : \sup_n \left\| \sum_{i=1}^n (-1)^i x_{m_i} \right\| \leq k \right\}$$

is closed in $\mathscr{P}_\infty(\mathbb{N})$.

Lemma 3′. *$\mathscr{A}_k$ as in Lemma 2′, if $M \in \mathscr{P}_\infty(\mathbb{N})$ and $\mathscr{P}_\infty(M) \subseteq \cup_{k=1}^\infty \mathscr{A}_k$, then there is an $M' \in \mathscr{P}_\infty(M)$ and a $k_M \geq 1$ so that*

$$\mathscr{P}_\infty(M') \subseteq \mathscr{A}_{k_M}.$$

Armed with these and proceeding similarly to the way we did in the proof of Lemma 4, we can find an $M \in \mathscr{P}_\infty(\mathbb{N})$ for which given $L = (l_i) \in \mathscr{P}_\infty(M)$, then

$$\sup_n \left\| \sum_{i=1}^n (-1)^i x_{l_i} \right\| < \infty.$$

If we now let $y_n = x_{m_{2n}} - x_{m_{2n+1}}$, then for any $L = (l_1) \in \mathscr{P}_\infty(M)$ we have

$$\sup_n \left\| \sum_{i=1}^n y_{l_i} \right\| < \infty,$$

and so (y_n) satisfies the hypotheses of Lemma 4 (with subsequences to spare). Thus, (y_n) has a subsequence equivalent to c_0's unit vector basis and $[x_n]$ contains a copy of c_0.

Now we are ready for the main act: the Elton-Odell theorem itself.

Let (x_n) be a normalized basic sequence in X satisfying the condition

$$\left\| \sum_{i=1}^n a_i x_i \right\| \leq (1+20^{-n}) \left\| \sum_{i=1}^\infty a_i x_i \right\|$$

for all n and all sequences (a_i) of scalars for which $\sum_{i=1}^\infty a_i x_i \in X$. A look at Mazur's method for constructing basic sequences will reassure you of the existence of such; also we might as well suppose (x_n) spans X.

If X contains an isomorphic copy of c_0, then we are done. So we assume X does not contain any copy of c_0. The Johnson criteria show us that (by passing to a subsequence, if necessary) we may assume that for each increasing sequence (m_n) of positive integers,

$$\sup_k \left\| \sum_{i=1}^k (-1)^i x_{m_i} \right\| = \infty.$$

Suppose α is any limit point of the real sequence $(\|x_n - x_{n+1} + x_{n+2}\|)_{n \geq 1}$; plainly, $1 \leq 3\alpha \leq 3$.

Let's establish a bit of notation especially useful for the present proof. Suppose $\delta > 0$. Call the vector b in X a *δ-block of* (x_n) (or simply a δ-block) if $\|b\| = 1$ and b is of the form

$$b = \beta \sum_{i=1}^{l} (-1)^{i+1} x_{m_i} = \beta(x_{m_1} - x_{m_2} + \cdots + x_{m_l}),$$

where $m_1 < m_2 < \cdots < m_l$, $|\alpha/\beta - 1| < \delta$, and l is an odd number ≥ 3. Let's agree to write $n < b_1 < b_2 < \cdots < b_k$ (where $b_1, b_2, \ldots, b_k$ are δ-blocks) if the b_i admit a representation

$$b_i = \sum_{j=p_i+1}^{p_{i+1}} a_j x_j,$$

where $n < p_1 < p_2 < \cdots < p_{k+1}$.

Notice that our choice of α ensures that given a $\delta > 0$ and $n \geq 1$ we can always find a δ-block b with $n < b$.

Our proof will focus attention on the following technical condition:

($*$) For each $\delta > 0$ and each $n \in \mathbb{N}$, there are δ-blocks $b_1, \ldots, b_k$ with $n < b_1 < \cdots < b_k$ so that if b is a δ-block with $b_k < b$, then one of the b_i lies within $1 + \delta$ of b.

Our interest in ($*$) derives from the following:

(not $*$) There exists a $\delta > 0$ and an $n \in \mathbb{N}$ so that for all δ-blocks $b_1, \ldots, b_k$ with $n < b_1 < \cdots < b_k$ there is a δ-block b with $b_k < b$ such that b is at least $1 + \delta$ from each of $b_1, \ldots, b_k$.

It is a short step from (not $*$) to the Elton-Odell theorem; we plan to show, on the other hand, that assuming ($*$) leads to nothing but mathematical trouble (i.e., a contradiction).

We assume that ($*$) holds. Letting $\delta_j = 20^{-j}(\delta_0 = 1)$, choose δ_j-blocks as follows

$$\underbrace{b_1^1 < b_2^1 < \cdots < b_{k_1}^1}_{\delta_1\text{-blocks}} < \underbrace{b_1^2 < b_2^2 < \cdots < b_{k_2}^2}_{\delta_2\text{-blocks}} < \cdots$$

in such a way that if b is a δ_j-block with $b_{k_j}^j < b$, then $\|b_i^j - b\| < 1 + \delta_j$ for some $1 \leq i \leq k_j$. We make the following claim.

Claim. We can select from each pack $\{b_1^j < \cdots < b_{k_j}^j\}$ of δ_j-blocks a term $b_{m_j}^j$,

$$b_{m_j}^j = \beta_{m_j}^j \sum_k (-1)^{k+1} x_{n_k},$$

so cleverly that if

$$d_{m_j}^j = \left(\frac{\alpha}{\beta_{m_j}^j}\right) b_{m_j}^j,$$

then

$$\sup_k \left\| \sum_{j=1}^{k} (-1)^j d_{m_j}^j \right\| < \infty.$$

Since X's containment of c_0 has already been ruled out, a look at Lemma 5 will hint at the trouble caused by the assumption of $(*)$.

The search for the appropriate $d_{m_j}^j$ will proceed in two stages.

First, we show how, given n, you can find $d_{m_1}^1, d_{m_2}^2, \ldots, d_{m_n}^n$ in such a way that

$$\left\| \sum_{j=1}^{n} (-1)^j d_{m_j}^j \right\| \le 3.$$

Our choice here of the n-tuple $(d_{m_1}^1, \ldots, d_{m_n}^n)$ will depend on n; it is possible that on passing from n to $n+1$, the $(n+1)$-tuple $(d_{m_1}^1, \ldots, d_{m_{n+1}}^{n+1})$ has little or nothing to do with the previous selection. A bit of patience on the part of the reader might be needed here—the fact that our notation hints that perhaps we are extending n-tuples to $(n+1)$-tuples is unfortunate, but if the notation is a bit troublesome, as it is, at least it is not sadistically cumbersome! Actually, in this first stage, we will choose $2n$-tuples of $d_{m_j}^j$, the idea being that each $d_{m_j}^j$ involves an odd number of x_k of which we can clip off a goodly number (say $n-1$) without seriously affecting the norms of interesting sums while achieving other goals. Starting with $2nd_{m_j}^j$ we are able to eliminate a few at the upper end and achieve the finitary condition

$$\left\| \sum_{j=1}^{n} (-1)^j d_{m_j}^j \right\| \le 3.$$

The second stage involves a diagonalization procedure the purpose of which is to show that under the assumption of $(*)$, we can select in one fell swoop a sequence $(d_{m_n}^n)$ so that

$$\sup_k \left\| \sum_{j=1}^{k} (-1)^j d_{m_j}^j \right\| \le 6.$$

Because X contains no copy of c_0, this consequence of $(*)$ plainly leads to a contradiction.

Start by looking at the δ_2-blocks: $b_1^2 < b_2^2 < \cdots < b_{k_2}^2$. Take any one of these, say $b_{i_2}^2$, where $1 \le i_2 \le k_2$. δ_2-blocks are δ_1-blocks; so there must be an i_1, $1 \le i_1 \le k_1$, such that

$$\left\| b_{i_1}^1 - b_{i_2}^2 \right\| \le 1 + \delta_1.$$

Correspondingly,

$$\left\|\left(d^1_{i_1}-d^2_{i_2}\right)-\left(b^1_{i_1}-b^2_{i_2}\right)\right\| \le \left\|d^1_{i_1}-b^1_{i_1}\right\|+\left\|d^2_{i_2}-b^2_{i_2}\right\|$$

$$\le \left|\frac{\alpha}{\beta^1_{i_1}}-1\right|+\left|\frac{\alpha}{\beta^2_{i_2}}-1\right| < \delta_1+\delta_2 < 2\delta_1,$$

and

$$\left\|\left(d^1_{i_1}-d^2_{i_2}\right)_{\text{less its last nonzero } x_k \text{ term}}\right\| \le (1+20^{-1})\left\|d^1_{i_1}-d^2_{i_2}\right\|$$

$$< (1+20^{-1})(1+3\delta_1).$$

Next, let's look at the δ_4-blocks: $b^4_1 < b^4_2 < \cdots < b^4_{k_4}$. Pick any one of these vectors, say b_{i_4}, where $1 \le i_4 \le k_4$. We know that there is an i_3, $1 \le i_3 \le k_3$, such that

$$\left\|b^3_{i_3}-b^4_{i_4}\right\| \le 1+\delta_3;$$

after all, δ_4-blocks and δ_3-blocks. Of course,

$$\left\|\left(d^3_{i_3}-d^4_{i_4}\right)-\left(b^3_{i_3}-b^4_{i_4}\right)\right\| \le \left\|d^3_{i_3}-b^3_{i_3}\right\|+\left\|d^4_{i_4}-b^4_{i_4}\right\|$$

$$\le \left|\frac{\alpha}{\beta^3_{i_3}}-1\right|+\left|\frac{\alpha}{\beta^4_{i_4}}-1\right| < \delta_3+\delta_4 < 2\delta_3,$$

forcing $\|d^3_{i_3}-d^4_{i_4}\| < 1+3\delta_3$, and

$$\left\|\left(d^3_{i_3}-d^4_{i_4}\right)_{\text{less its last nonzero } x_k \text{ term}}\right\| < (1+20^{-3})\left\|d^3_{i_3}-d^4_{i_4}\right\|$$

$$< (1+20^{-3})(1+3\delta_3)$$

$$< 1+\delta_2.$$

Further,

$$\left\|\left(d^3_{i_3}-d^4_{i_4}\right)_{\text{less its last nonzero } x_k \text{ term}}\right\| \ge \frac{\|d^3_{i_3}\|}{1+20^{-3}}$$

$$\ge \frac{1-\delta_3}{1+20^{-3}} \ge 1-\delta_2.$$

Notice that if we let

$$z_1 = \left(d^3_{i_3}-d^4_{i_4}\right)_{\text{less its last nonzero } x_k \text{ term}},$$

then $z_1/\|z_1\|$ is a δ_2-block having all its support to the right of b_{k_2}'s support. As before, there is an i_2, $1 \le i_2 \le k_2$, such that

$$\left\|b^2_{i_2}-\frac{z_1}{\|z_1\|}\right\| < 1+\delta_2.$$

Further,

$$\left\| \left(\sum_{j=2}^{4} (-1)^j d_{i_j}^j \right)_{\text{less its last nonzero } x_k \text{ term}} - \left(b_{i_2}^2 - \frac{z_1}{\|z_1\|} \right) \right\|$$

$$= \left\| \left(d_{i_2}^2 - b_{i_2}^2 \right) - \left(\left(d_{i_3}^3 - d_{i_4}^4 \right)_{\text{less its last nonzero } x_k \text{ term}} - \frac{z_1}{\|z_1\|} \right) \right\|$$

$$\le \left| \frac{\alpha}{\beta_{i_2}^2} - 1 \right| + \left\| z_1 - \frac{z_1}{\|z_1\|} \right\| < 2\delta_2.$$

Therefore,

$$\left\| \left(\sum_{j=2}^{4} (-1)^j d_{i_j}^j \right)_{\text{less its last two nonzero } x_k \text{ terms}} \right\|$$

$$\le (1+20^{-2}) \left\| \left(\sum_{j=2}^{4} (-1)^j d_{i_j}^j \right)_{\text{less its last nonzero } x_k \text{ term}} \right\|$$

$$< (1+20^{-2})(1+3\delta_2) < 1+\delta_1.$$

As before, if z_2 is given by

$$z_2 = \left(\sum_{j=2}^{4} (-1)^j d_{i_j}^j \right)_{\text{less its last two nonzero } x_k \text{ terms}},$$

then $\|z_2\|$ lies between $1-\delta_1$ and $1+\delta_1$. (*Notice* how few of the x_k terms we have clipped off the odd-lengthed $d_{i_j}^j$; at present we have wiped out only the last two nonzero x_k terms from $d_{i_4}^4$. There are lots of nonzero x_k terms left). In light of z_2's length we know that $z_2/\|z_2\|$ is a δ_1 block whose support lies to the right of $b_{k_1}^1$'s support. Consequently, there is an i_1, $1 \le i_1 \le k_1$, for which

$$\left\| b_{i_1}^1 - \frac{z_2}{\|z_2\|} \right\| \le 1+\delta_1.$$

It follows that

$$\left\| \left(\sum_{j=1}^{4} (-1)^{j+1} d_{i_j}^j \right)_{\text{less its last two nonzero } x_k \text{ terms}} - \left(b_{i_1}^1 - \frac{z_2}{\|z_2\|} \right) \right\|$$

$$= \left\| \left(d_{i_1}^1 - b_{i_1}^1 \right) - \left(\left(\sum_{j=2}^{4} (-1)^j d_{i_j}^j \right)_{\text{less its last two nonzero } x_k \text{ terms}} - \frac{z_2}{\|z_2\|} \right) \right\|$$

$$\le \left| \frac{\alpha}{\beta_{i_1}^1} - 1 \right| + \left\| z_2 - \frac{z_2}{\|z_2\|} \right\| < 2\delta_1.$$

Therefore,

$$\left\|\left(\sum_{j=1}^{4}(-1)^{j+1}d_{i_j}^j\right)_{\text{less its last three nonzero } x_k \text{ terms}}\right\|$$

$$\leq (1+20^{-1})\left\|\left(\sum_{j=1}^{4}(-1)^{j+1}d_{i_j}^j\right)_{\text{less its last two nonzero } x_k \text{ terms}}\right\|$$

$$\leq (1+20^{-1})(1+3\delta_1) < 1+\delta_0.$$

Since each $d_{i_j}^j$ has odd support size at least 3, we get

$$\left\|\sum_{j=1}^{2}(-1)^{j+1}d_{i_j}^j\right\| < 2(1+20^{-2}) < 3.$$

It should be clear that repeating the same kind of argument (starting with δ_{2n}-blocks, clipping away $(n-1)x_k$ terms, looking at $\sum_{j=1}^{n}(-1)^{j+1}d_{i_j}^j$), we can obtain, for any n, an n-tuple $(d_{i_1}^1,\ldots,d_{i_n}^n)$ such that

$$\left\|\sum_{j=1}^{n}(-1)^{j+1}d_{i_j}^j\right\| \leq 6.$$

Now to diagonalize, each step of our procedure produced an n-tuple $(d_{i_1}^1,\ldots,d_{i_n}^n)$ corresponding to and depending on that step. Let's commemorate the fact that $(d_{i_1}^1,\ldots,d_{i_n}^n)$ was selected as part of the nth step by tagging the indices $i_1,\ldots,i_n$ with an extra name tag $i_1(n),\ldots,i_n(n)$. We have then that for each n we've chosen $i_1(n), i_2(n),\ldots,i_n(n)$ so that $1\leq i_1(n)\leq k_1, 1\leq i_2(n)\leq k_2,\ldots,1\leq i_n(n)\leq k_n$, and

$$\left\|\sum_{j=1}^{n}(-1)^{j+1}d_{i_j(n)}^n\right\| \leq 6.$$

There are lots of n but only so many choices of i_1 (k_1 choices, in fact). It follows that there is some $\bar{i}_1$, $1\leq \bar{i}_1\leq k_1$ and an infinite set N_1 of n such that for each $n\in N_1$, $i_1(n)=\bar{i}_1$. There are lots of n in N_1 but only so many choices of i_2 (k_2 of them). It follows that there is some $\bar{i}_2$, $1\leq \bar{i}_2\leq k_2$, and an infinite subset N_2 of N_1 such that for each $n\in N_2$, $i_2(n)=\bar{i}_2$. Continuing in this fashion, we get a sequence $(\bar{i}_j)$ of indices $1\leq \bar{i}_j\leq k_j$ and a sequence (N_j) of infinite sets of positive integers satisfying $N_1\supseteq N_2\supseteq N_3\supseteq\cdots$, such that if $k\in\mathbb{N}$ and $j\leq k$, then $i_j(n)=\bar{i}_j$ for all $n\in N_k$. Let n_k be the kth element of N_k. Then the sequence $(d_{i_k(n_k)}^k)$ has the property that

$$\sup_k\left\|\sum_{j=1}^{k}(-1)^{j+1}d_{i_j(n_j)}^j\right\| \leq 6 < \infty.$$

Exercises

1. *Distorting l_1.* If X contains a subspace isomorphic to l_1, then for any $\delta > 0$ there is a sequence $(u_n) \subseteq B_X$ such that

$$(1-\delta)\sum_i |a_i| \le \left\| \sum_i a_i u_i \right\| \le \sum_i |a_i|$$

holds for any $(a_i) \in l_1$.

2. *The Banach-Saks property.*

   (i) Let (x_n) be a bounded sequence in the Banach space X. The set $A = \{(m_i) = M \in \mathscr{P}_\infty(\mathbf{N}) : \text{norm}\lim_k k^{-1}\Sigma_{i=1}^k x_{m_i} \text{ exists}\}$ is a Borel subset of $\mathscr{P}_\infty(\mathbf{N})$.

   (ii) If X has the Banach-Saks property, then each bounded sequence in X has a subsequence each of whose subsequences have norm-convergent arithmetic means with a common limit.

Notes and Remarks

Though we've belabored some details of the original, it is the proof of J. Elton and E. Odell that we follow religiously in the text. Quite frankly, we've gained some sense of satisfaction from finishing up with such a natural improvement of the principal result of our first chapter, Riesz's lemma.

Plainly our analysis revolves about recognizing a copy of c_0. The characterizations of c_0's unit vector basis due to W. B. Johnson and found in Lemmas 4 and 5 have been sharpened considerably. E. Odell and M. Wage, pursuing ideas quite similar to those in the text, have shown that *if (x_n) is a normalized weakly null sequence without any subsequence equivalent to the unit vector basis of c_0, then there is a basic subsequence (y_n) of (x_n) for which given any subsequence (z_n) of (y_n) and any sequence (ε_n) of 0's and 1's for which $(\varepsilon_n) \notin c_0$, then*

$$\lim_n \left\| \sum_{k=1}^n \varepsilon_k z_k \right\| = \infty.$$

In a devious departure from the usual combinatorial maneuvers but still calling on Ramsey's theorem at a crucial juncture, J. Elton has improved the Odell-Wage characterization to establish the following.

Theorem (Elton). *Let (x_n) be a normalized weakly null sequence having no subsequence equivalent to the unit vector basis of c_0. Then (x_n) has a basic subsequence (y_n) for which given any subsequence (z_n) of (y_n) and any $(a_k) \notin c_0$*

$$\lim_k \left\| \sum_{i=1}^k a_i z_i \right\| = \infty.$$

An interesting corollary of this was noted by E. Odell in his fast-paced survey of applications of Ramsey theorems to Banach space theory.

Theorem. *An infinite-dimensional Banach space without a copy of either c_0 or l_1 contains a subspace without the Dunford-Pettis property.*

The distortion theorem of R. C. James presented in the text as well as his companion theorem for l_1 (Exercise 1) are natural allies of anyone working on the detection of c_0 and l_1. It is an open question whether the corresponding theorem holds for l_p when $1 < p < \infty$. Here we might mention, too, the absence of any useful criterion for a Banach space to contain an isomorph of a fixed l_p $1 < p < \infty$ even in case $p = 2$. It is natural to suspect that any erstwhile answer will involve operator theoretic notions and so may, because of its intricacy, lose some of the elegance of statement enjoyed by Rosenthal's l_1 theorem or Elton's c_0 dilemma.

Returning momentarily to the Odell-Wage characterization of c_0, we would be remiss in our duties if we did not mention the crisp theorem of S. Kwapien bringing randomness to task in the affairs of c_0.

Theorem (S. Kwapien). *A Banach space X fails to contain a copy of c_0 if and only if given a series $\Sigma_n x_n$ in X for which $\Sigma_n \sigma_n x_n$ has bounded partial sums for almost all choices (σ_n) of signs ± 1, then $\Sigma_n \sigma_n x_n$ is norm convergent in X for almost all choices (σ_n) of signs.*

We take special pleasure in once again acknowledging that Kwapien's proof was the inspiration behind the first of our demonstrations of the Orlicz-Pettis theorem.

Ramsey theorems have played an important part in the Banach space theory of the past decade. Exercise 2, due to P. Erdös and M. Magidor (1976), gives but an inkling of the surprises that lay in store for the student when dealing with infinite-dimensional phenomena while in possession of a tool as powerful as Ramsey's theorem. The relevance of Ramsey theorems to this sort of averaging in Banach spaces is further borne out in the work of B. Beauzamy (1979, 1980), A. Brunel and L. Sucheston (1974), T. Figiel and L. Sucheston (1976), and J. R. Partington (1982) cited in the bibliography.

Other opportunities for the case of Ramsey theorems in Banach space theory are discussed by E. Odell in his aforementioned survey and it is to that survey that we direct the student.

Finally, we close with open problems.

Problem. *For which infinite-dimensional Banach spaces X is there an $\varepsilon > 0$ such that given any infinite-dimensional closed linear subspace Y of X, then one can find a $(1+\varepsilon)$-separated sequence in B_Y?*

Problem. *Which infinite-dimensional Banach spaces X have an equivalent norm $|||\cdot|||$ for which there is an $\varepsilon > 0$ such that if Y is an infinite-dimensional* *closed linear subspace of X, then* $B_{(Y, |||\cdot|||)}$ *contains a* $(1+\varepsilon)$-$|||\cdot|||$*-separated sequence?*

Bibliography

Beauzamy, B. 1979. Banach-Saks properties and spreading models. *Math. Scand.*, **44**, 357–384.

Beauzamy, B. 1980. Propriété de Banach-Saks. *Studia Math.*, **66**, 227–235.

Brunel, A. and Sucheston, L. 1974. On *B*-convex Banach spaces. *Math. Systems Theory*, **7**, 294–299.

Elton, J. and Odell, E. The unit ball of every infinite dimensional normed linear space contains a $(1+\varepsilon)$-separated sequence.

Erdös, P. and Magidor, M. 1976. A note on regular methods of summability and the Banach-Saks property. *Proc. Amer. Math. Soc.*, **59**, 232–234.

Figiel, T. and Sucheston, L. 1976. An application of Ramsey sets in analysis. *Advan. Math.*, **20**, 103–105.

Kwapien, S. 1974. On Banach spaces containing c_0. *Studia Math.*, **52**, 187–188.

Odell, E. 1981. *Applications of Ramsey Theorems to Banach Space Theory*. Austin: University of Texas Press.

Partington, J. R. 1982. Almost sure summability of subsequences in Banach spaces. *Studia Math.*, **71**, 27–35.

Index

Graduate Texts in Mathematics

continued from page ii

48 SACHS/WU. General Relativity for Mathematicians.
49 GRUENBERG/WEIR. Linear Geometry. 2nd ed.
50 EDWARDS. Fermat's Last Theorem.
51 KLINGENBERG. A Course in Differential Geometry.
52 HARTSHORNE. Algebraic Geometry.
53 MANIN. A Course in Mathematical Logic.
54 GRAVER/WATKINS. Combinatorics with Emphasis on the Theory of Graphs.
55 BROWN/PEARCY. Introduction to Operator Theory I: Elements of Functional Analysis.
56 MASSEY. Algebraic Topology: An Introduction.
57 CROWELL/FOX. Introduction to Knot Theory.
58 KOBLITZ. *p*-adic Numbers, *p*-adic Analysis, and Zeta-Functions.
59 LANG. Cyclotomic Fields.
60 ARNOLD. Mathematical Methods in Classical Mechanics.
61 WHITEHEAD. Elements of Homotopy Theory.
62 KARGAPOLOV/MERZLJAKOV. Fundamentals of the Theory of Groups.
63 BOLLOBÁS. Graph Theory.
64 EDWARDS. Fourier Series. Vol. I. 2nd ed.
65 WELLS. Differential Analysis on Complex Manifolds. 2nd ed.
66 WATERHOUSE. Introduction to Affine Group Schemes.
67 SERRE. Local Fields.
68 WEIDMANN. Linear Operators in Hilbert Spaces.
69 LANG. Cyclotomic Fields II.
70 MASSEY. Singular Homology Theory.
71 FARKAS/KRA. Riemann Surfaces.
72 STILLWELL. Classical Topology and Combinatorial Group Theory.
73 HUNGERFORD. Algebra.
74 DAVENPORT. Multiplicative Number Theory. 2nd ed.
75 HOCHSCHILD. Basic Theory of Algebraic Groups and Lie Algebras.
76 IITAKA. Algebraic Geometry.
77 HECKE. Lectures on the Theory of Algebraic Numbers.
78 BURRIS/SANKAPPANAVAR. A Course in Universal Algebra.
79 WALTERS. An Introduction to Ergodic Theory.
80 ROBINSON. A Course in the Theory of Groups.
81 FORSTER. Lectures on Riemann Surfaces.
82 BOTT/TU. Differential Forms in Algebraic Topology.
83 WASHINGTON. Introduction to Cyclotomic Fields.
84 IRELAND/ROSEN. A Classical Introduction Modern Number Theory.
85 EDWARDS. Fourier Series. Vol. II. 2nd ed.
86 VAN LINT. Introduction to Coding Theory.
87 BROWN. Cohomology of Groups.
88 PIERCE. Associative Algebras.
89 LANG. Introduction to Algebraic and Abelian Functions. 2nd ed.
90 BRØNDSTED. An Introduction to Convex Polytopes.
91 BEARDON. On the Geometry of Discrete Groups.
92 DIESTEL. Sequences and Series in Banach Spaces.

93 DUBROVIN/FOMENKO/NOVIKOV. Modern Geometry – Methods and Applications Vol. I.

94 WARNER. Foundations of Differentiable Manifolds and Lie Groups.

95 SHIRYAYEV. Probability, Statistics, and Random Processes.

96 ZEIDLER. Nonlinear Functional Analysis and Its Applications I: Fixed Points Theorem.